AF167648

Communications
in Computer and Information Science 2186

Rationale

The CCIS series is devoted to the publication of proceedings of computer science conferences. Its aim is to efficiently disseminate original research results in informatics in printed and electronic form. While the focus is on publication of peer-reviewed full papers presenting mature work, inclusion of reviewed short papers reporting on work in progress is welcome, too. Besides globally relevant meetings with internationally representative program committees guaranteeing a strict peer-reviewing and paper selection process, conferences run by societies or of high regional or national relevance are also considered for publication.

Topics

The topical scope of CCIS spans the entire spectrum of informatics ranging from foundational topics in the theory of computing to information and communications science and technology and a broad variety of interdisciplinary application fields.

Information for Volume Editors and Authors

Publication in CCIS is free of charge. No royalties are paid, however, we offer registered conference participants temporary free access to the online version of the conference proceedings on SpringerLink (http://link.springer.com) by means of an http referrer from the conference website and/or a number of complimentary printed copies, as specified in the official acceptance email of the event.

CCIS proceedings can be published in time for distribution at conferences or as post-proceedings, and delivered in the form of printed books and/or electronically as USBs and/or e-content licenses for accessing proceedings at SpringerLink. Furthermore, CCIS proceedings are included in the CCIS electronic book series hosted in the SpringerLink digital library at http://link.springer.com/bookseries/7899. Conferences publishing in CCIS are allowed to use Online Conference Service (OCS) for managing the whole proceedings lifecycle (from submission and reviewing to preparing for publication) free of charge.

Publication process

The language of publication is exclusively English. Authors publishing in CCIS have to sign the Springer CCIS copyright transfer form, however, they are free to use their material published in CCIS for substantially changed, more elaborate subsequent publications elsewhere. For the preparation of the camera-ready papers/files, authors have to strictly adhere to the Springer CCIS Authors' Instructions and are strongly encouraged to use the CCIS LaTeX style files or templates.

Abstracting/Indexing

CCIS is abstracted/indexed in DBLP, Google Scholar, EI-Compendex, Mathematical Reviews, SCImago, Scopus. CCIS volumes are also submitted for the inclusion in ISI Proceedings.

How to start

To start the evaluation of your proposal for inclusion in the CCIS series, please send an e-mail to ccis@springer.com.

Joe Tekli · Johann Gamper · Richard Chbeir ·
Yannis Manolopoulos · Salma Sassi ·
Mirjana Ivanovic · Genoveva Vargas-Solar ·
Ester Zumpano
Editors

New Trends in Database and Information Systems

ADBIS 2024 Short Papers, Workshops, Doctoral Consortium and Tutorials
Bayonne, France, August 28–31, 2024
Proceedings

 Springer

Editors
Joe Tekli 🆔
Lebanese American University
Beirut, Lebanon

Richard Chbeir 🆔
Univ Pau & Pays de l'Adour, Anglet
Anglet, France

Salma Sassi 🆔
Univ Pau & Pays de l'Adour, Anglet
Anglet, France

Genoveva Vargas-Solar 🆔
Délégation Rhône Auvergne
Villeurbanne, France

Johann Gamper 🆔
Free University of Bozen-Bolzano
Bolzano, Italy

Yannis Manolopoulos 🆔
Open University of Cyprus
Nicosia, Cyprus

Mirjana Ivanovic 🆔
University of Novi Sad
Novi Sad, Serbia

Ester Zumpano 🆔
University of Calabria
Arcavacata, Italy

ISSN 1865-0929   ISSN 1865-0937 (electronic)
Communications in Computer and Information Science
ISBN 978-3-031-70420-8   ISBN 978-3-031-70421-5 (eBook)
https://doi.org/10.1007/978-3-031-70421-5

Preface

This CCIS volume includes research papers presented at the 28th European Conference on Advances in Databases and Information Systems (ADBIS), plus research papers from four workshops accompanying ADBIS, as well research papers presented at three Doctoral Consortium (DC) sessions. The 28th ADBIS conference was held during August 28–31, 2024, in Bayonne, France.

Since 1997, ADBIS has been continuously organized as an annual event. Its previous editions were held in: Barcelona, Spain (2023); Torino, Italy (2022); Tartu, Estonia (2021); Lyon, France (2020); Bled, Slovenia (2019); Budapest, Hungary (2018); Nicosia, Cyprus (2017); Prague, Czech Republic (2016); Poitiers, France (2015); Ohrid, North Macedonia (2014); Genoa, Italy (2013); Poznan, Poland (2012); Vienna, Austria (2011); Novi Sad, Serbia (2010); Riga, Latvia (2009); Pori, Finland (2008); Varna, Bulgaria (2007); Thessaloniki, Greece (2006); Tallinn, Estonia (2005); Budapest, Hungary (2004); Dresden, Germany (2003); Bratislava, Slovakia (2002); Vilnius, Lithuania (2001); Prague, Czech Republic (2000); Maribor, Slovenia (1999); Poznan, Poland (1998), Saint Petersburg, Russia (1997). The official ADBIS portal at http://adbis.eu provides up to date information on all ADBIS Conferences, persons in charge, publications, and issues related to the ADBIS community.

The program of ADBIS 2024 included keynote talks, research papers, tutorials, workshops, a doctoral consortium (DC), as well as a diversity, equity and inclusion (DEI) panel and keynote. The main conference attracted 43 paper submissions. All the papers went through a process of rigorous single-blind reviewing by at least three reviewers. Eventually, the Program Committee selected 15 submissions as full contributions, and 7 short papers appearing in the present volume. The selected papers span a large spectrum of topics in the broader field of data management and were organized in seven sessions: (1) Algebra, Models, and Schemata; (2) Discovery and Data Analysis; (3) Algorithms and Optimization; (4) Access Methods and Query Processing; (5) Advanced Architectures; (6) Machine Learning; and (7) Large Language Models.

Following a successful tradition, selected best papers of ADBIS 2024 will be invited for special issues of the following journals: Information Systems (Springer), SN Computer Science (Springer), Digital (MDPI), and Computer Science & Information Systems. Therefore, the PC chairs would like to express their sincere gratitude to the Editors-in-Chief of the above journals for their approval regarding these special issues.

For this ADBIS event we had keynote talks by experts in different fields of the broader area of data management. In particular (in alphabetical order):

- Angela Bonifati (CNRS Liris research lab, Lyon 1 University, France) delivered a talk on *"The Future is Hybrid Graphs: Combining Graphs with Time Series"*.
- Christian S. Jensen (Center of Data Intensive Systems, Aalborg University, Denmark) dealt with the topic of *"Temporal and Spatio-Temporal Data Management and Analytics – A Personal Perspective"*, and

- Schahram Dustdar (TU Wien, Austria) covered the topic of *"Distributed Intelligence in the Computing Continuum"*.

In addition, two tutorials were also included in the program, which were delivered by:

- Zheying Zhang and Kostas Stefanidis (Tampere University, Finland) who presented a tutorial entitled *"Data-driven Analysis for Monitoring Software Evolution"*, and
- Witold Andrzejewski, Bartosz Bębel, Paweł Boiński and Robert Wrembel (Poznan University of Technology, Poland) who delivered a tutorial entitled *"On Customer Data Deduplication - Research vs. Industrial Perspective: Lessons Learned from a R&D Project in the Financial Sector"*.

The ADBIS conference program featured three DEI plenary sessions. Additionally, a data analytics activity with the DC Chairs gave Ph.D. students practical skills for designing equity-aware data-driven experiments. The DEI stream offered a keynote address by:

- Rita Bencivenga (University of Genova, Italy), who delivered a talk on *"Incorporating Gender+ and Intersectionality into Research Practices"*.

Furthermore, a panel discussion entitled *"More Than Muscles in the Lab: Expanding Masculinities in a Competitive Research Culture"* brought together distinguished researchers who study masculinity in the scientific community, exploring its various perspectives and impact. Our distinguished panelists were

- Rita Bencivenga (University of Genova),
- Jorge Alberto Hidalgo Toledo (University Anáhuac, Mexico),
- Amir Hossein Payberah (KTH Royal Institute of Technology, Sweden),
- Arun Kumar (University of California, San Diego, USA), and
- Romain Sabathier (COOP-EGAL, France).

The ADBIS 2024 conference hosted four affiliated workshops:

- 5th Workshop on Intelligent Data - From Data to Knowledge (DOING),
- 3rd Workshop on Knowledge Graphs Analysis on a Large Scale (K-GALS),
- 6th Workshop on Modern Approaches in Data Engineering and Information System Design (MADEISD), and
- 3rd Workshop on Personalization and Recommender Systems (PERS).

All four workshops addressed cutting-edge topics that call for extensive research from both a foundational and an application perspective to progress the state of the art. In total, 37 papers were submitted to these workshops, out of which 17 were selected for presentation and publication, yielding an overall acceptance rate of 46%. The accepted workshop papers are published in this volume. This volume also contains the contributions of the Doctoral Consortium stream; the latter call attracted 20 submissions, out of which 11 were selected for inclusion in the program. In particular, the following workshops were run at the ADBIS 2024 conference.

DOING 2024: 5th Workshop on Intelligent Data - From Data to Knowledge.

DOING 2024 was organized by Cristina Dutra de Aguiar (Universidade de São Paulo, Brazil), Mirian Halfeld Ferrari (Université d'Orléans, France), and Carmem Hara (Universidade Federal do Paraná, Brazil). The workshop focused on transforming data into information and then into knowledge, bringing together researchers from natural language processing, databases, and artificial intelligence. This workshop is associated with the DOING action of the French GDR MADICS. The workshop is the result of a collective effort. The organizers gratefully acknowledge the ADBIS conference chairs, the authors for submitting their work, and the PC members for their support and timely reviews:

- Cheikh Ba, Université Gaston Berger de Saint-Louis, Senegal
- Besim Bilalli, Universitat Politècnica de Catalunya, Spain
- Davide Buscaldi, Université Sorbonne Paris Nord, France
- Rogério Luís De Carvalho Costa, Polytechnic of Leiria, Portugal
- Laurent D'Orazio, Université de Rennes, France
- Sven Groppe, University of Lubeck, Germany
- Nicolas Hiot, Université d'Orléans, France
- Jixue Liu, University of South Australia, Australia
- Javam Machado, Universidade Federal do Ceará, Brazil
- Anne-Lyse Minard-Forst, Université d'Orléans, France
- Aurora Pozo, Universidade Federal do Paraná, Brazil
- Yves Rybarczyk, Dalarna University, Sweden
- Agata Savary, Université Paris-Saclay, France
- Wagner Zola, Universidade Federal of Paraná, Brazil

K-GALS 2024: 3rd Workshop on Knowledge Graphs Analysis on a Large Scale

K-GALS 2024 was organized by Mariella Bonomo, Simona E. Rombo, Ylenia Galluzzo (University of Palermo, Italy). Knowledge graphs are powerful models to represent networks of real-world entities, such as objects, events, situations, concepts, by illustrating the relationships between them. Information encoded by knowledge graphs is usually stored in graph databases, and visualized as graph structures. These models have been introduced in the Semantic Web context, and more recently they have been successfully applied also in other contexts, e.g., the analysis of medical data. Knowledge graphs often integrate datasets from various sources, which frequently differ in their structure. This, together with the increasing volumes of structured and unstructured data stored in a distributed manner, bring to light new problems related to data/knowledge representation and integration, data querying, business analysis and knowledge discovery. The ultimate goal of this workshop is to provide to participants the opportunity of introducing and discussing new methods, theoretical approaches, algorithms, and software tools that are relevant to the Knowledge Graphs based research, especially when it is focused on a large scale. To this regard, interesting open issues include how Knowledge Graphs may be used to represent knowledge, how systems managing Knowledge Graphs work, and which applications may be provided on top of a Knowledge Graph, in the distributed. The organizers gratefully acknowledge the PC members for their support and timely reviews:

- Leonardo Alexandre, INESC-ID, Instituto Superior Técnico, Universidade de Lisboa, Portugal
- Lorenzo Bellomo, University of Pisa, Italy
- Umberto Ferraro Petrillo, University of Rome "La Sapienza", Italy
- Valeria Fionda, University of Calabria, Italy
- Lorenzo Di Rocco, University of Rome "La Sapienza", Italy
- Blerina Sinaimeri, LUISS University, Italy
- Edoardo Serra, Boise University, USA
- Cristina Serrao, University of Calabria, Italy
- Filippo Utro, Computational Biology Center, IBM T. J. Watson Research, USA

MADEISD 2024: 6th Workshop on Modern Approaches in Data Engineering and Information System Design

MADEIS 2024 was organized by Ivan Luković (University of Belgrade, Serbia), Sonja Ristić (University of Novi Sad, Serbia), and Slavica Kordić (University of Novi Sad, Serbia). The main goal of the MADEISD workshop is to address open questions and real potentials for various applications of modern approaches and technologies in data engineering and information system design to develop and implement effective software services in support of information management in various organization systems. The intention was to address the interdisciplinary character of a set of theories, methodologies, processes, architectures, and technologies in disciplines such as Data Engineering, Information System Design, Big Data, NoSQL Systems, Data Streams, Internet of Things, Cloud Systems, and Model Driven Approaches in development of effective software services. The organizers gratefully acknowledge the PC members for their support and timely reviews:

- Moharram Challenger, University of Antwerp, Belgium
- Milan Čeliković, University of Novi Sad, Serbia
- Boris Delibašić, University of Belgrade, Serbia
- Vladimir Dimitrieski, University of Novi Sad, Serbia
- Dražen Drašković, University of Belgrade, Serbia
- João M. Fernandes, University of Minho, Portugal
- José E. Fernandes, Polytechnic Institute of Bragança, Portugal
- Krešimir Fertalj, University of Zagreb, Croatia
- Krzysztof Goczyła, Gdańsk University of Technology, Poland
- Ralf-Christian Härting, Aalen University, Germany
- Dušan Jakovetić, University of Novi Sad, Serbia
- Miklós Krész, InnoRenew CoE and University of Primorska, Slovenia
- Dragan Maćoš, Beuth University of Applied Sciences Berlin, Germany

- Sanda Martinčić-Ipšić, University of Rijeka, Croatia
- Manuel Mazzara, Innopolis University, Russia
- Nikola Obrenović, University of Novi Sad, BioSense Institute, Serbia
- Maxim Panov, Technology Innovation Institute, Abu Dhabi, UAE
- Rui Humberto Pereira, Polytechnic Institute of Porto, Portugal
- Aleksandar Popović, University of Montenegro, Montenegro
- Patrizia Poščić, University of Rijeka, Croatia
- Adam Przybyłek, Gdansk University of Technology, Poland
- Kornelije Rabuzin, University of Zagreb, Croatia
- Igor Rožanc, University of Ljubljana, Slovenia
- Nikolay Skvortsov, Russian Academy of Sciences, Russia
- William Steingartner, Technical University of Košice, Slovakia
- Vjeran Strahonja, University of Zagreb, Croatia
- Max Talanov, The Institute for Artificial Intelligence Research and Development, Serbia
- Valentino Vranić, Slovak University of Technology in Bratislava, Slovakia
- Slavko Žitnik, University of Ljubljana, Slovenia

PERS 2024: 3rd Workshop on Personalization and Recommender Systems (PERS)

PERS 2024 was organized by Aleksandra Karpus and Adam Przybyłek (Gdańsk University of Technology, Poland). Recommender systems are present in our everyday lives when we read news, log in to social media, or buy something in an e-shop. Therefore, it is not surprising that this domain is receiving more and more attention from researchers in academia as well as industry practitioners. However, the way in which they approach the same problem differs significantly. Personalization is an important element in novel recommendation techniques. Nonetheless, it is a wider topic that concerns also user modelling and representation, personalized systems, adaptive educational systems or intelligent user interfaces. The workshop on Personalization and Recommender Systems (PeRS) was founded in 2022 as a part of the FedCSIS multiconference. In 2023, PeRS has joined ADBIS with the aim to extend the state-of-the-art in Personalization and Recommender Systems by providing a platform where industry practitioners and academic researchers can meet and learn from each other. The organizers gratefully acknowledge the PC members for their support and timely reviews:

- Frederico Araujo Durao, Federal University of Bahia, Brazil
- Frantisek Babic, Technical university of Kosice, Slovakia
- Olga Cherednichenko, University of Lyon, France
- Mariam Dedabrishvili, International Black Sea University, Georgia
- Arpita Dutta, National University of Singapore, Singapore
- Tomasz Dziubich, Gdansk University of Technology, Poland
- Agata Filipowska, Poznan University of Economics, Poland
- Mouzhi Ge, Deggendorf Institute of Technology, Germany
- Krzysztof Goczyła, Gdansk University of Technology, Poland
- Tomasz Górecki, Adam Mickiewicz University, Poland

- Aleksander Jarzebowicz, Gdansk University of Technology, Poland
- Eyad Kannout, University of Warsaw, Poland
- Urszula Kużelewska, Bialystok University of Technology, Poland
- Maria Lencastre, Universidade de Pernambuco, Brazil
- Chalachew Muluken Liyew, University of Torino, Italy
- Ivan Luković, University of Belgrade, Serbia
- Mohsen Maraoui, LIDILEM, France
- Bartosz Marcinkowski, University of Gdansk, Poland
- Mirko Marras, University of Cagliari, Italy
- Jacek Maślankowski, University of Gdańsk, Poland
- Sanjay Misra, Institute For Energy Technology, Norway
- Durga Prasad Mohapatra, NIT, Rourkela, India
- Md. Saddam Mukta, LUT university, Finland
- Yen Ying Ng, Nicolaus Copernicus University, Poland
- Phuong T. Nguyen, University of L'Aquila, Italy
- Francesco Nocera, Polytechnic University of Bari, Italy
- Aneta Poniszewska-Maranda, Lodz University of Technology, Poland
- Piotr Szczuko, Gdansk University of Technology, Poland
- Adel Taweel, King's College London, UK
- Iacopo Vagliano, Amsterdam Universitair Medische Centra, Netherlands
- Mounir Zrigui, University of Monastir, Tunisia, Tunisia

ADBIS DC 2024: Doctoral Consortium

DC was organized by Mirjana Ivanovic (University of Novi Sad, Serbia), Genoveva Vargas-Solar (French Council of Scientific Research, France) and Ester Zumpano (University of Calabria, Italy). The ADBIS 2024 Doctoral Consortium was a forum for Ph.D. students to present their research projects to the scientific community and establish international collaborations with members and participants of the ADBIS community. The DC included eleven "ten minutes madness" oral presentations— one in person and ten online. PhD students at the DC engaged in a speed-dating format to enhance networking and effectively communicate their scientific identities. Following the presentations, the students were grouped thematically and received feedback from a panel of mentors on potential opportunities and strategies for advancing their research. Additionally, in line with the conference's commitment to Diversity, Equity and Inclusion (DEI), a data-driven activity was organized to promote awareness of inclusion in research, complementing the broader DEI conference agenda. The program for this DC edition featured a keynote dialogue titled "Text mining application in medicine" by Chiara Zucco, a junior research scientist at the University Magna Græcia of Catanzaro, Italy. The organizers gratefully acknowledge the PC members for their support and timely reviews:

- Judith Awiti, Université libre de Bruxelles, Belgium
- Andrea Calì, Birkbeck University of London, UK
- Barbara Catania, University of Genoa, Italy
- Jérôme Darmont, Université Lumière Lyon 2, France
- Marlon Dumas, University of Tartu, Estonia

- Abir Farouzi, LIAS/ISAE-ENSMA, France
- Sergio Flesca, University of Calabria, Italy
- Cristian Molinaro, University of Calabria, Italy
- Laura Po, University of Modena, Italy
- Nicolas Travers, Université Eiffel Paris, France
- Raquel Trillo, University of Zaragoza, Spain
- Domenico Ursino, Università Politecnica delle Marche, Italy
- Luciano Caroprese, University G. d' Annunzio of Chieti-Pescara, Italy

Acknowledgements. We would like to wholeheartedly thank all participants, authors, PC members, workshop organizers, session chairs, doctoral consortium chairs, workshop chairs, volunteers, and co-organizers for their contributions to making ADBIS 2024 a great success. We would also like to thank the ADBIS Steering Committee and all sponsors.

July 2024

Joe Tekli
Johann Gamper
Richard Chbeir
Yannis Manolopoulos
Salma Sassi
Mirjana Ivanovic
Genoveva Vargas-Solar
Ester Zumpano

Organization

General Chairs

Richard Chbeir   Université de Pau et des Pays de l'Adour, France
Yannis Manolopoulos   Open University of Cyprus, Cyprus

Program Co-chairs

Joe Tekli   Lebanese American University, Lebanon
Johann Gamper   Free University of Bozen-Bolzano, Italy

Workshops Chair

Salma Sassi   Université de Pau et des Pays de l'Adour, France

Special Issues Co-chairs

Dimitrios Katsaros   University of Thessaly, Greece
Oscar Romero   Universitat Politècnica de Catalunya, Spain

Doctoral Consortium Co-chairs

Ester Zumpano   Università della Calabria, Italy
Genoveva Vargas Solar   CNRS, France
Mirjana Ivanovic   University of Novi Sad, Serbia

Tutorials Co-chairs

Claudio Silvestri   Università Ca' Foscari Venezia, Italy
Goce Trajcevski   Iowa State University, USA

Diversion and Inclusion Co-chairs

Barbara Catania	University of Genova, Italy
Genoveva Vargas Solar	CNRS, France

Publicity Chairs

Khouloud Salameh	AURAK, UAE
Theodoros Tzouramanis	University of Thessaly, Greece

Proceedings Chair

Karam Bou-Chaaya	Expleo Group, France

Webmaster

Elie Chicha	Université de Pau et des Pays de l'Adour, France
Fouad Achkouty	Université de Pau et des Pays de l'Adour, France

International Program Committee Members

Agata Filipowska	Poznan University of Economics, Poland
Alberto Abello	Universitat Politècnica de Catalunya, Spain
Alberto Gutiérrez Torre	Barcelona Supercomputing Center, Spain
Alejandro Maté	University of Alicante, Spain
Aleksandra Karpus	Gdansk University of Technology, Poland
Alexandre Chanson	University of Tours, France
Alexandros Karakasidis	University of Macedonia, Greece
Andreas Behrend	University of Bonn, Germany
Andrzej Kozik	Opole University, Poland
Angelo Montanari	University of Udine, Italy
Anton Dignos	Libera Università di Bolzano, Italy
Antonio Corral	University of Almeria, Spain
Arpita Dutta	National University of Singapore
Audrone Lupeikiene	Vilnius University, Lithuania
Bentayeb Fadila	Lyon University, France
Bernd Amann	Sorbonne Université, France

Bernhard Thalheim	Christian-Albrechts-Universität zu Kiel, Germany
Besim Bilalli	Universitat Politècnica de Catalunya, Spain
Boris Novikov	National Research University, Russia
Boussaid Omar	Université Lyon 2, France
Carmem S. Hara	Universidade Federal do Parana, Brazil
Chalachew Muluken Liyew	University of Torino, Italy
Cheikh Ba	Université Gaston Berger de Saint-Louis, Senegal
Christos Doulkeridis	University of Piraeus, Greece
Claudia Steinberger	Universitaet Klagenfurt, Austria
Costin Badica	University of Craiova, Romania
Cristina D. Aguiar	Universidade de São Paulo, Brazil
Damiano Carra	University of Verona, Italy
Daniele Apiletti	Politecnico di Torino, Italy
Dejan Lavbic	University of Ljubljana, Slovenia
Dijana Oreški	University of Zagreb, Croatia
Drazen Brdjanin	University of Banja Luka, Bosnia and Herzegovina
Flavio Ferrarotti	Software Competence Centre Hagenberg, Austria
Francesco Guerra	Università di Modena e Reggio Emilia, Italy
Franck Ravat	Université de Toulouse, France
George Papastefanatos	ATHENA Research Center, Greece
Georgia Koloniari	University of Macedonia, Greece
Georgios Evangelidis	University of Macedonia, Greece
Giuseppe Polese	University of Salerno, Italy
Goran Velinov	Cyril and Methodius University in Skopje, North Macedonia
Gunter Saake	University of Magdeburg, Germany
Haridimos Kondylakis	Institute of Computer Science, Greece
Isabelle Comyn-Wattiau	ESSEC Business School, France
Ivan Lukovic	University of Belgrade, Serbia
Jaroslav Pokorny	Charles University in Prague, Czechia
Javier A. Espinosa-Oviedo	University of Lyon, France
Jérôme Darmont	Université Lyon 2, France
Joe Tekli	Lebanese American University, Lebanon
Johann Eder	Alpen Adria Universität Klagenfurt, Austria
José R. R. Viqueira	University of Santiago de Compostela, Spain
Josep Berral	Barcelona Supercomputing Center, Spain
Julius Köpke	Universität Klagenfurt, Austria
Kemal Polat	Abant Izzet Baysal University, Turkey
Khalid Belhajjame	Université Paris-Dauphine, France
Kjetil Nørvåg	Norwegian University of Science and Technology, Norway

Ladjel Bellatreche	University of Poitiers, France
Lorena Etcheverry	Universidad de la República, Uruguay
Marcos Sfair Sunye	Universidade Federal do Parana, Brazil
María del Mar Roldán	Universidad de Malaga, Spain
Mariella Bonomo	Università degli Studi di Palermo, Italy
Markus Endres	University of Applied Sciences Munich, Germany
Matteo Francia	University of Bologna, Italy
Miguel A. Martinez-Prieto	University of Valladolid, Spain
Milos Radovanovic	University of Novi Sad, Serbia
Milos Savic	University of Novi Sad, Serbia
Mirian Halfeld-Ferrari	Université d'Orléans, France
Moharram Challenger	University of Antwerp, Belgium
Mounir Zrigui	University of Monastir, Tunisia
Nicolas Labroche	Université de Tours, France
Nikolaos Polatidis	University of Brighton, UK
Panos Vassiliadis	University of Ioannina, Greece
Patrick Marcel	University of Orléans, France
Pawel Boinski	Poznan University of Technology, Poland
Petr Hajek	University of Pardubice, Czechia
Richard Chbeir	Université de Pau et des Pays de l'Adour, France
Robert Wrembel	Poznan University of Technology, Poland
Rogério Luís De Carvalho Costa	Polytechnic of Leiria, Portugal
Sandro Bimonte	INRAE, France
Sara Migliorini	Università degli Studi di Verona, Italy
Sergey Stupnikov	Russian Academy of Sciences, Russia
Sergi Nadal	Universitat Politècnica de Catalunya, Spain
Simona E. Rombo	Università degli Studi di Palermo, Italy
Slavica Kordic	University of Novi Sad, Serbia
Sonja Ristic	University of Novi Sad, Serbia
Spiros Skiadopoulos	University of Peloponnese, Greece
Stefan Jablonski	University of Bayreuth, Germany
Stefano Rizzi	University of Bologna, Italy
Sylvio Barbon	University of Trieste, Italy
Tadeusz Morzy	Poznan University of Technology, Poland
Tania Cerquitelli	Politecnico di Torino, Italy
Veronika Peralta	University of Tours, France
William Steingartner	Technical University of Kosice, Slovakia
Witold Andrzejewski	Poznan University of Technology, Poland
Yannis Manolopoulos	Open University of Cyprus, Cyprus
Ylenia Galluzzo	Università degli Studi di Palermo, Italy
Yves Rybarczyk	Dalarna University, Sweden
Zoubida Kedad	University of Versailles, France

Sponsors

Springer

Connected Environment & Distributed Energy Data Management Solutions
Research Group

Contents

**K-GALS 2024: 3rd Workshop on Knowledge Graphs Analysis on a
Large Scale**

**MADEISD 2024: 6th Workshop on Modern Approaches in Data
Engineering and Information System Design**

**PERS 2024: 3rd Workshop on Personalization and Recommender
Systems (PERS)**

ADBIS DC 2024: Doctoral Consortium

Tutorials

Short Papers

A Reproducibility Study of Subgroup Discovery Algorithms

Vadim Arzamasov$^{(\boxtimes)}$ ⓘ and Klemens Böhm ⓘ

Karlsruhe Institute of Technology, Karlsruhe, Germany
{vadim.arzamasov,klemens.boehm}@kit.edu

Abstract. Subgroup discovery is an important data mining technique that aims to uncover specific segments of data that exhibit noteworthy patterns or behaviors. Despite the existence of several subgroup discovery algorithms, there is no thorough comparative analysis of their effectiveness. This paper evaluates two recent subgroup discovery algorithms against the seminal PRIM method, demonstrating PRIM's superior accuracy, speed, and effectiveness in identifying interesting subgroups. These findings highlight the need for a comprehensive evaluation of subgroup discovery techniques to determine the state of the art.

Keywords: SuRF · DivExplorer · PRIM · Subgroup Discovery

1 Introduction

Subgroup discovery (SD) is a popular method in exploratory data analysis. It finds descriptions of data segments that behave unusually as defined by a target statistic. For example, an SD model might find [2]: "Young men with low levels of education tend to have higher unemployment rates." In contrast to association rule mining, SD methods belong to supervised learning. Exceptional model mining (EMM) [6,10] extends subgroup discovery to complex target statistics, often involving multiple attributes. An EMM model might find [6]: "For houses with driveways, recreation rooms, and at least two bathrooms, the relationship between lot size and price is very different from that of other houses sold." Following [2], we adopt a broad definition of SD that includes EMM.

Many SD algorithms have been proposed [2,9], but there is no comprehensive comparison or consensus on a state-of-the-art method. Hence, mature methods may outperform newer ones. We show this by comparing two recent SD algorithms from high-impact venues, SuRF [12] and DivExplorer [11], with an earlier method, PRIM [7]. Replicating their experimental settings, we found PRIM to be more accurate, faster, and as general as SuRF, and it tends to discover better subgroups than DivExplorer. Our code is openly available[1].

Paper outline: Sect. 2 describes the SD algorithms compared. Section 3 reproduces the experiments from [12] comparing PRIM with SuRF. Section 4 compares PRIM to DivExplorer using the setup from [11]. Section 5 concludes.

---

[1] https://github.com/Arzik1987/Prim_SuRF_DivExplorer.

© The Author(s), under exclusive license to Springer Nature Switzerland AG 2025
J. Tekli et al. (Eds.): ADBIS 2024, CCIS 2186, pp. 3–13, 2025.
https://doi.org/10.1007/978-3-031-70421-5_1

Algorithm 1. PRIM

1: **Input:** Training data, peeling parameter α.
2: **Output:** Desired number of subgroups
3: Calculate the bounding box that contains all data points.
4: Try contracting the bounding box by changing one of its sides to exclude a fraction α of the data for a numeric attribute X_i, or by removing a fraction of the data with a particular value of X_i if X_i is categorical. Select the contraction that yields the highest target statistic y.
5: Continue with Step 4 until the box contains a predefined number of points.
6: Extend the box along any dimension if this improves the target statistic y.
7: Exclude the data contained in the box found in Step 6 from the training data and repeat the process in Steps 3–6 to discover the desired number of subgroups.

---

2   Background

Over the past decades, dozens of subgroup discovery methods have emerged. Among other things, they differ in the search space and the types of attributes they can handle [2,9]. For our study, we chose PRIM as a reference point because (1) it is one of the earliest subgroup discovery methods, (2) it can handle both categorical and continuous data, (3) it searches a rich space, and (4) it is not tied to a particular target statistic. PRIM is also popular and has various applications, including finding multiple local maxima of a statistic in a data set [5] and analyzing the output of simulation models [3].

We reproduce experiments from [12] that compare SuRF to PRIM, and we also compare PRIM to DivExplorer [11]. We chose SuRF and DivExplorer because of their recency, their origins in high-impact venues, and the public availability of their code and data. In addition, the experimental settings in the respective papers are somewhat complementary: [12] uses simulated data with continuous attributes, while [11] uses real data with discrete attributes. The rest of this section describes PRIM, SuRF, and DivExplorer.

PRIM. The PRIM algorithm [7] (Algorithm 1) works by iteratively removing chunks of data while maximizing the target statistic over the remaining points. The algorithm is "patient" in the sense that it limits the proportion of points removed in each iteration. It works in "peeling" and "pasting" phases. The peeling phase (Steps 3–5) starts with a subgroup (box) containing the entire dataset and gradually shrinks it. Pasting (Step 6) works similarly, but in the opposite direction: It expands the candidate subgroup without reducing the target statistic. To find many subgroups, the covering procedure [7,8] (Step 7) is used, i.e. the peeling and pasting steps are repeated on data subsets that are not part of the previously found boxes.

SuRF. Algorithm 2 is the SuRF SD method [12]. It takes a training dataset, the threshold y_R for the target statistic y, and an auxiliary dataset D^{aux} containing the target statistic values from previous evaluations of different subgroups. If D^{aux} does not exist, it is created by evaluating random subgroups (Step 3). This auxiliary dataset allows fitting a surrogate model f (Step 4) that estimates

---

Algorithm 2. SuRF

1: **Input:** Training data, threshold y_R, parameters of the optimization procedure, optional: D^{aux} the data set of the historical evaluation of different subgroups.
2: **Output:** Subgroups where the target statistic exceeds y_R.
3: If not part of the input, create D^{aux} by computing the target statistic over the training data for random subgroups.
4: Train an ML model f using D^{aux} to the target statistic for any candidate subgroup.
5: Randomly create a starting set of candidate subgroups.
6: Evolve the candidate subgroups using a glowworm swarm optimization algorithm: Iteratively refine the identification of regions to increase the value of the target statistic. To speed up computation, use predictions of f instead of computing the target statistic from training data.
7: Repeat Step 6 a predefined number of times.
8: Return all subgroups where the target statistic exceeds y_R.

Algorithm 3. DivExplorer

1: **Input:** Training data with categorical attributes, support threshold s.
2: **Output:** Subgroups with support greater than s.
3: Define all possible subgroups. A subgroup description can include either a single value or all values of any attribute X_i.
4: Drop subgroups that have too few training data points (less than the share s).
5: Return the retained subgroups with their target statistic values.

---

the true statistic y over candidate subgroups during the search to speed up the computation. SuRF uses glowworm swarm optimization to identify regions in the dataset with $y > y_R$ (Step 6). It is unable to handle categorical attributes.

DivExplorer. Algorithm 3 is DivExplorer [11]. It works only with categorical data. Given a data set with categorical attributes, it exhaustively explores all subgroups in its search space and returns subgroups that contain more points than the pre-specified share along with the value of a target statistic. The subgroup description can include either all levels of a categorical attribute or a single level, in total $(|X_i| + 1)$ possibilities, where $|X_i|$ is the cardinality of attribute X_i. Thus, for a data set with categorical attributes $X_1, \ldots, X_n$, DivExplorer's search space has $(|X_1| + 1) \times \cdots \times (|X_n| + 1)$ elements.

In comparison, PRIM does not perform an exhaustive evaluation of all subgroups, but its search space is richer. For categorical attributes, PRIM allows any combination of their values. Let $n = |X_1| + \cdots + |X_n|$. The search space of PRIM then has 2^n elements. At iteration i, PRIM evaluates $n - i + 1$ candidate subgroups and thus does a maximum of $n(n+1)/2 - 1$ iterations for a subgroup.

3 PRIM Vs. SuRF

In this section, we reproduce the experiments from [12] that have proposed SuRF. We present the experimental setup and then report and discuss the results.

3.1 Experimental Setup

This section describes the datasets involved, the parameterizations of SuRF and PRIM, and the quality metric used. We largely reuse the experimental setup of SuRF[2] in Python, but use another PRIM implementation that allows for more control. We run the experiments on a Windows machine with AMD Ryzen 7 PRO 5850U with Radeon Graphics 1.90 GHz processor and 16 GB RAM.

Datasets. Using the source code of the SuRF experiments, we generated 20 synthetic datasets as described in [12]. Each dataset contains several ground truth (GT) regions — hyper-rectangles (boxes) that constrain a region in all dimensions. Each dataset corresponds to a variation of the following parameters: the number of GT regions $k \in \{1, 3\}$; the number of dimensions $d \in \{1, 2, 3, 4, 5\}$; and the type of the target statistic y, either (i) "density", which represents the number of data points in a region divided by its volume, or (ii) "aggregate", the average value of a target variable of the data points in a region.

Metrics. We do not change the procedure of [12] for the evaluation. An SD method should find subgroups corresponding to all ground truth regions in the data. Intersection over Union (IoU) measures the similarity between the regions identified by SuRF or PRIM and the ground truth regions. Specifically, we evaluate the IoU for each identified subgroup and the most similar GT region and choose the maximum IoU. Although we believe that taking the maximum is generally questionable, it is relatively fair for small numbers of boxes returned by a method (20 for SuRF, 1–6 for PRIM). Keeping it allows for easy comparison of our results with [12]. In the case of multiple ground truth regions, we compute the IoU for each identified region and the ground truth region with the highest overlap, and average the resulting IoU values.

SuRF. We did not modify the SuRF related code for the experiments. SuRF needs an auxiliary dataset D^{aux} of past evaluations of the target statistic y for different subgroups to train a model f, see Algorithm 2, which is XGBoost [4]. D^{aux} contains target statistic values for randomly generated subgroups, with center points and side lengths ranging from 1% to 15% of the data domain. As the number of dimensions increases, the number of these past evaluations also increases from 300 to 300K

PRIM. We use the PRIM implementation from the R package "prim"[3] unless otherwise noted. For the aggregate statistic, we use the same threshold $y_R = 2$ as in [12] and set the minimum support parameter to the volume of a GT region. For the density statistic, we set y_R slightly below the density of points within the GT region(s). This design makes it easier to automate the experiments. As we will explain in Sect. 3.2, it does not affect the usability of PRIM in real scenarios where the information about the GT regions is not known.

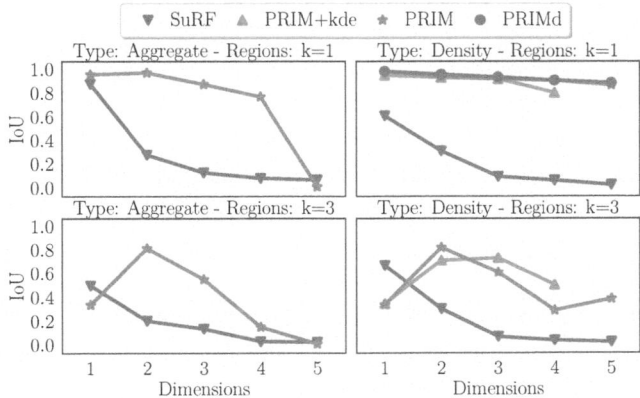

Fig. 1. Average IoU for different k and statistics.

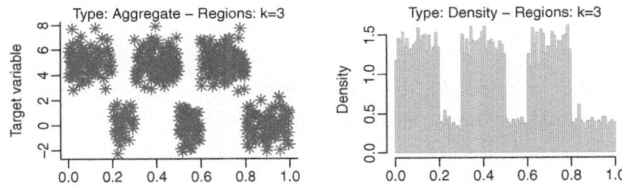

Fig. 2. 1D dataset with $k = 3$ and different statistics.

3.2   Results

First, we evaluate the quality of SuRF and PRIM for the "aggregate" and "density" target statistics. To handle the "density" statistic, one can either augment the data with an auxiliary target attribute and use the existing PRIM implementation that works with "aggregate", or edit the target statistic in the implementation of PRIM; we show both variants. We then compare the runtime of the algorithms. Finally, we highlight the added benefit of PRIM's interactivity.

Average. The left plots in Fig. 1 show the IoU of boxes found by SuRF and PRIM as a function of the number of dimensions d and the number of GT regions $k \in \{1, 3\}$. For $k = 1$ and $d = 1, 2, 3, 4$, the subgroups of PRIM are better than those of SuRF. This result is similar to the one in [12], except that in our experiments PRIM also found a high quality box for $d = 4$, while in [12] the reported quality for $d = 4$ was almost zero. We hypothesize that this is because PRIM had more iterations to explore smaller boxes, whereas in [12], PRIM probably stopped after a fixed number of iterations. For $k = 5$, both methods perform poorly due to the peculiarities of the data: Only a fraction 0.3^5 of 5000 points (i.e. about a dozen points) belong to a GT region, making it nearly impossible to locate this region.

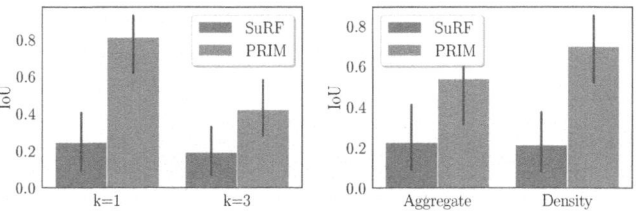

Fig. 3. Average IoU by the number of GT regions k (left); or target statistic (right).

Algorithm 4. Adaptation of PRIM to "density", variant 1

1: Create a new artificial data set of size 10^5, uniformly sampled at random from the domain of the original training data.

2: Label these points with the kernel density estimated from the original data or the negative distance to the first nearest neighbor in the original data.

3: Apply PRIM to this artificial data set.

For $k = 3$ and $d = \{2, 3, 4\}$, PRIM significantly outperforms SuRF, contradicting the consistently worse results reported in [12]. For $d = 5$ the poor result quality is due to the peculiarities of the data, with 1-2 points in each GT region. For $d = 1$ SuRF performs slightly better than PRIM. However, we find this case less important than for higher dimensions, especially $d \geq 3$. This is because for $d = 1$ one can visually locate interesting regions, see Fig. 2. We hypothesize that the new results favor PRIM because we used an implementation that includes the covering procedure (Algorithm 1, Step 7) to find multiple regions.

Density, Variant 1. A simple workaround to enable PRIM with the "aggregate" statistic to handle "density" is to create artificial data with a target attribute that reflects local density. We experiment with kernel density estimates (PRIM+kde) and negated nearest neighbor distance (PRIM) as such local density proxies. Algorithm 4 describes the procedure.

The right plots in Fig. 1 show the result. PRIMd is a modification that we will explain shortly. We do not use PRIM+kde for $d = 5$ because in this case kde takes more than 10 min to fit and the nearest neighbor distance gives quite accurate results. Except for the case $k = 3, d = 1$, the boxes found by PRIM are several times or even orders of magnitude better than those found by SuRF. This contrasts with the results of [12], which reports zero quality for PRIM. As before, we assume that the inferior quality of the PRIM result compared to SuRF at $k = 3, d = 1$ is not a problem, since one can detect interesting regions in the one-dimensional case by plotting the data, see Fig. 2 (right).

As in [12], Fig. 3 summarizes the performance of the algorithms as a function of the value of k or the type of statistic. In all cases, PRIM outperforms SuRF. It produces 2–4 times better results.

Density, Variant 2. [12] claims that PRIM does not support "density" statistics, but this limitation is in existing open source implementations, not in PRIM itself. These implementations iteratively remove fractions of points to

Table 1. Execution time in seconds.

Dimensions	SuRF					PRIMd				
	1	2	3	4	5	1	2	3	4	5
$N = 10^5$	2.52	2.28	2.68	2.85	2.84	0.01	0.03	0.04	0.05	0.06
$N = 10^6$	2.59	2.31	2.51	2.95	2.66	0.13	0.38	0.47	0.58	0.76
$N = 10^7$	2.66	2.33	2.43	2.81	2.85	1.55	4.02	5.45	6.64	8.79

maximize the mean of the target variable within the remaining box. This mean statistic can be replaced by any other [7], such as the density of the points. We have implemented a version of PRIM, PRIMd, to maximize the density of points within the box. In addition, PRIMd cuts off a predefined fraction of the box volume at each iteration, rather than a fraction of the points. We apply this solution to the case $k = 1$. Extending PRIMd to cases with larger k requires implementing a covering procedure, which is beyond the scope of this study. The plot in the upper right corner of Fig. 1 shows that PRIMd finds the boxes with almost perfect quality. This is remarkable since the volume of the GT region at $d = 5$ is only $0.3^5 = 0.243\%$ of the original volume.

Runtime. Table 1 compares the runtimes of SuRF and PRIM for datasets of different size (N) and dimensionality, following the procedure of [12]. For a fair comparison, we used PRIMd since it is implemented in Python like SuRF. PRIMd is an order of magnitude faster than SuRF for $N = 10^5$ and has comparable runtimes for $N = 10^7$. This comparison does not take into account the time SuRF spends building the auxiliary dataset D^{aux} of past subgroup evaluations and training the ML model with it.

One can get an idea of the relative time complexity of building the auxiliary dataset D^{aux} as follows. PRIMd runs 120 iterations, querying the data $2d$ times in each iteration, for a total of $240d$ times. In [12] and in our experiments, SuRF queries the database from 300 times at $d = 1$ to 300K times at $d = 5$ to build a database of past evaluations. I.e., for five-dimensional data, the time SuRF spends building a database is 250 times longer than the runtime of PRIM.

The time SuRF takes to train an ML model depends on the dimensionality of the training data and the size of D^{aux}. In our experiments, it took from 3–4 s for $d = 1$ and $|D^{aux}| = 300$ to 257 min for $d = 5$ and $|D^{aux}| = 300$K.

Additionally, as we will observe next, SuRF often requires multiple runs to find good subgroups, while PRIM finds each box in one pass. This suggests that SuRF can be much more time-consuming than PRIM in realistic scenarios.

Interactivity of PRIM. In the experiments so far, we used the setup of [12], setting the threshold to a target statistic close to its value within the GT regions. For automation, we matched the minimum box size in PRIM to the ground truth box size. Typically, good priors for these parameters are unknown and must be discovered experimentally. While SuRF requires multiple runs with different parameterizations, PRIM can determine these values in one pass. By iteratively applying the peeling procedure, PRIM constructs nested boxes that differ by a

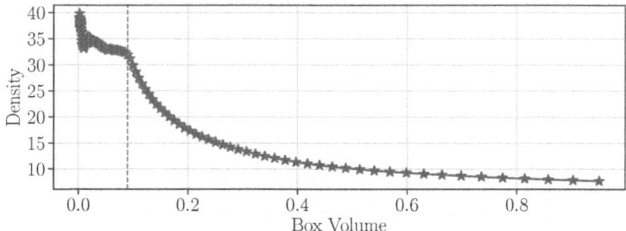

Fig. 4. Densities of boxes produces by PRIM in one pass contingent on their volumes.

single side. One can plot the statistic of interest against the box size and use the plot to infer the best parameter values and select the corresponding box. For example, Fig. 4 shows density versus box volume for PRIMd in a 2D data set with $k = 1$. The density plateaus at a box volume of about 0.09 (red dashed line), which corresponds to the ground truth region. This box can be selected as the PRIM output without re-running with different parameters.

4   PRIM Vs DivExplorer

In the previous section, we reproduced the experiments from [12] using synthetic datasets with continuous features. We will now look at a real dataset with categorical features and compare PRIM and DivExplorer. We first describe the experimental setup and then report and discuss the results.

4.1   Experimental Setup

We follow the experimental design from [11].[4] We describe the dataset and the PRIM configurations we use for the experiments.

Dataset. The COMPAS data [1] used for the case study in [11] includes demographic data and criminal history of defendants. The COMPAS score assesses the likelihood that a defendant will reoffend within two years. It uses a proprietary scoring model with undisclosed inner workings. The article [11] applies the DivExplorer SD algorithm to find the subsets of data where the accuracy (ACC), error rate (ER), false positive rate (FPR), or false negative rate (FNR) of this scoring model is significantly higher than the average of these statistics for the entire dataset. The original COMPAS dataset has four categorical and two continuous features. In [11], the continuous features were discretized before applying DivExplorer. We will use this modified dataset in the following.

PRIM Configurations. For categorical attributes of low cardinality, PRIM cannot control the proportion of points cut at each iteration (see Sect. 2). PRIM tends to favor such attributes because larger cuts often lead to greater potential improvement in the statistic of interest. With large cuts, PRIM would lose its

---

[4] https://github.com/divexplorer/divexplorer.

Table 2. Top subgroups in terms of accuracy (ACC), false positive rate (FPR), false negative rate (FNR), and error rate (ER) for the COMPAS dataset.

Method	Itemset	s	Δ_{ACC}
DivEx.	stay<week, #prior=0, race=Cauc	0.12	0.141
	charge=M, stay<week, #prior=0	0.15	0.133
PRIM	age≥25, race!='Afr-Am', #prior=0	0.16	0.171
	age≥25, #prior=0	0.24	0.143
		s	Δ_{FPR}
DivEx.	age=25-45, #prior>3, race=Afr-Am, sex=Male	0.13	0.22
	age=25-45, #prior>3, race=Afr-Am	0.15	0.211
PRIM	age≤45, race=Afr-Am, sex=Male, #prior>3	0.14	0.232
	age=25-45, race=Afr-Am, sex=Male, #prior>3	0.13	0.22
		s	Δ_{FNR}
DivEx.	age=25-45, stay<week, #prior=0	0.15	0.236
	charge=M, stay<week, #prior=[1,3]	0.10	0.233
PRIM	age≥25, charge=M, race!=Afr-Am, #prior≤3, stay<week	0.13	0.291
	age≥25, charge=M, race!=Afr-Am	0.17	0.242
		s	Δ_{ER}
DivEx.	age<25, stay<week, race=Afr-Am	0.10	0.098
	age<25, stay<week, sex=Male	0.13	0.095
PRIM	age<25, {race!=Asian, Native American, Other}, #prior>0, stay≤3M	0.10	0.123
	age<25, race!=Asian, #prior>0, stay≤3M	0.11	0.121

"patience" and stop after a few iterations, missing subgroups of potentially higher quality. To overcome this problem, the target statistic can be adjusted to give more weight to smaller cuts, as suggested by [7].

Let B be the current subgroup discovered by PRIM and b be the next smaller candidate subgroup. By default, PRIM makes the next cut to maximize the improvement in the target statistic $I(B,b) = y(b)-y(B)$, see Sect. 2. To prioritize smaller cuts, one can replace $I(B,b)$ by $J(B,b) = I(B,b)P(B,b)$, where $P(B,b)$ is some function. We consider two configurations of PRIM with different $P(B,b)$. In the first configuration we use $P(B,b) = \beta_b/(\beta_B - \beta_b)$, proposed in [7], where β_B, β_b are the numbers of points in B and b. For the second configuration, we use $P(B,b) = \beta_b^c$, where c is any non-negative number. We try different values for the parameter c from the set $\{0,0.1,\ldots,1\}$.

4.2   Results

We present the results in a form similar to [11]. Table 2 reports the two best subgroups for each method and different target statistics; s is the share of data in a subgroup. The last column is the increase in the statistic of interest in a subgroup compared to the entire dataset. Both configurations of PRIM discussed performed similarly well, so for brevity we only report the results for the second one. We observe that subgroups found by PRIM have higher values of the statistic of interest and often higher support than those found by DivExplorer. This is because PRIM has richer search space. For example, it finds descriptions like "age$\geq$45", which is a combination of the levels "age $>$ 45" and "age $=$ 25$-$ 45".

The article [11] makes several other contributions beyond proposing the subgroup discovery algorithm. For example, it proposes a method for quantifying and visualizing the role of each condition that defines a subgroup. Our analysis suggests that replacing DivExplorer with PRIM can improve the results. PRIM's ability to handle continuous attributes is particularly promising: It may allow for even more effective subgroup discovery than that just reported.[5]

5   Conclusions

Subgroup discovery is a popular data analysis technique used to identify distinct subsets of data with unusual values of some target statistic. We have compared the recent subgroup discovery algorithms SuRF and DivExplorer with the earlier PRIM algorithm using the experimental setups and datasets from the respective publications. PRIM shows superior efficiency and effectiveness in subgroup discovery. These results highlight the need for a comprehensive benchmark of subgroup discovery algorithms.

Disclosure of Interests. The authors have no competing interests to declare that are relevant to the content of this article.

References

1. Angwin, J., et al.: Machine bias. ProPublica (May 2016)
2. Atzmueller, M.: Subgroup Discovery. WIREs Data Min. Knowl. Discov. **5**, 35–49 (2015)
3. Bryant, B.P., Lempert, R.J.: Thinking inside the box: a participatory, computer-assisted approach to scenario discovery. Technol. Forecast. Soc. Change **77**, 34–49 (2010)
4. Chen, T., Guestrin, C.: XGBoost: a scalable tree boosting system. In: KDD (2016)
5. Dazard, J.E., Rao, J.S.: Local sparse bump hunting. J. Comput. Graph. Stat. **19**(4), 900–929 (2010)
6. Duivesteijn, W., Feelders, A.J., Knobbe, A.: Exceptional Model Mining. Data Min. Knowl. Discov. **30**(1), 47–98 (2015). https://doi.org/10.1007/s10618-015-0403-4

---

[5] Unfortunately, we do not have the dataset with undiscretized attributes to perform these experiments.

7. Friedman, J.H., Fisher, N.I.: Bump hunting in high-dimensional data. Stat. Comput. **9**(2), 123–143 (1999). https://doi.org/10.1023/A:1008894516817
8. Grosskreutz, H., Rüping, S.: On subgroup discovery in numerical domains. Data Min. Knowl. Discov. **19**(2), 210–226 (2009). https://doi.org/10.1007/s10618-009-0136-3
9. Herrera, F., et al.: An overview on subgroup discovery: foundations and applications. Knowl. Inf. Syst. **29**(3) (2011)
10. Leman, D., et al.: Exceptional model mining. In: ECML/PKDD (2008)
11. Pastor, E., et al.: Looking for trouble: analyzing classifier behavior via pattern divergence. In: SIGMOD Conference. ACM (2021)
12. Savva, F., et al.: Surf: identification of interesting data regions with surrogate models. In: ICDE (2020)

Interestingness Measures for Exploratory Data Analysis: a Survey

Alexandre Chanson[1] , Nicolas Labroche[1] , Patrick Marcel[2(✉)] ,
Verónika Perlata[1] , and Panos Vassiliadis[3]

[1] LIFAT - University of Tours, Blois, France
[2] LIFO - University of Orléans, Orléans, France
`patrick.marcel@univ-orleans.fr`
[3] University of Ioannina, Ioannina, Greece

Abstract. Exploratory Data Analysis (EDA) is the tedious activity of interactively analyzing a dataset to extract insights. Many approaches aiming at supporting EDA were recently proposed. They all rely on interestingness measures to score the importance of insights. This paper surveys and categorizes the different interestingness measures proposed in the literature for approaches aiming at automating EDA. The lessons learned from this survey allow to point out promising research directions.

Keywords: EDA · Insights · Interestingness measures

1 Introduction

Exploratory Data Analysis (EDA) is the notoriously tedious activity of Data Science consisting of interactively analyzing a dataset to gain insights, for "exposing the unanticipated" [34]. According to De Bie et al. [3] EDA poses the greatest challenges for automation, since background knowledge and human judgment are the keys to success. EDA is close to discovery-driven analysis [25–28] that guides the exploration of a datacube by providing users with interestingness values for measuring the peculiarity of the cells in a data cube, with the use of statistical models, e.g., based on the maximum entropy principle, and leveraging the intrinsic structure of multidimensional information. As we will see, EDA does not adhere to the multidimensional model, and the measures proposed go beyond the peculiarities of cube cells.

Recently many approaches were proposed to support EDA, including approaches to automatically generate EDA sessions, often defined as maximization problems (see, e.g., [5,10,33,36]). At the heart of each approach is the quantification of the importance of an *insight*, i.e., a piece of valuable information, for the user analyst, by means of one or more *interestingness measures*. While interestingness measures have been reviewed in several domains (see e.g., [6,13] for pattern mining or [17] for recommender systems), a survey and organization of interestingness measures tailored for EDA has not been done yet.

© The Author(s), under exclusive license to Springer Nature Switzerland AG 2025
J. Tekli et al. (Eds.): ADBIS 2024, CCIS 2186, pp. 14–24, 2025.
https://doi.org/10.1007/978-3-031-70421-5_2

This paper fills this gap. We review the measures proposed for EDA, and propose a classification using 6 dimensions from the literature (see e.g., [15]). This paper appeals to various readers and needs, including: the analyst interested in finding an off-the-shelf EDA system, the researcher looking to devise new EDA support approaches, or the system designer willing to combine measures from different approaches.

The paper is organized as follows. Section 2 details the organization of the survey. Section 3 reviews the measures proposed. Section 4 discusses combinations of interestingness measures. Section 5 discusses lessons learned and perspectives.

2   Categorization of Interestingness Measures

This section explains how we organize the survey. We define what insights are and explain that the interestingness of insights is inherently multidimensional.

2.1   What Are Insights?

Insights are properties or patterns of a subset of a dataset that signify the presence of an interesting relationship among the data participating to the insight. So, practically, an insight is:

- *scoped* by a set of data, typically a subset of a dataset,
- *defined* by the existence of a pattern, or property of the data (e.g., the existence of a peak in a distribution for a particular time point; the existence of seasonality, or the increase along a period of time, or drop in an otherwise steady series, in a time series),
- *computed* via an appropriate algorithm that verifies the presence or absence of the related property,
- *quantified* via a score that measures the degree of the presence of the pattern in the data (e.g., the support of an association rule, the Kendall τ score of the correlation of two measures, etc.).

To detect insights, query mechanisms are often used to isolate potential data subsets that are either (a) evaluated on their own, or (b) contrasted to each other for the fulfillment of the insight's defining property. Table 1 lists the insight data subset generation mechanisms most frequently found in the recent literature:

- The Group-by/filter form is the result of an aggregate query over an (often multidimensional) dataset. It was popularized with discovery-driven exploration of datacubes.
- The Sibling group form corresponds to the data of a one-dimensional slice in a multidimensional space.
- The Comparison form corresponds to two series of numbers being compared.

Examples of insights are: a rising trend in yearly sales [33], a factor being relevant to the difference on a given disease between two locations [19], a month

Table 1. Some popular insight forms

Insight forms	Salient contributions
Group-by / filter	[10, 12, 14, 26, 35]
Sibling groups	[8, 18, 33]
Comparison	[5, 11, 30, 37]

having minimum sales for some location [18]. We note that insights can be spurious, i.e., resulting from random data and a particular aggregation [37]. Therefore many works insist that insights should be statistically significant [8, 33, 37].

While any piece of data can be an insight, practically, the presence of an insight in a subset of the data, because of the existence of a pattern in these data, separates them from the rest of the dataset as interesting, or at least, potentially interesting for the analyst. What comes out as interesting for an analyst is, however, not immediately obvious. In general, the interestingness of an insight can be quantified via **an interestingness score**. Again, the semantics behind the interestingness score can be diverse; in the sequel, we try to organize these semantics along a principled framework.

2.2 Interestingness is a Multidimensional Notion

Two main approaches are used to capture the insights' interestingness: (i) the definition of heuristic measures and (ii) machine learning. Many heuristic measures were proposed, each capturing a different facet of the broad concept. However, as reported in [21], there is no single measure that consistently outperforms the others, interestingness being often subjective and changing dynamically [32]. This is why some works resort to machine learning (e.g., [10]) to dynamically select interestingness measures (and often combine them) or to model the users' interest with active-learning or learning-to-rank techniques. **We deliberately focus on heuristic definitions because they help understanding the nature of interestingness** in many ways – most importantly, as they are able to explain why a particular insight is proposed to the user.

Patil et al. [22] propose to evaluate EDA approaches using 3 categories of metrics: human, system and data.

- Human: quantitative and qualitative measures to evaluate user satisfaction (through questionnaires, tracking, etc.)
- System: measures evaluating the resources (memory, latency) consumed by the system. TPC benchmarks abound with this type of measures.
- Data: measures proposed to qualify an interesting property or pattern for a subset of the data in a dataset, often called insight, highlights, findings, discoveries, data facts, etc. [10, 14, 21, 26, 35].

In this Work We Focus on the Third One, Specific to EDA. Earlier works addressed the classification of these criteria [13, 15, 20], in particular contexts (pattern mining, data cube exploration) without actually reviewing and analyzing the measures proposed. We start by explaining what interestingness dimensions are.

We adopt a multidimensional view point, and propose 6 dimensions for characterising insight interestingness: *peculiarity, novelty, relevance, surprise, diversity* and *presentation* (they are defined in next paragraphs). These dimensions are inspired by the seminal work for cube exploration [13], where the authors review interestingness measures for results of OLAP queries, and by recent works [15,20] reworking such classification, and proposing interestingness aspects for datacubes, grounded by human behavior studies[1].

These dimensions are orthogonal, and have the advantage of clearly indicating what is needed to compute interestingness. We describe them hereafter, providing the signatures of the functions implementing their evaluation, and highlighting what is contrasted to generate interestingness.

- *Peculiarity*(i, D): The peculiarity of insight i indicates whether data of i is different and not in accordance to other data. An insight is contrasted to other data for commonalities or differences. Therefore, peculiarity depends on the dataset D where i comes from.
- *Novelty*(i, H): The novelty of novelty of insight i indicates whether i is new and previously unseen. Thus, an insight is contrasted to a user's history, and novelty depends on the history H of data seen before i.
- *Relevance*(i, g): The relevance of insight i indicates whether i is related to the overall analysis intention of the user, expressed in the user's exploration goal. Therefore, relevance depends on the user's goal g.
- *Surprise*(i, b): The surprise of insight i indicates whether i contradicts and revises the user's previous beliefs. Therefore, surprise depends on the belief b of the user.
- *Diversity*(i, C): The diversity of insight i indicates whether i covers various classes of the underlying data. Such classes may represent the user's targeted groups where a fair coverage is desirable (e.g., the values of a sensitive attribute like gender). Therefore, diversity depends on the coverage of user's targeted classes C;
- *Presentation*(i): The presentation of insight i indicates the difficulty for understanding i. This includes (but is not limited to its conciseness). Therefore, presentation depends on i itself.

Papers Selection. The papers reviewed in this survey were chosen based on the following considerations:

- we focus on approaches automating EDA proposed by the data management community;
- note that EDA approaches were already surveyed in the data management community [16]. The focus was slightly different (how to store and access data, how to interact with a data system to enable users and applications to

---

[1] [13] proposes *peculiarity, surprise, diversity* and *presentation* (some of them with different names) and [15,20] propose *relevance, novelty, peculiarity* and *surprise*.

quickly figure out which data parts are of interest). We mostly chose papers posterior to that survey, since they attach more importance to notions of interestingness and insights;

- some less recent but influential papers were included nonetheless (e.g., [7]) if they are key to understand important concepts;
- we chose papers pertaining to the most popular form of insights and pertaining to heuristic definitions of interestingness (see Sects. 2.1 and 2.2).

3   The Variety of Interestingness Measures for EDA

We review the measures proposed, according to the dimensions introduced in the previous section. For each dimension, we identify refinements based on the semantics of the measures proposed. We also indicate the importance of the dimension in helping building EDA explorations.

3.1   Peculiarity

Peculiarity allows to **quantify the importance of an insight among its peer data by evaluating how deviant, or, common the data of the insight are compared to the rest of the dataset.** The measures defined in this dimension concern either (i) the **outlierness**, or (ii) the **typicality** of the insights. Using this dimension, an analyst, or a recommendation system, can **steer the exploration to phenomena** (trends, outliers, etc.) or to **better represent the dataset.**

Outlierness. The *outlierness* of an insight quantifies its interestingness based on its difference with a broader set of data to which it is contrasted. Sintos et al. [31] measure the extent of the incorrectness of a value in a dataset (practically measuring the amount of false information of two values before and after a data cleaning procedure). Gkitsakis et al. [15] compute the outlierness of a newly posed cube query by aggregating the distances between the data of the query and past data retrieved. Many approaches consider the distribution of data [1,5,8,12, 33,37]. In [12], outlierness is measured using the difference in z-scores of the data obtained in two consecutive exploration steps. A recent trend is to turn insights into hypothesis testing [5,8,33,37], which allows to: (i) use the p-value for the insight significance, (ii) define false discoveries (type-1 errors, e.g., visualizations supporting a non-significant insight) and false omissions (type-2 errors, e.g., visualizations not supporting a significant insight), (iii) define credibility (e.g., percentage of visualizations supporting an insight). However, since the risk of type-1 error increases as more than one hypothesis are considered at once, a correction is needed to ensure reporting only non-spurious insights [37].

Typicality. Measuring the *typicality* of the insight consists of quantifying to what extent the subject of an insight can represent the *entire dataset* [8,18,33]. In most cases, anti-monotonic conditions are checked to prune insights. For instance: if the subject of insight A is a superset of the subject of insight B, then the impact of A should be no less than the impact of B. The *market share* measure used in [8,33] is defined as the ratio between the sum of values of the insights and the sum of all data.

3.2   Novelty

Novelty characterizes insights in terms of **being new observations** (or operations). Using this dimension allows to **make the exploration go further** or **make it focused**.

In its simplest expression, novelty is measured as a Boolean indicating whether some data have already been seen [12]. In [15], novelty is computed as the fraction of new data brought by a current query compared to the data retrieved by that query and the previous ones, either per se, or in different degrees of granularity. In [23], curiosity is measured as a function of the number of times a result is encountered (being inversely proportional to it).

3.3   Relevance

This dimension characterizes insights in terms of **fulfilling a user's goal** or being *familiar and coherent* to the user. The measures defined in this dimension concern (i) **goal fulfillment**, (ii) **familiarity** or (iii) **coherency**. Using this dimension allows to make explorations **connected to the analyst's interests**.

Goal Fulfillment. Gkitsakis et al. [15] distinguish two ways goals are declared: (a) explicit, directly stated by the user under the form of selection predicates over the dataset, or, (b) implicit, i.e., goal is approximated and estimated by the system. In case (a), the relevance of data computed by a query is measured as the fraction of data from the dataset it covers (i.e., the data used by the query) that overlaps with the user's goal. In case (b), the goal is inferred from the user's history (queries sent in the past), and relevance is measured as in case (a). The basic idea of the approach is openness: any other means of deriving a goal for the analyst can be plugged into the mechanisms, while retaining the fundamental essence of a goal, which is coarsely speaking, a "fence" that isolates the relevant subset of the data space (within the exploration goal) from the irrelevant one.

Familiarity. In [23], a familiarity measure is defined as the concentration ratio of target data in a set. It is implemented as a variant of the Jaccard index between data encountered during the exploration and a given target set of familiar data. This measure is expected to increase as the EDA session goes on, to avoid over-exploiting a set of familiar objects.

Coherency. The *coherency* of an insight contrasts the insight with other insights obtained in the *exploration session*, to check whether a given EDA operation is coherent at a certain point. For instance, in [10] heuristic classification rules are used to express general properties on the input dataset semantics (e.g., if the user focuses on flight delays, aggregating on the "departure-delay time" column is preferred). Some other works express coherency as a distance between exploration actions (separate from their definition of interestingness) and measure how coherent a sequence of actions is as a whole. For instance, in [5] a weighted Hamming distance of relational query parts is used.

3.4 Surprise

This dimension allows to characterize insights in terms of **how distant they are from the user's expectations**. The measures defined in this dimension express a **distance to expectations**. Using this dimension allows to **steer explorations to data showing unexpected values**.

A formal framework for defining measures of surprise has been introduced by De Bie for exploratory data mining [7]. Using an information-theoretic approach, the framework consists of quantifying the interactive exchange of information between data and user, accounting for the *user's prior belief state*. Approximating the belief that the user would attach to the result being expected is modeled as a background distribution, namely, a probability measure over the exploration results. This background distribution is updated after each result is presented to the user. Chanson et al. [4], propose a way to measure subjective interestingness for exploratory OLAP, inspired by De Bie's work [7]. The user belief is inferred based on the user's past interactions over a data cube, the cube schema and the other users' past activities. This belief is expressed by a probability distribution over all the query parts potentially accessible to the user. Surprise is then measured as in De Bie's work.

In the seminal work of Sarawagi [26], belief (i.e., expected values) is computed using maximum entropy principle, and Kullback-Leibler divergence is used to measure surprise. Gkitsakis et al. [15] distinguish two ways to account for beliefs: (a) expected values are provided by the user, or, (b) expectations are registered by annotating the expectation for a value to appear via a probability of appearance. In case (a), the surprise is measured using a distance function between actual data and expected data. In case (b), surprise for a given value is measured as the sum of the probabilities of all values that are different.

3.5 Diversity

Diversity characterizes insights in terms of **their coverage of population classes**. Using this dimension allows to **make the exploration more representative of the underlying dataset**.

Simple versions of diversity measures have been proposed. In [10], a diversity measure is introduced to encourage the analysis of different parts of the dataset. It is computed as the minimal Euclidean distance between the current observation and all the previous displays obtained. Francia et al. [12] also measure diversity[2] as the proportion of values that have not been seen frequently, presented in models (e.g., clustering) extracted from the insight. In [36], diversity

---

[2] Called surprise in [12], but reclassified here since it does not refer to a user's belief.

is measured as the pairwise difference between insights. In [24] the authors use a pairwise Jacquard similarity to measure diversity within their sub tables.

3.6 Presentation

This dimension characterizes either **how compact** the insight is when presented to a user, or **the amount if information** the insight displays. The measures defined in this dimension concern either (i) the **compactness**, or (ii) the **descriptional complexity** of the insights. Using this dimension allows to **favor insights being both informative and easy to understand**.

Table 2. What interestingness dimensions are combined (left part) and how they are combined (right part)

Contribution	Rel.	Nov.	Pec.	Sur.	Div.	Pre.	ratio	product	[weighted] sum
ATENA [9, 10]	✓		✓		✓	✓			✓
B.I.lief [4]			✓						
Calliope [29]			✓					✓	
Cube Query Int. [15]	✓	✓	✓	✓					
DataShot [35]			✓						✓
Describe [12]		✓	✓		✓				✓
DORA [23]	✓	✓							✓
EDA4Sum [36]		✓	✓		✓				✓
Forsied [7]				✓		✓	✓		
Metainsight [18]			✓			✓		✓	
Quickinsights [8]			✓					✓	
SubTab [24]			✓		✓				✓
TAP-Comparisons [5]			✓			✓		✓	
Top-k insights [33]			✓					✓	

Descriptional Complexity. Descriptional complexity measures how complex it is for a human to assimilate an insight [7]. For instance, the complexity of a set of values can be the number of elements in a set.

Conciseness. Conciseness measures how compact is an insight. For instance, when presenting aggregated results over a set of tuples, the ratio of tuples to groups or a function thereof can be used as a rough estimate of the chosen groups ability to summarize large quantity of information (tuples) [5, 10]. Conciseness can also be defined as a measure of entropy of the insight, acting for a proxy to the human effort necessary for its assimilation [18]. In this later form it also fits the definition of *descriptional complexity* of [7].

4   Combining Interestingness Dimensions

This section shows how interestingness dimensions are combined. Usually, insights are scored based on more than one dimension, to account for goal, history, or belief, or combinations thereof. Table 2 (left) indicates which dimensions of interestingness are commonly used together. Peculiarity is the most frequent dimensions used. Noticeably, there is no consensual approach as how dimensions are combined. For instance, a ratio is used in [7], a weighted sum is used in [10], and a product is used in [5]. This is summarized in Table 2 (right).

5   Conclusion

This paper surveys interestingness measures proposed to support Exploratory Data Analysis. The main lesson learned is that **no definitive measures or combinations of interestingness dimensions** have already been proposed. Some dimensions, like **peculiarity, attracted lots of attention** while others, like diversity, relevance, or surprise, that confront insights with the user's goals or beliefs, much less so.

This survey opens several research directions:

- development of new interestingness measures: the analyst is at the center of the data exploration activity, and measures tailored for personalized or collaborative EDA [2] are still to be proposed,
- formalizing the desirable properties of interestingness measures: in the spirit of what was done for pattern mining [13], the properties of interestingness measures will provide a fine understanding of how measures should be combined,
- contextualizing interestingness dimensions: a typology of EDA sessions is yet to be done. This will enable the characterization of what interestingness measures are required at what step of a given type of EDA session.

References

1. Abuzaid, F., Kraft, P., et al.: DIFF: a relational interface for large-scale data explanation. VLDB J. **30**(1), 45–70 (2021)
2. Amer-Yahia, S., Marcel, P., et al.: Data narration for the people: challenges and opportunities. In: EDBT, pp. 855–858. OpenProceedings.org (2023)
3. Bie, T.D., Raedt, L.D., et al.: Automating data science. Commun. ACM **65**(3), 76–87 (2022)
4. Chanson, A., Crulis, B., et al.: Profiling user belief in BI exploration for measuring subjective interestingness. In: DOLAP, CEUR Proceedings, vol. 2324 (2019)
5. Chanson, A., Labroche, N., et al.: Automatic generation of comparison notebooks for interactive data exploration. In: EDBT, pp. 2:274–2:284 (2022)
6. Dadvar, V., Golab, L., et al.: Exploring data using patterns: a survey. Inf. Syst. **108**, 101985 (2022)

7. De Bie, T.: Subjective interestingness in exploratory data mining. IDA **8207**, 19–31 (2013)
8. Ding, R., Han, S., et al.: QuickInsights: quick and automatic discovery of insights from multi-dimensional data. In: Proceedings of SIGMOD, pp. 317–332 (2019)
9. El, O.B., Milo, T., et al.: ATENA: an autonomous system for data exploration based on deep reinforcement learning. In: CIKM, pp. 2873–2876 (2019)
10. El, O.B., Milo, T., et al.: Automatically generating data exploration sessions using deep reinforcement learning. In: SIGMOD, pp. 1527–1537 (2020)
11. Francia, M., Golfarelli, M., et al.: Assess queries for interactive analysis of data cubes. In: EDBT (2021)
12. Francia, M., Marcel, P., et al.: Enhancing cubes with models to describe multidimensional data. Inf. Syst. Front. **24**(1) (2021)
13. Geng, L., Hamilton, H.J.: Interestingness measures for data mining: a survey. ACM Comput. Surv. **38**(3), 9 (2006)
14. Gkesoulis, D., Vassiliadis, P., et al.: CineCubes: aiding data workers gain insights from OLAP queries. Inf. Syst. **53**, 60–86 (2015)
15. Gkitsakis, D., Kaloudis, S., et al.: Cube query interestingness: novelty, relevance, peculiarity and surprise. Inf. Syst. **123**, 102381 (2024)
16. Idreos, S., Papaemmanouil, O., et al.: Overview of data exploration techniques. In: SIGMOD, pp. 277–281. ACM (2015)
17. Kaminskas, M., Bridge, D.: Diversity, serendipity, novelty, and coverage: a survey and empirical analysis of beyond-accuracy objectives in recommender systems. TiiS **7**(1), 2:1–2:42 (2017)
18. Ma, P., Ding, R., et al.: MetaInsight: automatic discovery of structured knowledge for exploratory data analysis. In: Proceedings of SIGMOD, pp. 1262–1274 (2021)
19. Ma, P., Ding, R., et al.: XInsight: explainable data analysis through the lens of causality. Proc. ACM Manag. Data **1**(2) (2023)
20. Marcel, P., Peralta, V., Vassiliadis, P.: A framework for learning cell interestingness from cube explorations. In: Welzer, T., Eder, J., Podgorelec, V., Kamišalić Latifić, A. (eds.) ADBIS 2019. LNCS, vol. 11695, pp. 425–440. Springer, Cham (2019). https://doi.org/10.1007/978-3-030-28730-6_26
21. Milo, T., Somech, A.: Automating exploratory data analysis via machine learning: an overview. In: SIGMOD (2020)
22. Patil, Y., Amer-Yahia, S., et al.: Designing the evaluation of operator-enabled interactive data exploration in VALIDE. In: HILDA@SIGMOD, pp. 4:1–4:7 (2022)
23. Personnaz, A., Amer-Yahia, S., et al.: DORA THE EXPLORER: exploring very large data with interactive deep reinforcement learning. In: CIKM (2021)
24. Razmadze, K., Amsterdamer, Y., et al.: SubTab: data exploration with informative sub-tables. In: SIGMOD, pp. 2369–2372 (2022)
25. Sarawagi, S.: Explaining differences in multidimensional aggregates. In: Proceedings VLDB, pp. 42–53 (1999)
26. Sarawagi, S.: User-adaptive exploration of multidimensional data. In: VLDB, pp. 307–316 (2000)
27. Sarawagi, S., Agrawal, R., Megiddo, N.: Discovery-driven exploration of OLAP data cubes. In: Schek, H.-J., Alonso, G., Saltor, F., Ramos, I. (eds.) EDBT 1998. LNCS, vol. 1377, pp. 168–182. Springer, Heidelberg (1998). https://doi.org/10.1007/BFb0100984
28. Sathe, G., Sarawagi, S.: Intelligent rollups in multidimensional OLAP data. In: Proceedings VLDB, pp. 531–540 (2001)
29. Shi, D., Xu, X., et al.: Calliope: automatic visual data story generation from a spreadsheet. TVCG **27**(2), 453–463 (2021)

30. Siddiqui, T., Chaudhuri, S., et al.: COMPARE: accelerating groupwise comparison in relational databases for data analytics. In: VLDB, vol. 14, no. 11, pp. 2419–2431 (2021)
31. Sintos, S., Agarwal, P.K., et al.: Selecting data to clean for fact checking: minimizing uncertainty vs. maximizing surprise. Proc. VLDB Endow. **12**(13), 2408–2421 (2019)
32. Somech, A., Milo, T., et al.: Predicting "what is interesting" by mining interactive-data-analysis session logs. In: EDBT (2019)
33. Tang, B., Han, S., et al.: Extracting top-k insights from multi-dimensional data. In: SIGMOD (2017)
34. Tukey, J.W.: Exploratory Data Analysis. Addison-Wesley (1977)
35. Wang, Y., Sun, Z., et al.: DataShot: automatic generation of fact sheets from tabular data. TVCG **26**(1), 895–905 (2020)
36. Youngmann, B., Amer-Yahia, S., et al.: Guided exploration of data summaries. Proc. VLDB Endow. **15**(9), 1798–1807 (2022)
37. Zgraggen, E., Zhao, Z., et al.: Investigating the effect of the multiple comparisons problem in visual analysis. In: Proceedings of CHI, p. 479 (2018)

A Compact and Efficient Data Structure for Line-Based Processing of Series of Raster Data

Luana Pereira dos Reis[✉] and Daniel S. Kaster

University of Londrina, Londrina, PR 86057-970, Brazil
{luana.pereira.reis,dskaster}@uel.br
https://www.uel.br

Abstract. Efficient management and processing of spatial data, especially in the context of map algebra, have become crucial for researchers. The challenge lies in the effectiveness of map algebra operations on voluminous series of raster data, which are essential for analyzing and extracting valuable insights. Many compact data structures have been developed to support such operations; however, they all suffer from data representation or performance issues. This work presents the Compressed Line Raster (CL-raster), a new compact data structure that performs a line-based compression of raster data. CL-raster stores raster data in a compressed format ready for processing in a line-by-line fashion, allowing fast processing of a series of raster data with reduced memory consumption. Experiments show that our approach is efficient, outperforming the state-of-the-art competitor in terms of processing time and memory requirements up to several times.

Keywords: Raster data series · Compact structure · Map algebra

1 Introduction

Efficient management and processing of spatial data have become increasingly critical as datasets grow in volume, particularly in fields like remote sensing and geographic information systems (GISs) [6,8]. Raster data, organized as multidimensional arrays of cells containing values, are fundamental for representing continuous fields such as temperature or precipitation over geographic regions [12]. Raster series, which consist of sequences of such arrays collected over time intervals [5], pose challenges in terms of storage and processing efficiency, especially for operations like map algebra that involve extensive data manipulation and analysis. Various compact data structures, including T-k^2-raster, have been developed to optimize storage and processing efficiency for both individual raster [2,6,9] and raster series [1,3,11]. However, T-k^2-raster's performance can diminish with longer raster series due to additional memory requirements for specific data structures like snapshots, crucial for its compression scheme.

© The Author(s), under exclusive license to Springer Nature Switzerland AG 2025
J. Tekli et al. (Eds.): ADBIS 2024, CCIS 2186, pp. 25–34, 2025.
https://doi.org/10.1007/978-3-031-70421-5_3

This work presents a new compact data structure for raster series processing called Compressed Line raster (CL-raster). Inspired by the run-length encoding (RLE) technique, CL-raster stores raster data lines without repeated values, preserving spatial and temporal resolution while capturing only distinct raster values, thereby eliminating redundancies. This method takes advantage of geographically close regions often sharing values, enabling efficient line-by-line processing of large raster datasets. This is particularly beneficial for applications such as raster map calculators, geospatial analysis, and data preparation for machine learning. Experimental results demonstrate that CL-raster outperforms T-k^2-raster, offering superior execution time and memory efficiency for typical map algebra operations over raster series. This makes it faster and more memory-efficient than the best T-k^2-raster variations.

2   Related Work

Map algebra operations on raster series, essential for GIS processing, analyzing, and visualizing, include local (cell-by-cell calculations), focal (neighborhood-based functions), and zonal (area-based aggregations) operations [13]. These are crucial in remote sensing, geology, climatology, and agriculture for understanding environmental and geographical changes. Efficiently handling large raster series is vital, and compact data structures enable processing and querying without decompression [7], optimizing storage and enhancing data access efficiency. Early optimization structures for raster data, like quadtrees, were initially used for image data compression and remain foundational for modern compact data structures in raster series [4].

The k^2-tree, a compact structure for representing graphs via adjacency matrices, is based on quadtrees and recursively divides space into quadrants, stopping when a quadrant is filled with 0 s or when cell values are reached [2]. Child nodes are represented by positive bits (indicating the presence of at least one 1 bit) or negative bits (indicating a quadrant filled with 0 bits). The k^2-tree uses two vectors: T (storing all tree bits) and L (storing leaf nodes). However, it has limitations, such as requiring square rasters (necessitating extension and padding with 0 values), binary rasters, and not considering the temporal sequence of raster data, which is crucial for sequential processing. The k^3-tree structure advances by adding a third dimension, useful for temporal rasters or non-binary rasters where the third dimension represents raster values. Despite these improvements, k^3-tree faces challenges like data precision loss due to value conversion for third-dimension indexing and performance issues with extreme outliers in raster data.

The 3D2D-mapping [9] and 4D3D-mapping [3] are compact structures designed for mapping non-binary raster data, with 4D3D-mapping being an evolution for raster series. Their main objective is to map non-binary raster data to a binary format, enabling efficient storage through the k^3-tree structure. These approaches advance the field by allowing raster data to handle continuous values instead of binary ones, enhancing applicability in various domains. Rasters must be read using Z-order to generate the structures, preserving spatial locality. However, 3D2D-mapping did not achieve satisfactory space efficiency, as columns rarely have more than one positive bit, resulting in relatively large mappings. Despite this, it was competitive and sometimes outperformed others in quadrant-specific search operations. The 4D3D-mapping structure shows significant memory space gains for series but has consistent data access times across different datasets, being approximately four times slower than the baseline.

The k^2-raster structure [6] efficiently compresses and indexes non-binary rasters by including spatial indices and a tree hierarchy for compact storage, recursively dividing space, and using differential encoding for additional compression. It uses three vectors: T for tree values, Lmax for maximum quadrant values, and Lmin for minimum values. When values are equal, it stores only the maximum. Despite advantages, k^2-raster's limitations include data precision issues from integer representation, handling non-square rasters, and potential restrictions on map algebra operations. The T-k^2-raster structure, an evolution of k^2-raster, handles temporal raster data using snapshots and logs to improve storage and processing efficiency [11]. Snapshots store a sample of the temporal sequence as a k^2-raster, while logs use differential encoding relative to snapshots. An additional vector indicates quadrant references to the snapshot, with binary values showing uniform differences or quadrant values. The construction process, illustrated in Fig. 1, involves two compression cases: when a quadrant has uniform cell values but is not identical to the snapshot Fig. 1(a), and when all subtractions are identical, allowing only the reference to be maintained in the log Fig. 1(b and c). Despite its space efficiency, T-k^2-raster faces challenges with extensive temporal series and limited memory scenarios, as operations like summation require loading both the snapshot and log into memory.

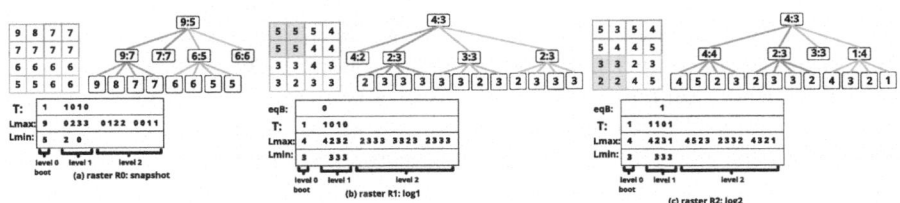

Fig. 1. T-k^2-raster structure construction process.

Raster series processing is often performed line by line in raster calculators to minimize memory usage, as this approach only requires loading the current line from each raster in the series. This strategy is beneficial for many applications since raster datasets are typically massive. However, all the mentioned structures have limitations in handling map algebra operations, particularly in line-by-line processing scenarios. Determining the actual value of a cell requires extensively traversing the tree and performing multiple operations. Furthermore, it's worth noting that these structures require the raster to be square. If the raster is not square, preprocessing the data to fit the structures is necessary, which can be computationally expensive.

3   The Compressed Line Raster Approach

The proposed Compressed Line Raster (CL-raster) performs line-based compression of raster data to enable fast processing and reduced memory consumption. CL-raster stores raster data in a compressed format that facilitates line-by-line processing, accommodating raster time series with varying sizes without needing to decide the series size in advance. Inspired by the RLE technique, CL-raster eliminates sequential repeated values in raster lines, leveraging the tendency of adjacent points to share similar values, thus reducing memory consumption while allowing direct processing of map algebra operations. Unlike related methods limited to binary or positive integer values, CL-raster supports floating-point values, providing greater precision in operations.

3.1   The Structure Organization

CL-raster uses a sparse array for each line in the raster, represented as a list of CL-rasters defined by the tuple $\tau\langle array\_of\_values,\ quantity,\ next\rangle$, where *array_ of_ values* stores distinct values in the raster line, *quantity* is the count of the last value's repetitions, and *next* points to the next CL-raster or is null if it is the last in the sequence. For example, a raster line with values $[1, 1.3, 1.5, 1.5, 1.5, 1.5, 1.5, 1.6, \ldots]$ is stored as $\langle[1,\ 1.3,\ 1.5],\ 5,\ *ptr\rangle$, with subsequent CL-raster starting at 1.6. Figure 2 shows the input raster representation, in Fig. 2(a) is represented the proposed structure, where Fig. 2(b) illustrates that lines l_1, l_2, l_3 contain one CL-raster each, while line l_4 needs two CL-rasters. To reduce memory consumption caused by memory pointers in conventional sparse arrays, CL-raster eliminates pointer chaining, using three vectors for each compressed line: a v vector storing distinct values from the raster line, a bit vector r indicating if each value in v repeats, and a vector n showing the number of repetitions. This approach minimizes space overhead while efficiently compressing and representing raster data.

Figure 2(c) illustrates the CL-raster implementation with three vectors for each raster line. For the first line, vector $v_1 = [9, 8, 7]$ contains non-repeating values, vector $r_1 = [0, 0, 1]$ indicates repetitions with true for 7, and vector $n_1 = [2]$ stores the repetition count for 7. Each line always has three vectors, regardless of the number of CL-rasters. For instance, line l_4 needs two CL-rasters represented by vectors $v_4 = [5, 6]$, $r_4 = [1, 1]$, and $n_4 = [2, 2]$, indicating that 5 and 6 each repeat twice. Vectors v and r are equal in size and no larger than the number of columns in the raster line, while vector n's size matches the number of true values in r. Lines l_2 and l_3 exemplify optimal compression, consisting only of repeated values. The worst-case scenario, with no repeated sequences, results in no compression as vector v stores all elements, using the same memory as the original line, plus minimal overhead from vector r. For example, a 512-column

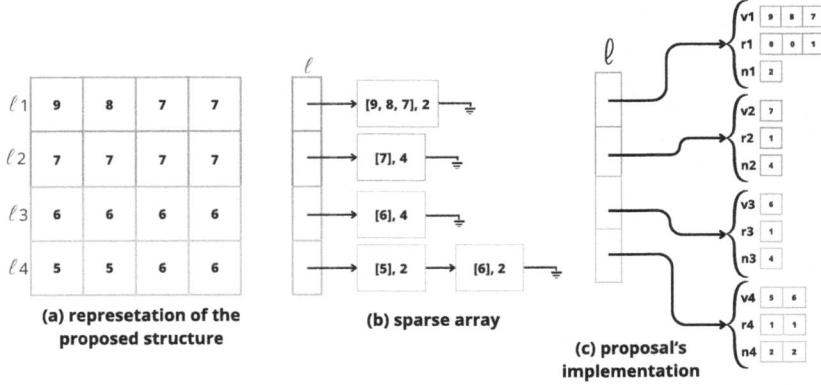

Fig. 2. Compressed raster using a CL-raster structure.

vector of 4 bytes per element (2,048 bytes) would need an additional 64 bytes for the bit vector, a 3% increase. With sequences of only two repetitions, memory use is equivalent to the original raster with the repetition count replacing the repeated value, plus bit vector overhead. Significant compression occurs with three or more contiguous repetitions, greatly reducing memory usage compared to the original raster size (see Sect. 4).

3.2   Algorithms

The encoding process of CL-raster ensures efficient data representation, as detailed in Algorithm 1. For each raster line, the first value is added to the v vector and marked as non-repeating in the r vector. Subsequent values are checked for repetition: if repeated, the count in the n vector is incremented, and if it is the last value, r is updated to true and the count recorded in n. If not repeated, any prior repetition is recorded in r and n, then the new value is added to v and marked as non-repeating in r. This process continues for all values in the line. The main strategy is to optimize raster processing operations, primarily through sequential scanning of line elements. CL-raster employs the iterator design pattern, as shown in Algorithm 2, to iterate over the vectors, returning the value at a specified line position (y) as it appears in the original raster.

The algorithm begins by using variable k to index the v y-line's n vector, which differs in size from vectors v and r Another variable, *iteratorPosition*, tracks the current iterator position in the original raster line, crucial for the algorithm *GetCell*. It iterates over vector v, returning its current value and updating the iterator position. Repetitions in the value are returned sequentially. The algorithm continues to the next v value or concludes if it reaches the end of the vector. This approach minimally impacts operations on v, r, and n when iterating over a raster line compared to an uncompressed vector. Accessing a specific cell $GetCell(x,y)$ and conducting partial scans involve linear time complexity

---

Algorithm 1. CL-raster Encoding

Require: M ▷ The input raster
 1: **for** $i \leftarrow 1$ **to** $num\_lines(M)$ **do**
 2:     $v[i].push\_back(M[i,1])$
 3:     $r[i].push\_back(false)$
 4:     $count\_repetitions \leftarrow 1$
 5:     **for** $j \leftarrow 2$ **to** $num\_columns(M)$ **do**
 6:         **if** $M[i,j] = v[i].back()$ **then**   ▷ Is the current value repeated?
 7:             $count\_repetitions \leftarrow count\_repetitions + 1$
 8:             **if** $j = num\_columns(M)$ **then**   ▷ Is the current value the last one?
 9:                 $r[i].back() \leftarrow true$
10:                 $n[i].push\_back(count\_repetitions)$
11:             **end if**
12:         **else**
13:             **if** $count\_repetitions > 1$ **then**   ▷ Was the previous value repeated?
14:                 $r[i].back() \leftarrow true$
15:                 $n[i].push\_back(count\_repetitions)$
16:                 $count\_repetitions \leftarrow 1$
17:             **end if**
18:             $v[i].push\_back(M[i,j])$ ▷ Adds the non-repeated value
19:             $r[i].push\_back(false)$
20:         **end if**
21:     **end for**
22: **end for**

---

for the first call on line y, with subsequent calls incrementing x operating in constant time due to saved iterator states per line. This method effectively spreads the cost of frequent partial line scans, which are more common than random cell accesses in raster data.

Algorithm 2. CL-raster Line Iterator

Require: v, r, n, y ▷ The CL-raster vectors and the line to scan
 1: $output \leftarrow \emptyset; k \leftarrow 0; iteratorPosition \leftarrow 0$
 2: **for** $i \leftarrow 0$ **to** $|v[y]|$ **do**   ▷ Scan on the cell values
 3:     $iteratorPosition \leftarrow iteratorPosition + 1$
 4:     $output \leftarrow v[y].at(i)$
 5:     **if** $r[y].at(i) = 1$ **then**   ▷ Does the value repeat?
 6:         **for** $n \leftarrow 0$ **to** $|n[y].at(k)|$ **do**   ▷ Add the repetitions to the output
 7:             $iteratorPosition \leftarrow iteratorPosition + 1$
 8:             $output \leftarrow v[y].at(i)$
 9:         **end for**
10:         $k \leftarrow k + 1$
11:     **end if**
12: **end for**
13: **return** $output$

---

4   Experimental Evaluation

The CL-raster structure significantly improves processing and managing spatial raster data, offering efficient handling of large datasets. This evaluation analyzes its effectiveness in map algebra operations using five cases for two raster datasets: CPTEC (Brazilian Center for Climate Study and Weather Forecasting) [10] with a 924 × 1001 grid, and NLDAS-2 (North American Land Data Assimilation System) [14] with a 224 × 464 grid. We compared CL-raster with T-k^2-raster, the state-of-the-art structure for raster series, evaluating encoding time, compression ratio, execution time, and memory consumption. Tests were conducted on a computer with an Intel® Core™ i5-9300H CPU@2.40 GHz × 8, 40 GB of RAM, and a 2.1 TB HDD. Results are detailed in the following sections.

4.1   Encoding Time and Compression Ratio

Figure 3 shows the wall clock time for encoding CL-raster and T-k^2-raster for increasing time series from the two datasets. The construction time increases with the number of rasters for both structures and datasets. However, CL-raster consistently shows a shorter construction time compared to T-k^2-raster. Notably, CL-raster performed significantly better for the CPTEC dataset, requiring only 40% of the time needed for T-k^2-raster encoding. This difference is attributed to the larger size of CPTEC rasters, which have almost nine times more cells than those in NLDAS-2.

Figure 4 compares compression ratios of T-k^2-raster and CL-raster on raster data. T-k^2-raster achieves superior compression by focusing on integers, while CL-raster maintains original real values for precision. In CPTEC data, CL-raster is effective but half as compressive as T-k^2-raster. In NLDAS-2, CL-raster faces challenges with high precision and larger cells, achieving lower compression. T-k^2-raster outperforms CL-raster significantly (87.9% vs. 5.1% compression) due to integer encoding. Preprocessing CL-raster with integer truncation improves compression to 58.7%, nearing T-k^2-raster's efficiency. Balancing compression

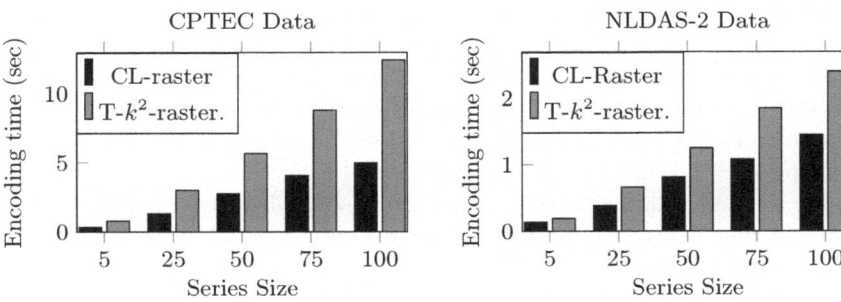

Fig. 3. Encoding time comparison for increasing sizes of series of rasters.

time and ratio is crucial, with higher compression requiring more computational resources.

4.2 Performance of Map Algebra Operations

Typical map algebra operations on raster series involve computing statistics like summations and averages over specific time windows, essential for analytics and machine learning tasks. In evaluating CL-raster's performance, we focused on a representative task of summing raster series and compared it with T-k^2-raster, which employs both cell-based (individual cell retrieval) and window-based (batch retrieval) access methods. Performance metrics, including execution time and memory usage, depicted in Fig. 5, illustrate CL-raster's consistent outperformance of T-k^2-raster across scenarios, as shown on a logarithmic scale. CL-raster was up to 60% faster for CPTEC and 68 s% faster for NLDAS-2 data compared to T-k^2-raster's window-based approach. Moreover, compared to T-k^2-raster's cell-based method, CL-raster showed remarkable speed gains, being up to 35 times faster for CPTEC and 20 times faster for NLDAS-2 datasets. Memory consumption evaluations using Valgrind confirmed CL-raster's efficiency in both execution time and memory usage during typical map algebra operations.

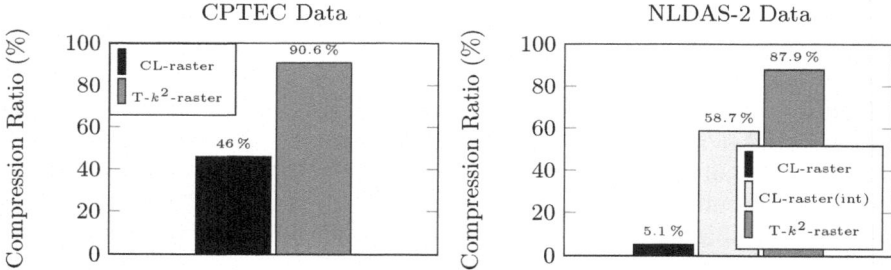

Fig. 4. Compression ratio between the compact structures and the original grid.

Regarding memory efficiency in raster series processing, CL-raster demonstrated superior performance compared to both cell-based and window-based T-k^2-raster variants as depicted in Fig. 5. The window-based T-k^2-raster exhibited faster execution times but required up to 10 times more memory than its cell-based counterpart. In contrast, the cell-based T-k^2-raster, while slower, consumed up to 7 times more memory than CL-raster for CPTEC data and 3 times more for NLDAS-2 data. As series sizes increased, CL-raster showed a modest increase in memory consumption (1.3 to 1.8 times for CPTEC and NLDAS-2, respectively), whereas T-k^2-raster variants experienced significantly larger increases (6.6 to 14.5 times and 4 to 11.3 times, respectively). In summary, CL-raster's evaluation underscores its critical role in efficiently processing and analyzing raster data series, offering a promising solution for managing extensive

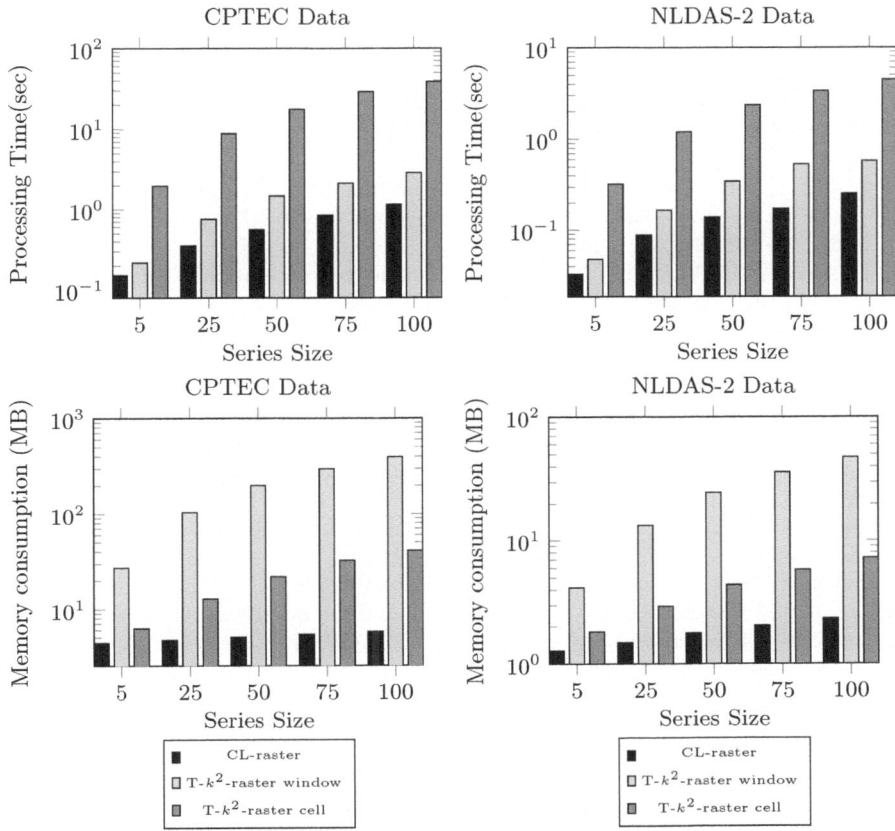

Fig. 5. Performance of executing the sum of a raster series.

geospatial datasets with superior performance in processing time and memory efficiency.

5 Conclusion

This work introduces CL-raster, a compact data structure designed for line-based processing of raster data series. CL-raster enables efficient compression of raster data while facilitating map algebra operations in a memory-efficient manner. Experimental analysis demonstrates its superiority over existing methods in terms of both execution time and memory requirements. CL-raster offers a promising alternative for storing and processing raster data, pushing the boundaries of current technology. Future directions include extending CL-raster to support various map algebra and spatial analysis operations, integrating it into raster calculators for seamless processing of raster series, and enhancing functionalities such as cell and region value retrieval.

Acknowledgements. This work has been supported by the Brazilian funding agencies Araucaria Foundation, CNPq, and CAPES.

References

1. Brisaboa, N.R., Cerdeira-Pena, A., de Bernardo, G., Navarro, G.: Óscar Pedreira: extending general compact querieable representations to GIS applications. Inf. Sci. **506**, 196–216 (2020)
2. Brisaboa, N.R., Ladra, S., Navarro, G.: Compact representation of web graphs with extended functionality. Inf. Syst. **39**, 152–174 (2014)
3. Cruces, N., Seco, D., Guitérrez, G.: A compact representation of raster time series. In: 2019 Data Compression Conference (DCC), pp. 103–111 (2019)
4. Gonzalez, R., Woods, R.: Digital Image Processing. Prentice Hall (2008)
5. Han, D., Nam, Y.M., Kim, M.S., Park, K., Han, S.: SciDFS: an in-situ processing system for scientific array data based on distributed file system. In: 2018 IEEE International Conference on Big Data and Smart Computing (BigComp), pp. 375–382. IEEE (2018)
6. Ladra, S., Paramá, J.R., Silva-Coira, F.: Compact and queryable representation of raster datasets. In: Proceedings of the 28th International Conference on Scientific and Statistical Database Management, pp. 1–12 (2016)
7. Ladra, S., Paramá, J.R., Silva-Coira, F.: Scalable and queryable compressed storage structure for raster data. Inf. Syst. **72**, 179–204 (2017). https://doi.org/10.1016/j.is.2017.10.007
8. de Oliveira, S.S., Rodrigues, V.J., Martins, W.S.: Smart: Machine learning approach for efficient filtering and retrieval of spatial and temporal data in big data. J. Inform. Data Manage. **12**(3), 273–289 (2021)
9. Pinto, A., Seco, D., Gutiérrez, G.: Improved queryable representations of rasters. In: 2017 Data Compression Conference (DCC), pp. 320–329 (2017)
10. Rozante, J.R., Gutierrez, E.R., Fernandes, A.d.A., Vila, D.A.: Performance of precipitation products obtained from combinations of satellite and surface observations. Int. J. Remote Sens. **41**(19), 7585–7604 (2020)
11. Silva-Coira, F., Paramá, J.R., de Bernardo, G., Seco, D.: Space-efficient representations of raster time series. Inf. Sci. **566**, 300–325 (2021)
12. Tomlin, C.D., et al.: Geographic Information Systems and Cartographic Modeling, vol. 249. Prentice Hall Englewood Cliffs, Englewood Cliffs (1990)
13. Tomlin, C.: Map algebra: one perspective. Landscape and Urban Planning **30**(1), 3–12 (1994). special Issue Landscape Planning: Expanding the Tool Kit
14. Xia, Y., et al.: Continental-scale water and energy flux analysis and validation for the north american land data assimilation system project phase 2 (NLDAS-2): 1. intercomparison and application of model products. J. Geophys. Res. Atmos. **117**(D3) (2012). https://doi.org/10.1029/2011JD016048

FedeM: Federated Learning-Based Privacy-Preserving Record Matching

Michail Zervas and Alexandros Karakasidis$^{(\boxtimes)}$ (ID)

Department of Applied Informatics, University of Macedonia, Thessaloniki, Greece
{ics20015,a.karakasidis}@uom.edu.gr

Abstract. Privacy Preserving Record Linkage is the task of identifying the same real-world entities (usually humans) in databases originating from different dataholders, without revealing to any of them any other information apart from their matching records. In this paper, we focus on Privacy-Preserving Record Matching, the stage of Privacy-Preserving Record Linkage where records are checked for matching between data owners and we present FedeM: Federated Learning-based Privacy-Preserving Record Matching. FedeM is generic, relying on Federated Learning, without requiring a Linkage Unit, achieving high matching quality. Using Support Vector Machines, FedeM performs equivalently to a non-Federated Learning SVM-based setup for plain-text record matching, with only 1% decrease in Precision and 5% decrease in Recall.

Keywords: Record Linkage · Privacy · Federated Learning

1 Introduction

Record linkage is a fundamental process of identifying records referring to the same entity. As these records do not usually share common unique identifiers, combinations of attributes (usually names, surnames, addresses etc.), called *quasi-identifiers*, which together are able to uniquely identify an individual are used. These attributes may exhibit low quality due to typos so approximate string matching methods are required.

When these records originate from distinct data owners and refer to humans, privacy is an additional concern due to a variety of reasons, such as recent developments in privacy-oriented legislation and business competition. This implication led to developing methods for *Privacy-Preserving Record Linkage* (PPRL). To this end, the goal of PPRL is to detect data describing the same real world entities, i.e. individuals, across different databases while preserving their privacy through de-identification.

In this work, we focus on Privacy-Preserving Matching, the stage of PPRL where records are matched, utilizing a *Federated Learning* (FL) [13] approach, not requiring a Linkage Unit. We propose a generic approach, compatible both with server-based and server-less [17] FL, potentially usable with any Machine

J. Tekli et al. (Eds.): ADBIS 2024, CCIS 2186, pp. 35–45, 2025.
https://doi.org/10.1007/978-3-031-70421-5_4

Learning algorithm and we showcase its efficiency in an implementation relying on Support Vector Machines, which have proved their performance in addressing the original version of the Record Linkage problem [4].

The rest of this paper is organized as follows. Section 2 discusses related work. In Sect. 3 we provide the details of our method and the required background. Section 4 contains our empirical evaluation. Finally, we discuss our conclusions and our next steps in Sect. 5.

2   Related Work

For the problem of Privacy-Preserving Record Linkage, recent advancements may be found in [9]. Bloom filters [19] comprise a very popular approach in this area. Bloom filters are combined with *n-grams* and the resulting bit vectors are ANDed to determine whether they match in a separate server referred to as the *Linkage Unit*. It has been shown that such solutions are vulnerable, requiring additional hardening measures [8].

Bloom filters have recently been enhanced with differential privacy guarantees also using FL and deep learning for enhancing the quality of the results [16]. Another approach that exploits FL is presented in [11]. Yao et al. [24] combine Bloom filters with Siamese Neural Networks for increased security and performance. In [6] authors use autoencoders for hardening the Bloom filters. However, all of these methods rely on the availability either of a Linkage Unit [6,16] or of a Trusted Third Party [11,24]. FedeM does not exhibit such limitations as it does not use a Linkage Unit and it might operate either in a server-based or in server-less FL setup.

In other techniques, Smith [20] proposes encoding sensitive data into bit vectors and applying Locality Sensitive Hashing, with the drawback of increased computational cost [5]. In [12], Soundex is incorporated in a privacy-preserving big data setup. However, this method also requires the use of a Linkage Unit.

Furthermore, in the last few years, methods based on cryptographic primitives have emerged. These methods include homomorphic encryption [7], known, however, for its high computational cost [2] and susceptibility to certain types of attacks [10] and garbled circuits [3], needing to be further investigated in terms of execution time, size and reusability in this context [18]. PPRL using Fuzzy Vaults [14] falls in this category of methods, not requiring a Linkage Unit but also exhibiting high computational complexity. FedeM, does not exhibit the limitations of such methods, as it features low computational cost, both for training and matching.

3   Methodology

In this section, we start by presenting the building blocks for our approach and then we lay out our methodology

3.1 Building Blocks

Problem Formulation. Without loss of generality, we consider two data sources, Alice (A) and Bob (B), who respectively hold r^A and r^B records each. Let r_i^A and r_i^B be the i-th record of Alice and Bob, respectively. The j-th attribute of these records is represented as $r_i^A.j$ and $r_i^B.j$. *Privacy-Preserving Record Matching* is the problem of matching all pairs of r^A and r^B records that refer to the same entity, so that no more information is disclosed to either A, B or any other party involved in the process besides the identifiers of the linked r^As and r^Bs.

As Alice and Bob are expected to use different schemas in their databases, their records have different attributes not sharing any common candidate keys. We assume that in these schemas, m of the attributes are common between the two sources being quasi-identifiers, forming a composite key. We refer to these attributes as *matching attributes*. The composite key is used to determine when two records *match*, by comparing the respective attributes, relying on a similarity or distance function, as our data is often dirty.

Reference Set. A *Reference Set* (RS) is a data corpus exploited as a means of indirect comparison when matching sensitive information. This corpus, or a subset of it, which may be publicly available, is pre-agreed among the matching parties. Then, each of them performs an embedding of their data through some pre-agreed mapping function, which relates each of the values in their dataset to one or more values in RS. Then, they compare the results of the embedding. Such a construct has been used before [15] requiring, however, a Linkage Unit.

Federated Learning. Federated Learning (FL) is a technique for training machine learning models when data is distributed across multiple clients [13], making it ideal for privacy-oriented applications. In its general formulation, clients, who hold their local data, train local models and send them to an aggregator for calculating a global model. Thus, only the parameters of the particular model being trained are transferred. Federated Learning can be categorized as follows [23]. Depending on the existence or not of a server, there is the Centralized and the Decentralized architecture. Depending on data distribution, there is Horizontal FL, with the clients' data residing in the same feature space and a small overlap of the sample space (different rows), vertical federated learning, where data in clients have similar sample spaces, but different feature spaces and finally, there is federated transfer learning, where data in each client have different sample and feature spaces. Our approach may be implemented using either centralized or decentralized FL. We examine the case of a horizontal FL with the peculiarity that we are only interested in the full name of each record, which is a common column in each client, so we are in the same feature space.

3.2　FedeM

The Protocol. FedeM's protocol contains the steps taken for federated record matching. Initially, all dataholders agree on a common RS and the matching attributes. The RS should either originate from a different domain from the datasets or not contain data residing in the datasets. Local data are then associated to the RS creating a *Reference Representation* (RR) for each tuple. Then, each dataholder generates from its local dataset its training samples. Local models are trained at each dataholder and the parameters of these models will be federated, either in a centralized or decentralized manner. For this to happen, dataholders exchange their RRs. Using the global model derived from the FL procedure, each dataholder matches RRs. Finally, each dataholder delivers the matched data to the other. In our protocol, all communications take place using secure channels.

Creating Reference Representations. For Privacy-Preserving Record Matching, we will consider the case of using Support Vector Machines [21]. Training data should reflect candidate record pairs, either matching, or non-matching. As opposed to the case of classical record linkage, considering the distances between corresponding record fields is not suitable as, to preserve privacy, data cannot be transferred among dataholders. For this purpose, we need to calculate distances between matching data and RS data first. These are the RRs mentioned before. To calculate RRs, a mapping P between dataset and RS attributes is used and Edit Distance is calculated between them for each record in a dataholder's dataset, resulting, in a tensor TD of vectors with distances. To prevent ties, i.e. more than two dataset records having the same distances with RS rows, we introduce *Feature Augmentation*: In mapping P, a matching attribute is related with multiple RS attributes. Considering that for each record in RS there are k attributes in RS, a vector can have up to $P = k * m * |RS|$ values.

Algorithm 1: Build Feature Vector.

Input:
- **RS**: Reference Set
- **r**: Dataset with alphanumeric fields
- **P**: Mapping between r & RS fields

Output:
- **TF**: Tensor of feature vectors

1　$r' \leftarrow$ Corrupt (r);
2　$RR \leftarrow$ CreateRR (r, RS, P);
3　$RR' \leftarrow$ CreateRR (r', RS, P);
4　$TF \leftarrow$ [];
5　**foreach** $r_i \in r$ **do** // for each row
6　　TF.Append $([d\ (RR_i, RR'_i), 1])$;
7　　TF.Append $([d\ (RR_i, RR'_{\forall k \neq i}), 0])$;
8　**end**
9　**return** TF;

Algorithm 2: Data Matching with Federated SVM.

Input:
- RR^A: Ref. Representation of A
- RR^B: Ref. Representation of B
- $SVM()$: The federated model

Output:
- MA: A matching array

1　$MA \leftarrow$ [];
2　**foreach** $i \in RR^A$ **do** // A's rows
3　　**foreach** $j \in RR^B$ **do** // B's rows
4　　　MA.Append $(\text{SVM}(d\ (i, j)))$;
5　　**end**
6　**end**
7　**return** MA;

Data Generation and Training. In order to use an SVM for privately classifying data, training data should be generated first. To ensure no data exchange
among clients, training data are generated locally, within each dataholder using
the dataset r, the Reference Set RS and the mapping P. This method is illustrated in Algorithm 1. Initially (line 1), a corrupted version of r, r' is created by
performing Edit Distance operations (e.g. character additions). Then, Reference
Representations are built for r and r' (lines 2–3) as described just before. Next, a
feature vector is created by calculating the distances between records of RR and
RR', being labeled as *matching* (or 1) when comparing a record's RR with its
corrupted version RR' (line 6) or non-matching (or 0), when comparing a record
with the corrupted version of another record (line 7). Having created the feature
vectors for representing training pairs, training begins. Initially, a local SVM
model is trained using the created data. When training concludes, an aggregate
model is created through federation which is delivered to the dataholders.

Data Matching. Now, dataholders exchange their RRs so that matching takes
place locally. The matching process involves the steps illustrated in Algorithm 2.
For every item of the Reference Representation RR^A from Alice, its distance is
calculated against every item of Bob's RR^B. Then, the resulting feature vector
is labeled as matching or non-matching using the federated trained SVM (line
4). The resulting array of matches MA is eventually returned (line 7), so that
each party delivers to the other the actual matching data.

3.3 Privacy Discussion

FedeM assumes a Honest But Curious without Collusion model, according to
which, all participating entities try to infer as much information as possible but
without deviating from the protocol and without cooperating, as FedeM may
be used by multiple parties, or an aggregator may be employed. To prove the
privacy-preserving characteristics of out method, we shall consider a series of
attack types as described in [22] and discuss how it performs in such situations.

First, there are *Dictionary Attacks*, where an adversary attempts to identify
a sensitive value using a publicly available dictionary and encoding its values so
as to match a dataset's encoded values. FedeM is invulnerable to these attacks
since each party receives from the other an array of distances. As these functions
are non-bijective (more than one pairs of strings may produce such a value), the
calculated distances between the dictionary entries and the RS cannot uniquely
identify a real record.

For the *Frequency Analysis Attack*, where a public dataset is employed to
study its distribution and identify quasi-identifier attributes, FedeM's vectors
contain distances resulting from non-bijective functions. So, there is no relation
between the frequency of a dictionary entry and its distance from RS's.

In *Similarity Attacks*, an adversary may exploit that distributions of similarities between encoded and plain text fields are maintained to relate plain text
and encoded values. FedeM uses Edit Distance, which is neither limited by the

bounds of similarity functions, nor it is bijective. Additionally, these distances are calculated against a custom made RS, unknown to an attacker, making a similarity attack infeasible, without the attacker knowing either of the strings these distances refer to.

For *Linkage and Ciphertext-only Attacks*, where publicly available information is linked to reveal the quasi-identifiers. The adversary analyses ciphertexts to recover plain texts. Let us consider the case where a dataholder tries to perform a brute-force attack. First of all, an external attacker will not know the agreed RS. Also considering Edit Distance's non-bijectivity and that the intersection between the RS and the recordset is empty, even if an attacker manages to create a vector of distances identical to the recovered one, there is no evidence of the strings compared that will result in such a vector.

Finally, FedeM may be deployed using an aggregator. In this case, the aggregator gains no knowledge over client data as she receives no other information besides the support vectors of the clients local models.

4   Empirical Evaluation

In this section, we provide empirical proof of FedeM's high performance.

4.1   Experimental Setup and Datasets

Using Python 3, scikit-learn and NVFLare[1], NVIDIA Federated Learning Application Runtime Environment, we implement our algorithms[2]. All experiments have been conducted on an Core i5-10600K host with 16GB of main memory running Ubuntu 22.04. Without harming the general case, we assumed the use of a centralized aggregator for calculating the federated SVM model.

We employed the North Carolina voters database[3], having used three samples with 2000, 5000 records and 10000 records each. In each sample, we have used the attributes: 'last name', 'first name' and 'middle name'. We assume that the attributes chosen comprise a candidate key. Thus, we have deduplicated the dataset using the respective attribute combination. Then, we took samples and we generated two databases, belonging to Alice and Bob, respectively. We corrupted Bob's records using [1], applying one Edit Distance operation randomly on a matching attribute of each record in Bob's dataset, so that a join operation using these quasi-identifiers yields an empty result set. For RS creation, we used actor names from Wikipedia[4]. We have created two RSs, one with 200 records and one with 2000 records not containing any attribute value from the dataset samples we are using.

---

[1] https://github.com/NVIDIA/NVFlare.

[2] Implementation available at: https://github.com/mikez3/FL-record-linkage.

[3] https://dl.ncsbe.gov/?prefix=data.

[4] https://en.wikipedia.org/w/index.php?title=Category:20th-century_American_male_actors.

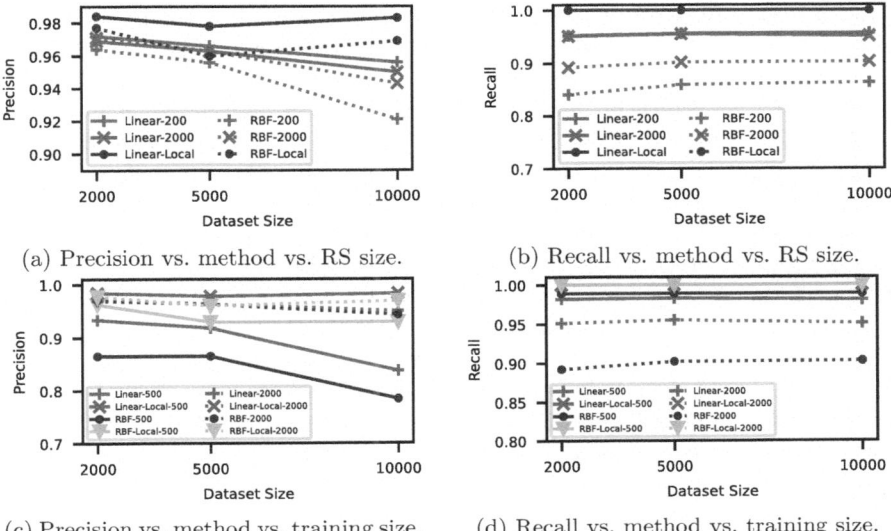

(a) Precision vs. method vs. RS size.   (b) Recall vs. method vs. RS size.

(c) Precision vs. method vs. training size.   (d) Recall vs. method vs. training size.

Fig. 1. Experimental evaluation results.

To evaluate matching performance, we measure Precision and Recall. Precision is defined as the fraction of the relevant elements among the retrieved elements, while Recall is defined as the fraction of the retrieved relevant elements divided by the total relevant elements: $Precision = \frac{TP}{TP+FP}$ and $Recall = \frac{TP}{TP+FN}$. In our case, elements are matching record pairs. We then evaluate efficiency by means of overall processing time, which comprises of data transformation time, training time and matching time. To have a broader view of our method, we considered two setups for our SVM classifier. The first setup employs a linear SVM kernel with a hyperparameter C=100 at both the clients and the server. The second setup employs an RBF kernel with C=0.01 at the clients and C=0.0001 at the server. Finally, to be able to have a basis of comparison, we also compare against a simple local SVM classification as described in [4].

4.2 Evaluation of Matching Performance

To evaluate the matching performance and the behavior of FedeM, we will consider the size of the reference set used and the size of the training set used. We will vary these parameters and examine how FedeM behaves with respect to the dataset sample size used for matching in terms of Precision and Recall.

Impact of Reference Set Size. First, we will vary the size of RS, while using a 2000 records training set. The results of these experiments are illustrated in Fig. 1a and Fig. 1b. The horizontal axis represents the size of the dataset we attempted to match, while the vertical axis the respective measure, i.e. Precision or Recall. We report results both for the SVM and RBF kernels. In the case of the federated setup, we illustrate results for both reference set sizes, while the case of simple SVM, without privacy characteristics is designated by "Local".

Starting with Precision, we can discern that the highest value is for the Local Linear SVM with values above 0.98. Then the RBF SVM follows with a performance in the area of 0.97. We may observe that these classifiers are not affected by the dataset sizes to be matched. Then, we proceed to the federated classifiers, where the Linear classifier's performance is superior than all other cases in terms of Precision, for both dataset sizes. Also, different RS sizes do not affect performance. Nevertheless, Precision is affected by the size of datasets to be matched. Even there, the difference from the Local Linear classifier is near 0.01, which is the cost to be paid to achieve privacy-preserving characteristics. RBF kernels do not seem to behave towards the desired direction. Using a smaller RS, the RBF kernel lacks up to 0.05 compared to the Local RBF kernel for the 10000 records dataset and approximately 0.06 from the Local Linear kernel. However, when increasing the size of RS, the situation significantly improves and this difference halves to 0.02 and 0.03 respectively.

Examining Recall (Fig. 1b), both Local SVM classifiers achieve perfect Recall, without being affected by matching set sizes. The Linear federated classifier is not affected by dataset size either, nor by the RS size used. Nevertheless, it performs lower than the Local classifiers by 5%. The RBF kernels, again, come last. However, in this case their performance is not affected by the dataset size they are used at. On the contrary, their performance is affected by the RS size used to train them, with RBF's Recall vastly improving from 0.85 to 0.9 when increasing RS size.

Impact of Training Set Size. Here, we keep the RS size fixed at 2000 records. The results of this part of our evaluation for Precision and Recall are illustrated in Fig. 1c and Fig. 1d respectively. The setup of these Figures are similar with Fig. 1a and Fig. 1b. For Precision, and considering the case of Linear Local SVM, we may discern that it exhibits the highest performance of all methods again, reaching 0.98, regardless of the Training Set size. The RBF Local SVM then follows when trained with the 2000 records dataset, exhibiting Precision between 0.96 and 0.975, without being affected by the dataset size to be matched. Its behavior is similar when the 500 records training dataset is used but with lower performance, ranging between 0.93 and 0.96, without seeming to be affected by the matching dataset size either. Moving to FedeM's implementations, we may deduce both for the Linear and the RBF SVMs that the training set size significantly affects their performance. With the 2000 records dataset, both methods reach the performance of the local RBF SVM with a score near 0.95. Nevertheless, using a smaller training set significantly affects the performance of both of these approaches which is now also affected by the dataset size to be matched. Thus, achieving for the 10000 records dataset Precision of 0.84 for the Linear SVM and 0.78 for the RBF SVM. Given these, we may conclude that increasing the Training Set size positively affects FedeM's Precision performance.

For Recall, both local SVM implementations perform perfectly. For FedeM, its behavior is reversed compared to its Precision performance. For the 500 records Training Set, both Linear and RBF SVMs achieve a score over 0.98. Increasing the Training set size, Recall drops for Linear to 0.95, while for RBF it drops to 0.90. This leads us to the conclusion that increasing the Training Set size leads to a drop of Recall. In any case, however, the Linear kernel performs better. Thus, a good practice is to use a Linear kernel, increasing the size of the training set, as in the privacy-preserving context Precision is more important.

4.3 Evaluation of Time Performance

FedeM's consumed time may be discerned in Data Preparation time, Training Time and Matching Time. Data Preparation time is proportional to the sizes of the Reference Set and the size of the dataset to be matched. However, it is independent of the used kernel. To this end, for a RS of 200 records, and for the three datasets, 6, 18 and 57 secs. are required, exhibiting a superlinear behavior. This is more evident with the 2000 reference rows. Then, approximately 28, 136 and 495 secs are respectively required. Training time requires for the Local Linear and RBF SVMs 16.4 and 17.7 secs on average for 500 training records and 19.8 and 22.5 for 2000 training records. For the federated case these times are, as expected, elevated. For 500 records 19.6 and 19.8 secs are required for Linear and RBF training and 32.7 and 44.7 for 2000 records. Finally, Matching time is also affected, due to the increased representation size of each feature vector. As such, while the Local Linear SVM may require for matching 6, 9 and 21 secs, FedeM requires 7, 19 and 57 secs respectively.

5 Conclusions & Future Work

In this paper, we presented FedeM, a method for performing Privacy-Preserving Record Matching using Federated learning and SVMs without requiring a Linkage Unit and with a minimal performance overhead compared to the use of SVMs for plain text record linkage. Our next steps are directed towards further improving the performance of FedeM in terms of matching and tine performance and studying the efficient extension of our method to multiple parties.

References

1. Bachteler, T., Reiher, J.: A test data generator for evaluating record linkage methods. Tech. rep., German RLC Work. Paper No. wp-grlc-2012-01 (2012)
2. Bonomi, L., Huang, Y., Ohno-Machado, L.: Privacy challenges and research opportunities for genomic data sharing. Nat. Genet. **52**(7), 646–654 (2020)
3. Chen, F., et al.: Perfectly secure and efficient two-party electronic-health-record linkage. IEEE Int. Comput. **22**(2), 32–41 (2018)

4. Christen, P.: Automatic record linkage using seeded nearest neighbour and support vector machine classification. In: ACM SIGKDD, pp. 151–159 (2008)
5. Christen, P., Ranbaduge, T., Schnell, R.: Linking Sensitive Data - Methods and Techniques for Practical Privacy-Preserving Information Sharing. Springer (2020). https://doi.org/10.1007/978-3-030-59706-1
6. Christen, V., Häntschel, T., Christen, P., Rahm, E.: Privacy-preserving record linkage using autoencoders. Int. J. Data Sci. Anal. **15**(4), 347–357 (2023)
7. Essex, A.: Secure approximate string matching for privacy-preserving record linkage. IEEE Trans. Inf. Forensics Secur. **14**(10), 2623–2632 (2019)
8. Franke, M., Sehili, Z., Rohde, F., Rahm, E.: Evaluation of hardening techniques for privacy-preserving record linkage. In: 24th International Conference on Extending Database Technology, pp. 289–300. OpenProceedings.org (2021)
9. Gkoulalas-Divanis, A., Vatsalan, D., Karapiperis, D., Kantarcioglu, M.: Modern privacy-preserving record linkage techniques: an overview. IEEE Trans. Inf. Forensics Secur. **16**, 4966–4987 (2021)
10. Goodrich, M.T.: The mastermind attack on genomic data. In: 30th IEEE Symposium on Security and Privacy, pp. 204–218. IEEE Computer Society (2009)
11. Heidt, C.M., Hund, H., Fegeler, C.: A federated record linkage algorithm for secure medical data sharing. In: German Medical Data Sciences: Bringing Data to Life, pp. 142–149. IOS Press (2021)
12. Karakasidis, A., Koloniari, G.: Efficient privacy preserving record linkage at scale using apache spark. In: 2022 IEEE International Conference on Big Data (Big Data), pp. 402–407. IEEE (2022)
13. McMahan, B., Moore, E., Ramage, D., Hampson, S., y Arcas, B.A.: Communication-efficient learning of deep networks from decentralized data. In: Artificial Intelligence and Statistics, pp. 1273–1282. PMLR (2017)
14. Mullaymeri, X., Karakasidis, A.: Using fuzzy vaults for privacy preserving record linkage. In: The 23rd International Workshop on Design, Optimization, Languages and Analytical Processing of Big Data. CEUR Workshop Proceedings, vol. 2840, pp. 101–110. CEUR-WS.org (2021)
15. Pang, C., Gu, L., Hansen, D., Maeder, A.: Privacy-preserving fuzzy matching using a public reference table. In: McClean, S., Millard, P., El-Darzi, E., Nugent, C. (eds.) Intelligent Patient Management, Studies in Computational Intelligence, vol. 189, pp. 71–89. Springer, Berlin Heidelberg (2009). https://doi.org/10.1007/978-3-642-00179-6_5
16. Ranbaduge, T., Vatsalan, D., Ding, M.: Privacy-preserving deep learning based record linkage. IEEE Trans. Knowl. Data Eng. (2023)
17. Roy, A.G., Siddiqui, S., Pölsterl, S., Navab, N., Wachinger, C.: BrainTorrent: a peer-to-peer environment for decentralized federated learning. arXiv preprint arXiv:1905.06731 (2019)
18. Saleem, A., Khan, A., Shahid, F., Alam, M., Khan, M.K.: Recent advancements in garbled computing: how far have we come towards achieving secure, efficient and reusable garbled circuits. J. Netw. Comput. Appl. **108**, 1–19 (2018)
19. Schnell, R., Bachteler, T., Reiher, J.: Privacy-preserving record linkage using bloom filters. BMC Med. Inform. Decis. Mak. **9**, 41 (2009)
20. Smith, D.: Secure pseudonymisation for privacy-preserving probabilistic record linkage. J. Inf. Secur. Appl. **34**, 271–279 (2017)
21. Vapnik, V.N.: The Nature of Statistical Learning Theory. Springer, New York, NY (2000). https://doi.org/10.1007/978-1-4757-3264-1

22. Vidanage, A., Ranbaduge, T., Christen, P., Schnell, R.: A taxonomy of attacks on privacy-preserving record linkage. J. Priv. Confidentiality **12**(1) (2022)
23. Yang, Q., Liu, Y., Chen, T., Tong, Y.: Federated machine learning: concept and applications. ACM Trans. Intell. Syst. Technol. (TIST) **10**(2), 1–19 (2019)
24. Yao, S., Ren, Y., Wang, D., Wang, Y., Yin, W., Yuan, L.: SNN-PPRL: a secure record matching scheme based on siamese neural network. J. Inf. Secur. Appl. **76**, 103529 (2023)

Estimating *MPdist* with *SAX* and Machine Learning

Mihalis Tsoukalos[(✉)] , Pantelis Chronis , Nikos Platis ,
and Costas Vassilakis

Department of Informatics and Telecommunications, University of the Peloponnese,
Tripolis, Greece
{mtsoukalos,chronis,nplatis,costas}@uop.gr

Abstract. *MPdist* is a distance measure which considers two time series
to be similar if they share many similar subsequences. However, computing *MPdist* can be slow, especially for large time series. We propose a
technique for the approximate computation of *MPdist* that uses the *SAX*
representation of the time series to quickly estimate the Nearest Neighbor
(NN) distance of each subsequence, and then applies a Machine Learning model to correct the accuracy loss incurred. Our method is orders
of magnitude faster than the exact computation of *MPdist*; at the same
time, our best approximation computes the NN of a time series with high
accuracy. A thorough evaluation of our technique is provided.

Keywords: Time series · SAX · Distance metric · MPdist · Machine
Learning · Random Forest

1 Introduction

Nowadays, data frequently comes in the form of time series [13,14], in domains
such as healthcare, finance, computing, and marketing. The most common task
is *comparing* time series, in order to perform data mining tasks [1] such as classification, clustering, anomaly detection and outlier analysis, and motif discovery.
Similarity search is at the epicenter of all these tasks as it allows the identification of similar and dissimilar time series to the one at hand. The selection of
an appropriate distance metric is often the most important part of similarity
search. Traditional distance metrics include the Euclidean Distance (ED), which
is a basic and widely used metric that cannot capture complex patterns well, and
Dynamic Time Warping (DTW) [2], which is particularly useful when comparing
time series that may have different temporal resolutions or phase shifts.

The *MPdist* [9] distance metric takes a different approach when comparing time series: instead of considering a time series as a whole, like ED and
DTW, *MPdist* breaks the time series into subsequences using a *sliding window*,
and makes the comparisons at the subsequence level. This usually leads to more
accurate results than ED and DTW for tasks such as classification and clustering.
However, *MPdist* is slow to compute, as it requires the retrieval of the Nearest

J. Tekli et al. (Eds.): ADBIS 2024, CCIS 2186, pp. 46–57, 2025.
https://doi.org/10.1007/978-3-031-70421-5_5

Neighbor (NN) among a large number of subsequences. To decrease its computational complexity, approximate computation of *MPdist* may be employed.

Contributions. In this work, we use the *SAX* representation to group similar subsequences during the calculation of *MPdist*. This reduces the number of calculations required to find the NN of each subsequence at the cost of accuracy. Afterwards, we use a Machine Learning (ML) model to compensate for that loss of accuracy. We experiment with four ML models; among them, Random Forest had the best accuracy. We use the most accurate ML model to perform NN computations and show that it produces accurate results. Finally, we show that the size of the sliding window plays a key role in the performance of the presented techniques.

Paper Structure. The rest of the paper is structured as follows: In Sect. 2 we outline related work, explaining the operation of *MPdist* and its $O(n^2)$ complexity. In Sect. 3 we present our technique along with the features used for the ML models. In Sect. 4 we evaluate the performance and the accuracy of our technique, and compare them to *MPdist*. In Sect. 5 we present our conclusions and paths for future work.

2   Preliminaries and Related Work

2.1   Definitions

A *time series* $T = (t_1, \ldots, t_n)$ is an ordered list of n *data points* $t_1, \ldots, t_n$. A *sliding window* of size w decomposes a time series into all its subsequences of size w. Specifically, using a sliding window of size w, T is decomposed into $n - w + 1$ subsequences: $(t_1, \ldots, t_w)$, $(t_2, \ldots, t_{w+1})$, $\ldots$, $(t_{n-w+1}, \ldots, t_n)$.

The *similarity* between two time series is determined using *distance functions* [2,5–7,9,17]. The Euclidean distance (ED) is defined for time series of the same size n, as $\mathcal{D}(T, T') = \sqrt{\sum_{i=1}^n (t_i - t_i')^2}$.

2.2   *MPdist*

The *MPdist* distance [9] considers two time series to be similar if they have many similar subsequences, regardless of their order. To compute the distance between two time series T and T' (possibly of different sizes), *MPdist* uses a sliding window of size w to split T and T' into all their subsequences of size w. Then, for each subsequence of T (respectively T'), *MPdist* finds the ED to its NN among the subsequences of T' (resp. T) and stores this distance in a list $\mathcal{L}$. In the end, $\mathcal{L}$ contains the distances between all subsequences of size w of T and their NN among the subsequences of T', and vice-versa for the subsequences of T'. $\mathcal{L}$ contains $|T| + |T'| - 2 \cdot w + 2$ values, where $|T|$ and $|T'|$ are the lengths of the time series.

The creators of *MPdist* propose to use a value in $\mathcal{L}$ to measure the similarity of T and T'. Smaller values in $\mathcal{L}$ correspond to subsequences of T (resp. T') for which T' (resp. T) contains some subsequence that has high similarity; and inversely, the larger values correspond to subsequences of T (resp. T') for which T' (resp. T) does not contain any subsequence of high similarity. Selecting a small value (resp. large) in $\mathcal{L}$ as *MPdist* means that T and T' are considered similar when they have few (resp. many) similar subsequences. Typically, the 5% lowest value in $\mathcal{L}$ is selected as the *MPdist* value. Computing *MPdist* requires $\mathcal{O}(n^2)$ time, where n is the size of the input time series (we consider both time series of length n without loss of generality). In more detail, T and T' have $\mathcal{O}(n)$ subsequences and for each subsequence, we need to find its nearest neighbor in the other time series, which requires $\mathcal{O}(n)$ comparisons. This bound may prohibit *MPdist* usage for very large time series. The reference computation of *MPdist* contains several optimizations and is performed using the Matrix Profile [20]. Nevertheless, computing the *MPdist* distance of two time series with 3 million points each on a single current home-class CPU core using the Stumpy Python library (https://pypi.org/project/stumpy/) requires more than 18 h.

2.3   The *SAX* Representation

The Symbolic Aggregate Approximation (*SAX*) [12,15] is a symbolic representation of a time series which transforms $T = (t_1, \ldots, t_n)$ into a representation with m elements where $m << n$. This is referred to as *dimensionality reduction* and offers a way of representing a time series in summary form, thus saving space and accelerating related computations.

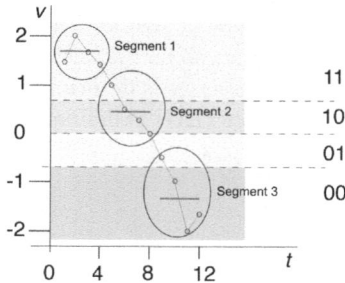

Fig. 1. The SAX representation of a time series T: $SAX(T) = \{\mathbf{11}, \mathbf{10}, \mathbf{00}\}$.

The computation of the *SAX* representation of a time series T is illustrated in Fig. 1, where T contains 12 points (illustrated as circles). To compute $SAX(T)$, the data value space is partitioned into a number of areas. Each area is labeled with a binary number (e.g., the topmost is labeled with 11, etc.), called a *SAX word*. The number of areas used is called the *cardinality* of the SAX representation. To compute $SAX(T)$, we divide T into *segments* that contain the same number of data points (in our example, 3 segments of 4 points each). For each segment, we compute the average value of its data points (this is its PAA [11]) and note the label of the area that this average value falls in; all these labels form $SAX(T)$. In our example, $SAX(T) = \{\mathbf{11}, \mathbf{10}, \mathbf{00}\}$ (average values illustrated with the red lines).

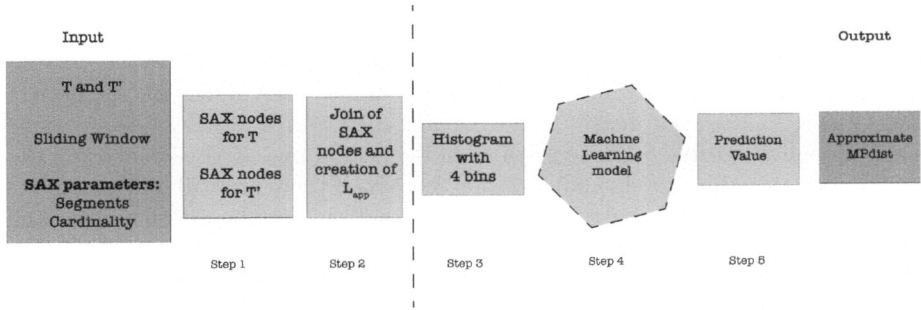

Fig. 2. The logical steps of the method.

3   Method Description

We aim at reducing the algorithmic complexity of *MPdist* by decreasing the time needed to find the NN of each subsequence. The core idea behind the proposed technique is to match each subsequence from one time series with subsequences having the same *SAX* representation in the other time series in search for its NN, thus excluding all other subsequences. The intuition behind the use of *SAX* is that there is high probability that the NN of a subsequence s has the same *SAX* representation with s. As expected, this will not always be true, therefore the proposed method will overestimate the distance, resulting in accuracy loss.

At this point, we considered approaches to compensate for the loss of accuracy. Initial attempts for the approximate calculation of *MPdist* involved analytical methods for choosing a value from the list $\mathcal{L}_{app}$ of the estimated NN distances, including techniques for predicting the rank (index) of the approximate value in the list or the approximate value itself. These efforts were unsuccessful, leading to estimations with high error margins, highly complex formulas which produced acceptable results only for a subset of the time series, and model overfitting. Finally, we resorted to the use of ML models, which are fed with values and features from both the time series and the $\mathcal{L}_{app}$ list. The method finally adopted comprises the steps illustrated in Fig. 2.

3.1   Working with *SAX*

In the *first step*, we compute the *SAX* representation of each subsequence of the input time series T and T' using the provided sliding window, and group the subsequences into *SAX nodes* based on their *SAX* representation. This creates two separate sets of *SAX* nodes, one for each of the input time series.

In the *second step*, the nodes with matching *SAX* representations from T and T' are joined, and the minimum ED for each subsequence is put into a list $\mathcal{L}_{app}$ of Euclidean distances. It is possible, however, that some nodes of T have a *SAX* representation that does not match the *SAX* representation of any node of T'. For these nodes, a fallback matching approach is followed: we consider that, in the absence of exact matches, the best NN candidates are probably found in *partially matching SAX* nodes, i.e. those that have $(number\_of\_segments - 1)$ common *SAX* words *regardless of order* (for instance, *SAX* nodes $(00, 11, 00, 01)$ and $(00, 01, 00, 10)$); not insisting on matching *SAX* words at the same positions likely includes multiple *SAX* nodes of T' in the partial match, thus increasing the probability of finding good candidates for the NN. If there are still nodes without a match, either exact or partial, these are ignored, under the rationale that the subsequences in such nodes are probably too distant to all subsequences in the other time series. It is worth noting that, in our experiments, the vast majority of *SAX* nodes have an exact match on the other time series, while very few need to be partially matched and only rarely is it needed to ignore a node. The process continues until all nodes have been examined. Thus, we end up with $\mathcal{L}_{app}$ containing an approximate NN distance for each subsequence in each time series.

3.2   Working with the Machine Learning Model

The *third step* involves the computation of the histogram of the lower 5% of the values of $\mathcal{L}_{app}$. Recall from Sect. 2 that the value of *MPdist* is the 5% lowest value in $\mathcal{L}$, and since the elements of $\mathcal{L}_{app}$ may overestimate the corresponding elements of $\mathcal{L}$, the actual value of *MPdist* will fall in the range defined by the lower 5% of $\mathcal{L}_{app}$. The rationale behind the use of a histogram is to create a summary of the lower 5% of $\mathcal{L}_{app}$ and feed the ML model with this information. We have found that dividing the values of the lower 5% of $\mathcal{L}_{app}$ into 4 histogram bins is efficient: the first two histogram bins can essentially be discarded as the values contained in them are too few and small, whereas the last two bins, and the distribution of values therein, are important since the real *MPdist* value is usually found there. We note that the fourth bin usually has the most values. The usage of more bins has been found to hinder the determination of an accurate *MPdist* value. Descriptive statistics regarding the bin bounds and bin population are used to construct input features to the ML models in the next step, in order to improve the *MPdist* value prediction accuracy.

The *fourth step* is about feeding the ML model with data based on the chosen features. For all ML models we use five features, denoted as *F1–F5*. **F1** is the sliding window size w, which is selected because larger window sizes imply larger subsequence ED values; due to this fact, w is expected to affect the prediction of the ML model. **F2** is the sum of the lengths of T and T' and allows us to describe the lengths of the two input time series; in general, the bigger the time series lengths are, the more values the histogram will contain. **F3** is the fraction

$\frac{MiddleVal}{MaxVal}$, where $MiddleVal$ is the middle and $maxVal$ is the highest of the 5 equally spaced values bounding the 4 histogram bins. **F4** is the number of values in the fourth bin divided by the total number of values in the histogram. Features *F3* and *F4* provide an indication of the distribution of values in the histogram and help the ML model make better predictions. Last, **F5** is the $MiddleVal$ value, which complements the descriptive statistics provided by features *F3* and *F4* by contributing a concrete, anchored value.

The output of the ML model is a fraction value, named $PredictionValue$ and is equal to $\frac{MiddleVal}{Real\ MPdist}$. We opted to predict this fraction instead of the distance itself so that the ML models are not restricted to the range of values already present in the training dataset.

The *last step* concerns the use of $PredictionValue$ to compute the approximate $MPdist$ value, which is equal to $\frac{MiddleVal}{PredictionValue}$.

The Machine Learning Models. We considered four ML models, and all were fed with the same features *F1–F5* described above. **Random Forest** (RF) [4] is a model that consists of a collection of decision trees, each fitted on a different random sample of the dataset. RF is a non-parametric model, which means it can represent arbitrary relations that are not limited to a single form (e.g., linear). RF is resistant to overfitting, which makes it suitable for small datasets. **SVR-Linear** [8] is a regularized linear model, trained using an ϵ-sensitive loss function, meaning that during training it ignores errors smaller than ϵ. SVR-Linear can only model linear relations and can be effectively trained on small datasets. **SVR-RBF** [8] is also regularized and trained using an ϵ-sensitive loss function, but is not limited to linear relations. Instead it transforms the features of the dataset using the Radial Basis Function (RBF), which allows it to model arbitrary relations. SVR-RBF can model more complex relations that SVR-Linear, but it requires a larger training dataset. **Multilayer Perceptron** (MLP) [10] is a model composed of layers of multiple nodes; each node calculates a linear combination of its input, applies an activation function and produces an output that is passed to the next layer of nodes. MLP has several parameters that need to be specified in order to be effectively trained. It can represent arbitrarily complex relations but it, generally, requires a larger training dataset.

4   Experimental Evaluation

In this section we report on the experimental evaluation performed to assess the effectiveness of the proposed algorithm, regarding the (i) result accuracy, (ii) execution time and (iii) scalability. Regarding the accuracy, we consider both the relative error in the magnitude of the *MPdist* metric and the use of the estimated *MPdist* metric to identify the NN of a given time series, a very common use case in time series processing.

All experiments were conducted on an Intel i9 machine with 64Gb of RAM and a 2TB NVMe SSD disk running Arch Linux. The software was developed in Python v3.10, using the Stumpy Python package implementation of *MPdist* distance computation, while all ML models were implemented using Scikit-learn (https://scikit-learn.org/). taskset(1) was used for limiting the available CPU cores to 1 when recording *MPdist* execution times.[1]

We used three datasets, comprising time series used in medical applications, each with different characteristics. We refer to them as **EEG** (Electroencephalogram), **ECG** (Electrocardiogram) and **EOG** (Electrooculogram). Out of the original datasets, we used subsets of 225 k, 450 k and 900 k elements for EEG, and 500 k, 1 M and 1.5 M o=l EOG and ECG. In all experiments we used a segments value of 4 and a cardinality value of 32. The sliding windows used were 128, 512, 1024 and 2048. Choosing the optimal *SAX* parameters is an open research problem [16].

Our technique stores the generated $\mathcal{L}_{app}$ lists in plain text files in order to avoid their recalculation during our experiments. As the time to pick the distance from $\mathcal{L}_{app}$ is negligible compared to the rest of the process, the task that is being timed is the generation of these files.

Machine Learning Setup. The training set has 28 entries for EEG and 164 entries for each one of ECG and EOG. The test set has 112 entries for EEG and 44 for each one of ECG and EOG. EEG has a smaller training set as we found out that it produced accurate *MPdist* results with only this training; this is likely due to the even distribution of its values.

Apart from testing the accuracy of each model on a per dataset basis, we also merge all training datasets into a single set, which we call the *Unified Dataset* and we assess the performance of the models trained with the Unified Dataset as well. In this case, the training set contains 356 entries and the test set contains 200 entries. A model trained with a Unified Dataset could be used for input time series from any domain, without the need for additional specific training.

Model Parameters. Each ML model requires the selection of various parameters. For RF we used n_estimator=25, max_depth=10 and absolute_error as the splitting criterion. For MLP we used one hidden layer of size 10, early stopping with 0.2 validation size, and learning_rate=0.01. For SVR-RBF we used nu=0.1. We kept all other parameters at the defaults of Scikit-learn.

4.1   Accuracy Tests

Figure 3 shows the accuracy error of each individual ML model per dataset. The lowest accuracy error is achieved on the EEG dataset with all ML models. Random Forest achieves the lowest accuracy error on all datasets.

ML models trained with the *Unified Dataset* are more generic and thus less accurate than models trained for a specific dataset. Figure 4 shows the accuracy error of each model, trained on the Unified dataset, and tested on each dataset separately. Tables 1 and 2 show the relative error of the computed approximate

---

[1] The source code and test data are available upon request.

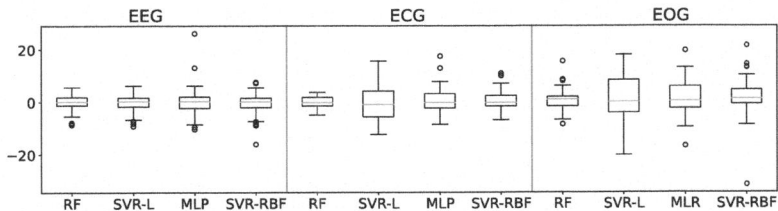

Fig. 3. Accuracy error of each ML model per dataset.

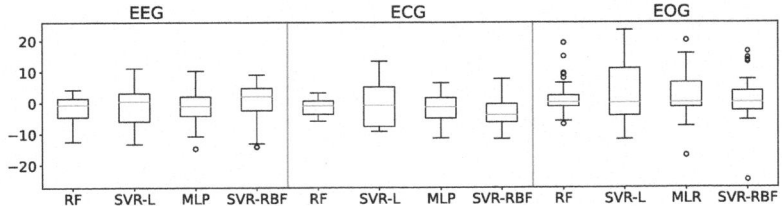

Fig. 4. Accuracy error of each ML model trained on the Unified Dataset.

MPdist values for all the ML models trained with the Unified Dataset, compared to the real *MPdist* values, for time series pairs from the evaluation set and various sliding window sizes. Once again, the RF model is the most accurate. Increasing the sliding window size or the time series length has no major effect on the accuracy of the ML model. However, the dataset type plays a key role in the accuracy of the ML model. EEG performs well in most cases, whereas ECG is the most difficult dataset, primarily because many of its values are close to 0, which means that they are unevenly distributed among *SAX* nodes.

Table 1. Accuracy of the Unified Random Forest and SVR-Linear models.

Dataset	Unified Random Forest				Unified SVR-Linear			
	w 128	w 512	w 1024	w 2048	w 128	w 512	w 1024	w 2048
EEG 225k	0.88%	−5.92%	−0.30%	0.39%	5.35%	2.51%	−11.01%	−7.33%
450 k	3.09%	−1.22%	−5.92%	−10.73%	5.01%	0.39%	−7.75%	−3.66%
450 k	−1.49%	−4.56%	−1.14%	−6.72%	−1.88%	3.03%	1.84%	−1.60%
ECG 500 k	−0.31%	−0.30%	−1.37%	5.24%	−0.18%	0.15%	−5.60%	16.06%
1M	−3.11%	−4.86%	1.07%	−1.82%	−1.57%	16.71%	5.96%	−2.17%
1.5M	−5.86%	−6.78%	0.73%	−7.70%	−11.77%	13.31%	4.28%	−7.43%
EOG 500 k	−0.04%	0.12%	−2.79%	−4.89%	−5.49%	7.30%	−8.02%	−7.26%
1M	−1.28%	1.09%	3.81%	−0.28%	−7.51%	11.23%	9.40%	−4.75%
1.5M	0.35%	−5.99%	−6.35%	−5.38%	−9.15%	−0.84%	−2.24%	−5.62%

Table 2. Accuracy of the Unified MLP Regressor and SVR-RBF models.

Dataset	Unified MLP Regressor				Unified SVR-RBF			
	w 128	w 512	w 1024	w 2048	w 128	w 512	w 1024	w 2048
EEG 225k	1.10%	2.63%	−2.72%	−7.90%	2.37%	−1.79%	2.36%	−0.07%
450k	2.73%	−2.75%	−7.48%	−4.79%	1.01%	−6.43%	−4.12%	−2.48%
450k	6.30%	−2.56%	2.06%	6.15%	0.52%	−4.32%	−1.13%	3.94%
ECG 500k	**0.3%**	**−0.30%**	−3.16%	9.53%	−2.85%	1.99%	−0.84%	7.44%
1 M	14.12%	7.49%	1.42%	−3.41%	1.90%	3.36%	−0.94%	2.24%
1.5 M	−4.55%	4.95%	−3.17%	−7.94%	−24.85%	3.79%	0.91%	−4.36%
EOG 500 k	−1.27%	−0.44%	−8.94%	−4.48%	−6.97%	0.64%	−9.60%	−7.42%
1 M	−2.91%	4.02%	3.58%	−0.42%	−5.16%	7.65%	2.53%	3.32%
1.5 M	−0.35%	−6.01%	−7.22%	−2.00%	−6.50%	−9.82%	−11.54%	−0.73%

4.2 Feature Importance Analysis for the Random Forest Model

Figure 5 presents an analysis concerning the importance of the features used in the ML models —the feature importance is computed as the mean of accumulation of the impurity decrease within each tree. For conciseness, we focus on the RF model, which has demonstrated the highest performance. It can be seen that *F5* is consistently ranked first, across all datasets, followed by *F3* and *F4*, in this order.

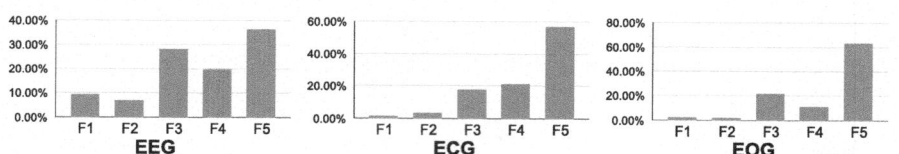

Fig. 5. Feature importance analysis for the Random Forest models.

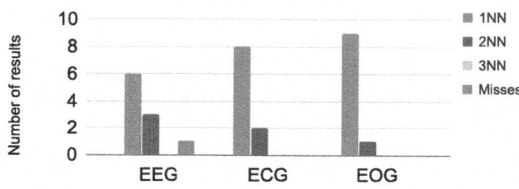

Fig. 6. Accuracy of NN computations per dataset.

4.3 Use Case: Nearest Neighbor Accuracy

We examine the effectiveness of the approximate *MPdist* value when computing the NN of a time series. As presented in Fig. 6, in 77% of all the cases, the NN was identified correctly using the approximate *MPdist* value; in 20% of the cases the use of the approximate *MPdist* value has resulted in selecting the 2-NN as nearest, and only in 3% of the cases the proposed approach has chosen an NN not in the top-3 NNs of the target time series. For the cases where the actual NN was not identified correctly, the divergence between the *MPdist* value of the chosen NN and the *MPdist* value of the actual NN ranges from 0.07% to 2.22%, being 1.12% on average; therefore the selected NN is, even in these cases, very close to the actual one.

4.4 Scalability Tests

We compared the scalability of our technique to that of the original *MPdist* method. We limited our scalability experiments to the EOG and ECG time series because the EEG time series were processed very fast in all cases by our method. For these tests we used subsets of the original datasets of increasing size, from 1.750 M to 3 M elements in steps of 250 k elements. For each one of the 6 different time series sizes, we created a pair of time series from each one of the two datasets, resulting in 24 time series in total. A selection of results concerning execution times are depicted in Figs. 7 and 8, and Table 3 presents the accuracy results from the experiments. The proposed approach achieves an execution speedup ranging from 22.4% to 95.4%; for small window sizes, the proposed approach increases execution speed by one order of magnitude.

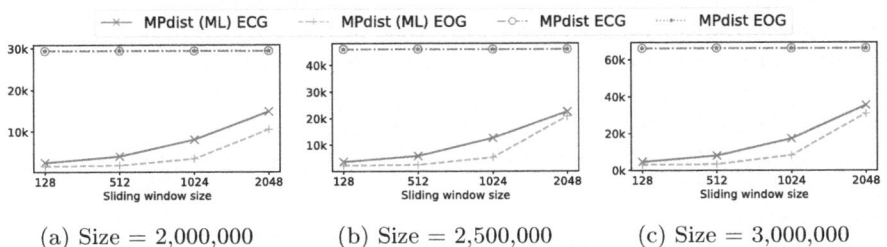

(a) Size = 2,000,000   (b) Size = 2,500,000   (c) Size = 3,000,000

Fig. 7. Execution times (seconds) for the scalability tests per time series length.

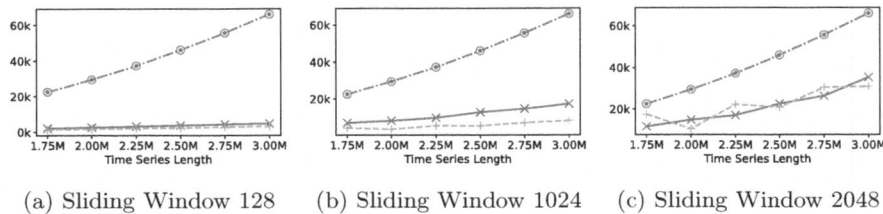

(a) Sliding Window 128   (b) Sliding Window 1024   (c) Sliding Window 2048

Fig. 8. Execution times for the scalability tests per sliding window size.

Table 3. Accuracy results for the scalability experiments.

| | EOG | | | | ECG | | | |
Dataset size	w 128	w 512	w 1024	w 2048	w 128	w 512	w 1024	w 2048
1,750,000	−0.82%	−4.63%	6.92%	0.61%	−7.71%	−2.73%	2.62%	−0.17%
2,000,000	−9.53%	−8.85%	0.84%	0.55%	−7.81%	−3.58%	2.68%	**0.04%**
2,250,000	−9.31%	−8.30%	0.90%	0.38%	−7.70%	−3.78%	2.57%	5.26%
2,500,000	−9.59%	−8.84%	0.90%	−0.36%	−6.83%	−3.11%	2.06%	−0.17%
2,750,000	−9.83%	−8.99%	−0.23%	−0.35%	−7.10%	−3.38%	1.42%	5.83%
3,000,000	−10.10%	−9.26%	−0.55%	−0.71%	−7.06%	−4.53%	2.27%	4.63%

5 Conclusions and Future Work

In this paper, we presented a technique that combines the *SAX* representation with a ML model for approximating *MPdist*. Our method results in significantly faster execution times than the original method while achieving high accuracy in NN computations. Among the four ML models tried, the Random Forest had the best accuracy. In the future, we plan to experiment with summarization techniques, like [3], that offer a more even distribution of subsequences among nodes, parallelize our method, take advantage of modern GPUs, and use time series indexes, such as [19] and [18]. Last, we plan on trying more ML models.

Acknowledgments. This research was partly funded by the SODASENSE project (https://sodasense.uop.gr/) under grant agreement No. MIS 5060275 (co-financed by Greece and the EU through the European Regional Development Fund).

Disclosure of Interests. All authors have no conflicts of interest.

References

1. Aggarwal, C.C.: Data Mining. Springer, Cham (2015). https://doi.org/10.1007/978-3-319-14142-8
2. Berndt, D.J., Clifford, J.: Using dynamic time warping to find patterns in time series. In: Proceedings of the 3rd ICKDDM, pp. 359–370 (1994)

3. Bountrogiannis, K., Tzagkarakis, G., Tsakalides, P.: Distribution agnostic symbolic representations for time series dimensionality reduction and online anomaly detection. IEEE Trans. Knowl. Data Eng. **35**(6), 5752–5766 (2023)
4. Breiman, L.: Random forests. Mach. Learn. **45**(1), 5–32 (2001)
5. Cai, Y., Ng, R.: Indexing spatio-temporal trajectories with chebyshev polynomials. In: ACM SIGMOD, p. 599–610 (2004)
6. Chan, K., Fu, A.W.: Efficient time series matching by wavelets. In: ICDE (1999)
7. Chatzigeorgakidis, G., Skoutas, D., Patroumpas, K., Palpanas, T., Athanasiou, S., Skiadopoulos, S.: Twin subsequence search in time series. arXiv:2104.06874 (2021)
8. Drucker, H., Burges, C.J.C., Kaufman, L., Smola, A., Vapnik, V.: Support vector regression machines. In: Advances in Neural Information Processing Systems, vol. 9 (1996)
9. Gharghabi, S., Imani, S., Bagnall, A., Darvishzadeh, A., Keogh, E.: Matrix profile XII: MPdist: a novel time series distance measure to allow data mining in more challenging scenarios. In: ICDM, pp. 965–970 (2018)
10. Haykin, S.: Neural Networks: A Comprehensive Foundation, 1st edn. Prentice Hall PTR, USA (1994)
11. Keogh, E.J., Chakrabarti, K., Pazzani, M.J., Mehrotra, S.: Dimensionality reduction for fast similarity search in large time series databases. Knowl. I. S. **3**, 263–286 (2001). https://doi.org/10.1007/PL00011669
12. Lin, J., Keogh, E., Wei, L., Lonardi, S.: Experiencing SAX: a novel symbolic representation of time series. Data Min. Knowl. Discov. **15**(2), 107–144 (2007)
13. Palpanas, T.: Data series management: the road to big sequence analytics. SIGMOD Rec. **44**(2), 47–52 (2015)
14. Palpanas, T.: Data series management: the next challenge. In: ICDM, pp. 196–199 (2016)
15. Tsoukalos, M.: Time Series Indexing. Packt Publishing (2023)
16. Tsoukalos, M., Platis, N., Vassilakis, C.: Estimating iSAX parameters for efficiency. In: ADBIS, pp. 3–12 (2023)
17. Wang, X., Mueen, A., Ding, H., Trajcevski, G., Scheuermann, P., Keogh, E.: Experimental comparison of representation methods and distance measures for time series data. Data Min. Knowl. Disc. **26**(2), 275–309 (2013)
18. Wang, Y., Wang, P., Pei, J., Wang, W., Huang, S.: A data-adaptive and dynamic segmentation index for whole matching on time series. Proc. VLDB Endow. **6**(10), 793–804 (2013)
19. Wang, Z., Wang, Q., Wang, P., Palpanas, T., Wang, W.: Dumpy: a compact and adaptive index for large data series collections. Proc. ACM Manage. Data **1**(1), 1–27 (2023)
20. Yeh, C.M., et al.: Matrix Profile I: all pairs similarity joins for time series: a unifying view that includes motifs, discords and shapelets. In: ICDM (2016)

Entity Matching with Large Language Models as Weak and Strong Labellers

Diarmuid O'Reilly-Morgan$^{(\boxtimes)}$, Elias Tragos , Erika Duriakova ,
Honghui Du , Neil Hurley , and Aonghus Lawlor

Insight Centre for Data Analytics, University College Dublin, Dublin, Ireland
{diarmuid.oreillymorgan,elias.tragos,erika.duriakova,honghui.du,
neil.hurley,aonghus.lawlor}@insight-centre.org

Abstract. A number of recent studies have shown that pre-trained large language models (LLMs) display highly competitive performance on entity matching tasks while acting in a zero shot manner, thus reducing the need for labelled training data. However, the use of the most capable LLMs (e.g. GPT-4) comes at a strongly prohibitive cost, whilst weaker LLMs (e.g. GPT-3) are significantly cheaper yet still provide reasonable performance without training data. In this study, we consider a scenario in which a budget constrained data practitioner with limited access to training data attempts to perform LLM based inference for an entity matching task. Within budget, they can afford to have only some portion of the data labelled by a stronger LLM. We cast this problem as that of deciding when to defer to a stronger LLM. In this scenario, we assume that a weak LLM provides cheap but noisy labels, while a strong LLM provides labels with better precision, but greatly increased cost. To address this, we develop a technique capable of intelligently allocating labels between strong and weak LLM labellers to maximise performance within the given budget. We show that, given a small amount of already labelled data, it is possible to reliably learn when the weak LLM is likely to be inaccurate, and thus defer to the stronger more expensive LLM. Employing GPT-3 and GPT-4 on four popular entity matching benchmarks we find that, given a specific budget to spend on the strong LLM, our approach generally performs better than a random labelling allocation, and outperforms an in-context-learning strategy. As such, this work outlines a simple yet effective methodology whereby real-world practitioners can employ LLMs in entity matching tasks.

Keywords: Entity Matching · Large Language Models · Natural Language Processing

1 Introduction

The entity matching task concerns determining which pairs of examples drawn from one or more datasets refer to the same entity. Despite many years of algorithmic advances, this has remained a challenging computational task,

J. Tekli et al. (Eds.): ADBIS 2024, CCIS 2186, pp. 58–67, 2025.
https://doi.org/10.1007/978-3-031-70421-5_6

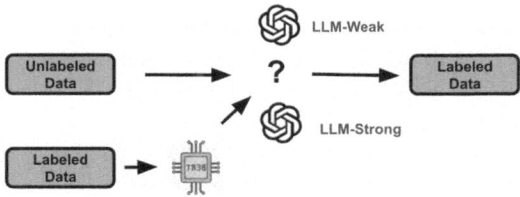

Fig. 1. Depiction of our proposed system. A small language model finetuned on a portion of labelled data allows us to decide, when given unlabelled pairs, whether to label that data with a weak or strong LLM.

requiring large amounts of training data to reach desirable performance levels on many datasets. In particular, many past works rely upon fine tuning pre-trained language models (LMs) such as BERT [5] with the downside that they require numerous labelled examples, and often under-perform on out of distribution samples [2]. At the same time, entity matching and the related task of data deduplication still consume a large proportion of the time of data practitioners, often with the need to source human labellers for large numbers of examples [12]. However, lately, a number of studies have shown that large language models (LLMs) [1], trained on web scale data, actually achieve state of the art performance on entity matching tasks out of the box, without the need for practitioners to provide any training data at all [14,16,17]. This ability can likely be ascribed to their capacity for semantic understanding of text, demonstrated in numerous areas of natural language processing [3]. However, though the performance attained by state of the art LLMs such as GPT-4 on entity matching benchmarks is certainly impressive, it comes at a prohibitive expense in terms of computation and API fees, while at the same time less expensive LLMs such as GPT-3 still provide decent performance without training data [17].

In this work, we consider the example of a data practitioner confronted with an entity matching task. The practitioner has limited labelled training data, and wants to employ an out of the box LLM method to classify some large unlabelled set of blocked pairs. We also assume that the practitioner has a limited budget, enough to cover annotating all of the pairs with a weaker LLM, with some leftover to annotate a carefully selected limited number of pairs with a stronger LLM. This seems quite a realistic scenario moving forward, where for instance, in our experiments, GPT-4 costs us around $3.5 to label 1200 entity matching pairs, whereas the weaker GPT-3.5 turbo costs a fraction (about 1/20th) of this cost, with open-source LLMs costing less provided one has the technical ability and infrastructure to run them. As such, we frame the problem within the well established framework of choosing between strong (expensive) and weak (inexpensive) labellers [22]. While it often appears in the context of reducing labelling cost for active learning, in our example we are primarily concerned with the cost of inference, where human annotators are swapped out for LLMs directly completing the entity matching task.

In this paper, we develop a technique capable of intelligently allocating labels between strong and weak LLM labellers, maximising the performance while adhering to a given budget. We show that (see Fig. 1), given a small amount of already labelled training data, it is feasible to learn a local model that can discern with sufficient accuracy as to when labels provided by a weak LLM (GPT-3) are inaccurate, so that we can defer to a stronger LLM (GPT-4) for those labels. We apply this approach to 4 publicly available datasets. Our results show that this approach is highly effective, with F1-score at any budget level limit being better than relabelling randomly chosen pairs, while also outperforming an in-context-learning strategy.

2 Related Work

2.1 Entity Matching with Language Models

An Entity Matching system generally features two components - a blocker and a matcher. The blocker consists of a low cost component that can efficiently retrieve likely match candidates, whether by simply indexing the data on some attribute, or applying e.g. TF-IDF similarity retrieval. For matching, the widely acknowledged state of the art consists of fine-tuning a pre-trained language model (LM) such as BERT on pairs of products, to output a classification score [9,15]. Though successful, this approach requires a large amount of labelled training data, and can lack robustness on out of distribution samples [2] - for instance when applied to entities unseen in the training data. Over the past few years, however, an increasing number of studies have shown that LLMs pretrained on web scale data excel out of the box at the entity matching task [4,6,14,17]. Peeters & Bizer [16,17] have established a new state of the art for entity matching, relying entirely upon LLMs such as GPT-3 and GPT-4 and show that on many benchmarks LLMs exceed fine-tuned LMs by a wide margin [16]. Additionally, they show that the use of "in-context" or "few-shot" learning, whereby a number of examples are given to the LLM as demonstrations, can often improve performance. Although one might reasonably question the extent to which the publicly available entity matching benchmarks may have appeared within the LLMs' training data, Peeters & Bizer offset these objections by evaluating solely on a benchmark which postdates GPT's training data cutoff point [17]. As the results are consistent with those observed on other benchmarks, it seems reasonable to assume that they are driven by the LLMs' semantic capabilities derived from size and web scale training, rather than from data contamination.

2.2 Weak and Strong Labellers

The trade-off between the use of different human labellers is well established in machine learning research [13,21,22], especially in the field of active learning. Often, the scenario explored is something akin to a group of data scientists building a cancer imaging model, who have access to, on the one hand, experienced but expensive radiologists (strong labellers), and a group of inexpensive students (weak labellers) on the other [22]. A significant reduction in labelling cost can be achieved by using the students to annotate the 'easy' examples in the dataset, while deferring to an expert for those that are more difficult. One prominent example in the literature [22] explores this trade off in the context of active learning, by building a difference classifier for predicting when the weak and strong labellers will disagree in their predictions. A number of recent papers have looked at treating LLMs as annotators/labellers within the wider NLP domain [20,23]. One very recent approach [8] explores the trade off between human and LLM annotation on a number of NLP tasks, utilising multiple prompts to the LLM to estimate the entropy and thus uncertainty of its responses, and then querying a human expert annotator according to the LLM uncertainty estimate and a given budget. As the use of multiple prompts (7 in their case) increases the cost of querying the weak LLM, this is not applicable to our case where we are operating within a limited cost margin - GPT-4 annotation is still significantly cheaper than human expert annotation [8], while repeated calls to GPT-3 would push its base cost towards half of GPT-4's. Another recent approach [19] directly queried GPT-3 for it's confidence level on annotation tasks, and explored setting thresholds as a heuristic for when to switch to few shot learning, also switching to human labelling given low enough confidence, but as a component of active learning rather than inference.

3   Methodology

3.1   Problem Formulation

In the entity matching problem, there is a blocked dataset of pairs $D_{unlabelled} \subseteq E \, X \, E'$ where E and E' are two collections of entity descriptions, so that each item in D is a tuple $x = (e, e')$, where the two descriptions have a reasonable chance of referring to the same entity. The matching task, then, requires producing a label $y = \{0, 1\}$ indicating whether the pair of descriptions refer to the same entity. In this paper we assume access to two LLM labellers LLM_{weak} and LLM_{strong}, each of which can label the dataset but with a cost/performance trade-off, where the labels given by LLM_{strong} have higher precision but also greater expense. In addition we have a small correctly labelled dataset D_{true}. We assume that our budget allows us to relabel k pairs with LLM_{strong}. Given $D_{unlabelled}$ and the two labellers, we want to decide in a principled manner which samples to relabel with LLM_{strong}, so as to increase performance as much as possible whilst remaining within budget.

3.2   Prompt Construction

Peeters & Bizer [17] present a comprehensive evaluation of a number of different prompting strategies across all of the datasets employed in this study. They primarily distinguish between "free" and "forced" prompts, where the forced prompts instruct the LLM to specifically answer with 'yes' or 'no', while with the free prompts the LLM might generate more tokens as part of its answer. To reduce token cost, here we select what are shown to be the most competitive GPT-3 and GPT-4 zero shot prompts on average. Specifically, for GPT-4 we employ the "domain-complex-force" prompt, and for GPT-3 the "domain-simple-force" prompt. In our evaluation we also employ the best average in-context-learning strategy with varied numbers of examples as a baseline, utilising the "domain-simple-force-handpicked" strategy, whereby a number of examples handpicked from the WDC-Products dataset are included as examples to the LLM.

Using these prompts, we can acquire a series of weak labels:

$$\hat{y}_i = LLM_{weak}(x_i) \tag{1}$$

3.3 Estimating the Accuracy of the Weak Labeller

For estimating the accuracy of LLM_{weak}, we seek to assign a probability to the labels that it generates. Thus, we adopt the entity-matching approach common in the literature, wherein a pre-trained model F from the BERT family is fine-tuned on the entity matching task [9,10,15]. Specifically, the model F is presented with strings of text in which the descriptions of two entities are concatenated to give input x_i, while a linear head layer is added to the pre-trained model, and trained via cross-entropy loss to output a softmax probability score of a match. Fine tuning this model on a small sample of expert labelled training data D_{true}, we can then obtain an estimate as to the probability (given the labelled training data) of a label $\hat{y}$ provided by LLM_{weak}.

$$p(\hat{y}_i|x_i, D_{true}) \approx \hat{y}_i \, F(x_i) + (1 - \hat{y}_i)(1 - F(x_i)) \tag{2}$$

3.4 Label Repair

Given an approach for detecting erroneous labels, we can rank samples in accordance with estimates of their label probability. As such, we can then select, in accordance with our budget, the set of samples to send to LLM_{strong} for possible relabelling. We define a cutoff c, being the maximum estimated probability p at which a sample will be labelled by the LLM_{strong} given the budget. Specifically, given a sorted list of probabilities, and if we are able to relabel k samples within budget, c will be the k-th probability in the list.

For these examples we can then obtain strong LLM labels:

$$y_i^* = LLM_{strong}(x_i) \tag{3}$$

Thus we define a function:

$$\delta(x_i) := \begin{cases} \hat{y}_i & \text{if } p(\hat{y}_i|x_i, D_{true}) >= c, \\ y_i^* & \text{if } p(\hat{y}_i|x_i, D_{true}) < c. \end{cases} \tag{4}$$

such that we get the resulting labelled dataset:

$$D_{labelled} = \{\delta(x_i) \mid x_i \in D_{unlabelled}\} \tag{5}$$

4 Experimental Setup

4.1 Datasets

We select 4 datasets that compare products from different domains and are popular in the entity matching literature, using the train and test splits provided by Peeters & Bizer [16]. These datasets are WDC-Products [18], Abt-Buy [7], Amazon-Google [7] and Walmart-Amazon [7]. The total size of the training datasets is 5000, 7659, 9167 and 8193 respectively. For testing, we use the same splits of the datasets employed by Peeters & Bizer [16], which are constructed so as to contain a significant number of unseen and hard negative pairs. The size of these test sets is 1239, 1206, 1234 and 1193 respectively. As we only query the LLM APIs on the test sets, the use of these small testing splits cuts down on the overall expense incurred during this study. In our experiments we randomly select small portions of the training sets to train the underlying small LM.

4.2 Language Models

We adopt RoBERTa [11] as the local language model used to estimate the probability of labels from LLM_{weak}. In our experiments, we conduct five training runs, each for fifty epochs, taking the model from the final epoch. We then conduct five relabelling and testing operations, reporting the aggregated results. Following Peeters & Bizer [16], we use GPT-4-0613 as the strong labeller, and GPT-3.5-turbo-0301 as the weak labeller. At the time of experimentation, GPT-4-0613 had a cost of \$30/\$60 for 1 million input/output tokens, while GPT-3.5-tubro-0301 had a cost of \$1.50/\$2. As a baseline, we report the metrics attained by Ditto [9] on each test dataset in [16], where it has been trained using the full training sets. We also demonstrate in-context-learning with 1,2,4,8 and 10 pairs of positive/negative examples, using the prompt from Sect. 3.2.

4.3 Metrics

To evaluate the downstream performance of our approach, we employ the commonly used F1-score, which provides a harmonic mean of precision and recall. To gauge how well our method for allocating examples to LLM_{strong} performs above a random selection, we calculate normalised performance difference:

$$\frac{\sum_k (model(c_k) - random(c_k))}{K} \tag{6}$$

where $model(c_k)$ is performance attained by a model based approach at the k-th cost cutoff, and $random(c_k)$ shows performance given a random selection.

5 Evaluation

5.1 Cost/performance Tradeoff

In this section, we show how the methods compare when a fixed number of labelled examples are used to train the local RoBERTa model. In this instance, we use 500 labelled examples. The results are depicted in Fig. 2. We report the performance of a Ditto model trained on all labelled examples in the training set. We also present cost/performance trade-off results for in-context-learning, where we increase the number of examples added as part of the input prompt to

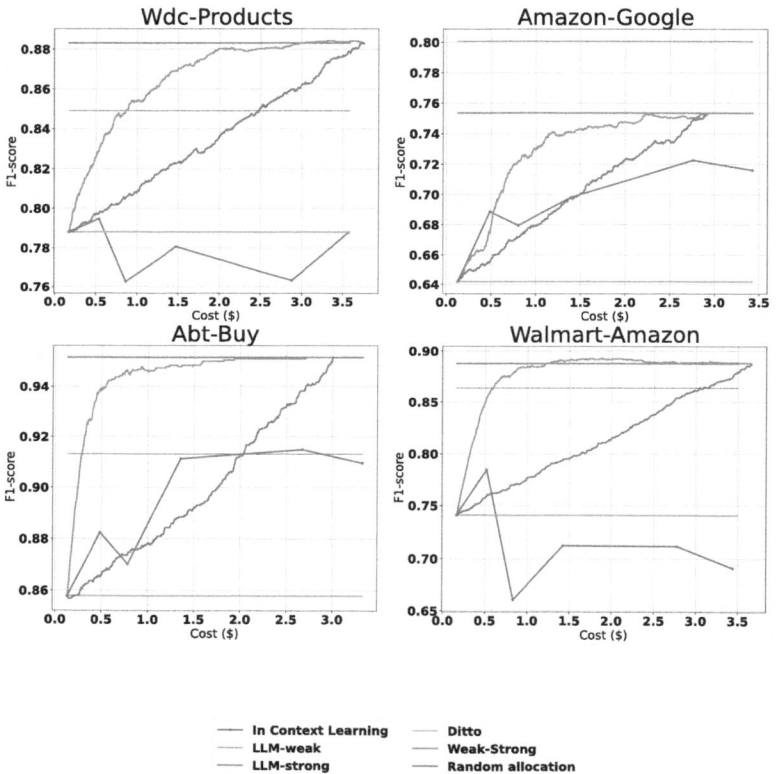

Fig. 2. The figure displays the overall F1-score for the tested methods with additional cost in dollars displayed on the x-axis. The red line shows the performance attained by Ditto when trained on the entire training dataset. In this figure we have used 500 samples to train the local RoBERTa model. The orange and green lines show, respectively, the performance attained by LLM_{weak} and LLM_{strong} out of the box. LLM_{strong} outperforms Ditto on all but one dataset (Amazon-Google). The blue line shows in-context-learning with different numbers of examples. While in-context-learning scales poorly with additional examples, our method Weak-Strong (purple line) shows a steep cost benefit over randomly relabelling samples (brown line).

LLM_{weak}. LLM_{strong} outperforms LLM_{weak} by a large margin on all datasets, and outperforms Ditto by a significant margin except on Amazon-Google, where the locally trained baseline outperforms it. Our method demonstrates a large improvement over random allocation on two datasets, Abt-Buy and Walmart-Amazon, where a steep increase in F1-score is noted with only a small increase in labelling cost. On Wdc-Products and Amazon-Google, it still outperforms both in-context-learning and random allocation, though the cost saving is less pronounced, especially when only a small number of samples are relabelled with LLM_{strong}. In-context-learning is seen to show inconsistent results, as on two datasets it effectively decreases the overall F1-score attained, whereas on the other two it gives a small increase, but is far from being preferable to randomly relabelling examples with LLM_{strong}. We would however caution against interpreting this as a strong result against the use of in-context-learning. As Peeters & Bizer [16] argue, in-context-learning can be inconsistent, and its use should be determined experimentally, perhaps by comparative evaluation of multiple prompts on a validation set.

Table 1. The normalised performance difference of our method vs random allocation when the size of labelled data used to train the RoBERTa model is varied.

Dataset	50	100	200	500	750	1000
Wdc-Products	0.0145	0.016	0.0157	0.0312	0.0248	0.0291
Abt-Buy	0.0334	0.0359	0.0387	0.0403	0.0378	0.0382
Amazon-Google	0.0091	0.0129	0.014	0.028	0.036	0.0372
Walmart-Amazon	0.053	0.0564	0.0584	0.0627	0.0628	0.0668

5.2 Varying the Amount of Training Data

In this section, and in Table 1, we show how performance above random scales with the amount of training data allocated to training the base RoBERTa model. We observe that the normalised performance difference (see Sect. 4.3) generally trends upwards as the number of examples is increased, though not consistently. Though we exclude the results for want of space, we note that the RoBERTa model showed highly varied results on the test set between runs, and assume that there is some instability when training a large model on small numbers of examples, which would have a downstream effect on the performance of our proposed method. The fact that our method continues to perform very well on both Abt-Buy and Walmart-Amazon, even when only 50 samples are used to train the local model, warrants further explanation. A brief analysis showed that this could be explained by the data. Specifically, while GPT-3 predicts matches for roughly 20% of the data, on both datasets the majority of its disagreements with GPT-4 were for samples where GPT-3 had predicted a match. The models learned at 50 samples in both cases gave $F(x_i) < 0.5$ for the vast majority of samples, with the effect being that GPT-3's match predictions were corrected first. While it is unclear if this behaviour would extend beyond the single test splits used in this study, it does suggest that even simple heuristics (e.g. 'check matches first') may prove useful in correcting GPT-3's predictions.

6   Conclusion and Future Work

In this work we have outlined a simple method for employing state of the art LLMs in entity matching workflows while being constrained by a budget. We have shown that, with only a small amount of labelled training data, we can effectively decide when to defer to a strong LLM labeller, which can significantly boost performance on the entity matching task. Although our method still requires labelled data, in future work we would like to explore obtaining this through active learning. Furthermore, as the labels obtained from GPT-4 have high precision, it might be possible to sample them as the initial training data. Though this study was limited to single test splits, and a stronger evaluation of in-context-learning is required, this still provides a promising first step for further research in this area. The future is likely to bring larger, more performant models, with increased costs, and as such being able to intelligently discern when to utilise less expensive models will become ever more pressing.

Acknowledgments. This work has been supported and funded by the EU Horizon Europe SEDIMARK (SEcure Decentralised Intelligent Data MARKetplace) project (Grant no. 101070074) and by the Science Foundation Ireland through the Insight Centre for Data Analytics under grant number SFI/12/RC/2289_P2.

References

1. Achiam, J., et al.: GPT-4 technical report. arXiv preprint arXiv:2303.08774 (2023)
2. Akbarian Rastaghi, M., Kamalloo, E., Rafiei, D.: Probing the robustness of pre-trained language models for entity matching. In: Proceedings of the 31st ACM International Conference on Information & Knowledge Management, pp. 3786–3790 (2022)
3. Brown, T., et al.: Language models are few-shot learners. Adv. Neural. Inf. Process. Syst. **33**, 1877–1901 (2020)
4. CHen, Z., et al.: SEED: simple, efficient, and effective data management via large language models. arXiv preprint arXiv:2310.00749 (2023)
5. Devlin, J., Chang, M.W., Lee, K., Toutanova, K.: BERT: pre-training of deep bidirectional transformers for language understanding. preprint arXiv:1810.04805 (2018)
6. Jaimovitch-López, G., Ferri, C., Hernández-Orallo, J., Martínez-Plumed, F., Ramírez-Quintana, M.J.: Can language models automate data wrangling? Mach. Learn. **112**(6), 2053–2082 (2023)
7. Köpcke, H., Thor, A., Rahm, E.: Evaluation of entity resolution approaches on real-world match problems. Proc. VLDB Endowment **3**(1–2), 484–493 (2010)

8. Li, M., Shi, T., Ziems, C., Kan, M.Y., Chen, N., Liu, Z., Yang, D.: Coannotating: uncertainty-guided work allocation between human and large language models for data annotation. In: Proceedings of the 2023 Conference on Empirical Methods in Natural Language Processing, pp. 1487–1505 (2023)

9. Li, Y., Li, J., Suhara, Y., Doan, A., Tan, W.C.: Deep entity matching with pre-trained language models. Proc. VLDB Endowment **14**(1), 50–60 (2020)

10. Li, Y., Li, J., Suhara, Y., Wang, J., Hirota, W., Tan, W.C.: Deep entity matching: challenges and opportunities. J. Data Inf. Qual. (JDIQ) **13**(1), 1–17 (2021)

11. Liu, Y., et al.: RoBERTa: a robustly optimized BERT pretraining approach. arXiv preprint arXiv:1907.11692 (2019)

12. Meduri, V.V., Popa, L., Sen, P., Sarwat, M.: A comprehensive benchmark framework for active learning methods in entity matching. In: Proceedings of 2020 ACM SIGMOD International Conference on Management of Data, pp. 1133–1147 (2020)

13. Mozannar, H., Sontag, D.: Consistent estimators for learning to defer to an expert. In: International Conference on Machine Learning, pp. 7076–7087. PMLR (2020)

14. Narayan, A., Chami, I., Orr, L., Ré, C.: Can foundation models wrangle your data? Proc. VLDB Endowment **16**(4), 738–746 (2022)

15. Peeters, R., Bizer, C.: Supervised contrastive learning for product matching. In: Companion Proceedings of the Web Conference 2022, pp. 248–251 (2022)

16. Peeters, R., Bizer, C.: Entity matching using large language models. arXiv preprint arXiv:2310.11244 (2023)

17. Peeters, R., Bizer, C.: Using chatgpt for entity matching. In: In: Abelló, A., et al. (eds.) European Conference on Advances in Databases and Information Systems, pp. 221–230. Springer, Cham (2023). https://doi.org/10.1007/978-3-031-42941-5_20

18. Peeters, R., Der, R.C., Bizer, C.: WDC products: a multi-dimensional entity matching benchmark. arXiv preprint arXiv:2301.09521 (2023)

19. Rouzegar, H., Makrehchi, M.: Enhancing text classification through LLM-driven active learning and human annotation. In: Proceedings of The 18th Linguistic Annotation Workshop (LAW-XVIII), pp. 98–111 (2024)

20. Wang, S., Liu, Y., Xu, Y., Zhu, C., Zeng, M.: Want to reduce labeling cost? GPT-3 can help. In: Findings of the Association for Computational Linguistics: EMNLP 2021, pp. 4195–4205 (2021)

21. Whitehill, J., Wu, T.f., Bergsma, J., Movellan, J., Ruvolo, P.: Whose vote should count more: Optimal integration of labels from labelers of unknown expertise. In: Advances in Neural Information Processing Systems, vol. 22 (2009)

22. Zhang, C., Chaudhuri, K.: Active learning from weak and strong labelers. In: Advances in Neural Information Processing Systems, vol. 28 (2015)

23. Zhang, R., Li, Y., Ma, Y., Zhou, M., Zou, L.: LLMAAA: making large language models as active annotators. arXiv preprint arXiv:2310.19596 (2023)

LLMClean: Context-Aware Tabular Data Cleaning via LLM-Generated OFDs

Fabian Biester[1], Mohamed Abdelaal[2(✉)], and Daniel Del Gaudio[1]

[1] University of Stuttgart, Stuttgart, Germany
{fabian.biester,daniel.gaudio}@ipvs.uni-stuttgart.de
[2] Software AG, Darmstadt, Germany
mohamed.abdelaal@softwareag.com

Abstract. Data cleaning tools, particularly those that exploit functional dependencies within ontological frameworks or context models, are instrumental in augmenting data quality. Nevertheless, crafting these context models is a demanding task, both in terms of resources and expertise, often necessitating specialized knowledge from domain experts. This paper introduces an innovative approach, called LLMClean, for the automated generation of context models, utilizing Large Language Models to analyze and understand various datasets. LLMClean encompasses a sequence of actions, starting with categorizing the dataset, extracting or mapping relevant models, and ultimately synthesizing the context model. To demonstrate its potential, we have developed and tested a prototype that applies our approach to three distinct datasets from the Internet of Things, healthcare, and Industry 4.0 sectors. The results of our evaluation indicate that LLMClean can achieve data cleaning efficacy comparable with that of context models crafted by human experts and with state-of-the-art data cleaning tools.

1 Introduction

Ontological Functional Dependencies (OFD)-based data cleaning tools have proven effective in improving error detection and correction. However, manually constructing context models for extracting OFDs is inefficient and impractical, especially for real-time applications [2]. This inefficiency arises from the need for extensive domain expertise to interpret complex and evolving data relationships, the overwhelming volume of data, and the necessity for rapid model adaptation to environmental changes. Manual methods are further hindered by human error, limited scalability as system complexities grow, and the challenges of maintaining consistency during model updates. Therefore, automating this process is essential to maintain the precision and reliability of context models while ensuring their scalability and adaptability in rapidly changing data environments.

Proposed Solution. In this paper, we introduce a novel method, designated as LLMClean, which automatically generates context models from real-world data without requiring supplementary meta-information. The data-cleaning process

J. Tekli et al. (Eds.): ADBIS 2024, CCIS 2186, pp. 68–78, 2025.
https://doi.org/10.1007/978-3-031-70421-5_7

begins with creating a context model from the dataset, outlining vital relationships and attributes, which serves as the base for the cleaning procedure (cf. Figure 1). LLMClean detects and uses OFDs within this model to detect invalid entries. Error correction tools, e.g., Baran and HoloClean [1], then utilize the detections to produce data repairs.

LLMClean utilizes the capabilities of Large Language Models (LLMs) to adapt to dynamic data patterns effectively. It involves several steps, including dataset classification, model extraction or mapping, and context model generation. By automatically generating OFDs, LLMClean establishes a

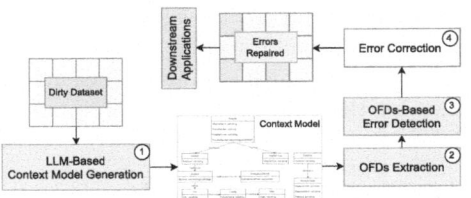

Fig. 1. Architecture of LLMClean

robust framework for data cleaning and analysis, addressing the complexities of real-world data, such as IoT datasets. Additionally, LLMClean introduces Sensor Capability Dependencies and Device-Link Dependencies for precise error detection. Evaluations indicate that LLMClean matches the data cleaning effectiveness of manually curated context models while providing enhanced efficiency and scalability.

Summary of Contributions. The paper introduces several key contributions to the field of error detection in tabular data: (1) It presents a novel three-stage architectural framework that integrates LLM models, context models, and data-cleaning tools, significantly enhancing the effectiveness and efficiency of error detection. (2) It proposes a method to autonomously generate context models from real-world data using LLM models such as Llama-2, GPT-3.5, and GPT-4. (3) The paper introduces a prompt ensembling technique to improve the stability of LLM models and develops an error detection tool that leverages OFD dependencies extracted from the generated context models. (4) The paper provides extensive experimental evaluations, comparing the proposed method, referred to as LLMClean, against baseline methods using datasets from IoT, Industry 4.0, and healthcare domains. Notably, LLMClean is claimed to be the first method to effectively utilize LLM models to enhance data cleaning tools through automatically generated context models.

2   Automated Context Modeling

In this section, we present a systematic approach for the automated generation of context models leveraging LLM models. We categorize data into two principal classes, namely *IoT datasets* and *non-IoT relational datasets*, each with distinct requirements for the context model. Non-IoT datasets do not generally adhere to most OFDs as these dependencies are tailored to the architectural patterns of IoT sensors. Instead, only Matching and Denial dependencies are pertinent to non-IoT data. Conversely, IoT datasets comprise data from a network of interconnected sensors. For such datasets, dependencies unique to IoT, like Device-Link, Temporal, Locality, Monitoring, and Capability Dependencies, are relevant. These dependencies, which imply certain relationships within the data, inform the structure of the context model. Figure 2 depicts the workflow that encompasses a sequence of steps tailored for both IoT and non-IoT relational datasets. The steps delineated in green signify the tasks where LLM models are employed to yield specific outcomes. The procedure commences with a dirty dataset, from which column names are extracted. These column names form the basis for the classification of the input dataset into respective types. Below, we elaborate on the steps specifically designed for each type of dataset.

Handling IoT Datasets. The workflow of IoT datasets initiates with column mapping. In this step, associations between the dataset's columns and the corresponding entities within a meta-context model are established. This mapping, critical for the subsequent generation of a context model, is executed via exploiting LLM models. The designated LLM model undertakes a systematic review of the predefined concepts, such as `ssn:System`, `ssn:Device`, `ssn:SensingDevices`, `ssn:Sensor`, `iot-lite:Location`, `iot-lite:Attribute`, `ssn:ActuatingDevice`, `iot-context:Measurement`, and `iot-list:Metadata`, to determine their relevance to the

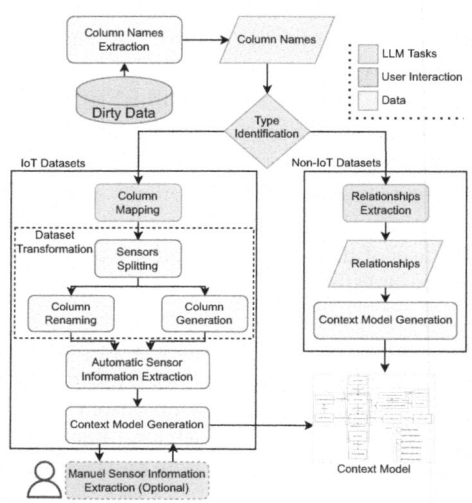

Fig. 2. Generation of context model

columns at hand. In instances where a concept lacks a corresponding column, synthetic generation is employed to ensure completeness. Following successful column-to-concept correlation within the meta-model, the dataset undergoes a transformation phase to facilitate the creation of an actionable context model compatible with data-cleaning tools. This step is partitioned into three sub-steps, including sensor splitting, column renaming, and column generation.

Sensor Splitting. This sub-step is initiated upon the identification of multiple sensor readings within a single row in the input dataset during the column mapping step. In this sub-step, a composite dataset with multiple sensor readings per row is restructured into a singularized format. To illustrate, consider an initial dataset where each row is a tuple composed of temperature (T), CO2 concentration (C), location (L), and timestamp (t). The outcome of the sensor splitting sub-step is a dataset where each tuple's sensor readings are disaggregated into distinct rows. For instance, a row (T1, C1, L1, t1) in the original dataset is divided into two separate rows in the transformed dataset: one for temperature,

(Temp, T1, L1, t1), and another for CO2 concentration, (CO2, C1, L1, t1). The location and timestamp for each sensor reading are replicated to maintain the integrity of the data, ensuring that each sensor value is contextualized by its original spatial and temporal information.

Column Generation. During the mapping step, there is a possibility that certain concepts may not be present in the input dataset or might not be recognized in the previous step. Such missing data may include parameters like the minimum and maximum sensor values, indicative of Capability Dependency, or data relating to the system's structural components, such as the device and sensor network details. The column generation phase is designed to resolve these gaps by introducing the requisite columns and populating them with synthetically derived values. However, it is pertinent to note that not all concepts or dependencies can be synthetically generated, leading to the potential exclusion of some concepts during this phase. Consider an example where the input dataset comprises only columns for "value", "location", and "timestamp". Here, if "System", "Device", "SensingDevice", and "Sensor" are requisite entities within the meta-context model, the absence of these columns necessitates their creation. Synthetic values are then assigned to these new columns to simulate system configuration. Moreover, to meet the requirements of Capability Dependency, additional columns like "MinValue" and "MaxValue" might be introduced, with default or synthetic ranges specified for sensor capacities.

Column Renaming. Upon successful assignment of columns to each concept, a validation ensures column titles align with the necessary naming conventions for OFD generation. Discrepancies are rectified through systematic renaming according to a predefined schema, converting original identifiers like "Sensor_name", "temperature", "place", and "time" are systematically converted to "sensor", "value", "location", and "timestamp", respectively. This standardization facilitates seamless integration with the OFD extraction process, avoiding errors from inconsistent naming during data cleaning. The automatic sensor information extraction step then identifies sensor types and establishes capability dependencies by specifying operational values, using queries to resources such as LLM models, Wikipedia, and Wikidata. LLMClean also allows end-users to contribute sensor information directly, enhancing context model accuracy and comprehensiveness. The final phase generates a concrete instance of the context model, starting with initial data-cleaning using statistical methods to rectify errors, yielding sanitized entities. This refined dataset is then structured into an RDF graph, where each row becomes an RDF triple network, capturing sensor data relationships and properties in a semantic construct. This transformation semantically augments the raw data, enhancing its utility for applications requiring semantically enriched, high-quality datasets.

Handling Non-IoT Datasets. Constructing context models from non-IoT relational datasets necessitates a distinct approach from that employed for IoT data, utilizing only Matching and Denial dependencies. In this context, the workflow comprises three steps. First, all possible pairs of column names within the dataset are extracted. These pairs are subjected to analysis by LLM models to ascertain the presence of semantic relationships between the columns. Second, if a relationship between two columns is identified, the concept of the two columns is determined. Finally, LLMClean assesses whether either column functions as an attribute of the other or stands as an independent concept. This critical evaluation serves to clarify the relationship between the columns, distinguishing whether they form part of a hierarchical arrangement or represent distinct concepts. Leveraging the data extracted from relationship extraction and concept mapping, the process stores column names as discrete concepts within the RDF graph. Relationships are then methodically established, linking less-encompassing concepts to more comprehensive ones. This approach facilitates the clear definition of hierarchical structures among the concepts, ensuring an organized and semantically meaningful RDF graph representation.

3   Prompt Ensembling

This section discusses the prompt engineering setup for querying LLM models in LLMClean, focusing on utilizing their reasoning capabilities for context model generation. Due to the tendency of LLM models to produce misleading results, LLM-Clean employs *Prompt Ensembling* to enhance accuracy (cf. Figure 3). This involves determining effective prompt combinations through a consensus method with various thresh-

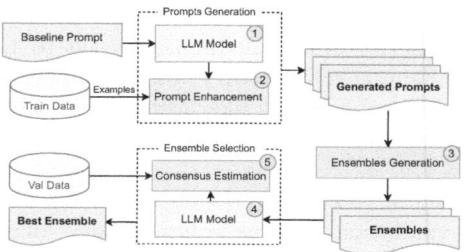

Fig. 3. Prompt ensembling method

olds. An input dataset is split into a training set, used to identify accurate prompt ensembles and their thresholds via metrics like the F1 score, and a validation set to find the optimal ensemble and threshold. The process begins with crafting a baseline prompt, generating enhanced prompts through few-shot learning, and evaluating each prompt on the training and validation datasets. LLM-Clean explores all prompt combinations, assessing ensembles by aggregating their results and computing the F1 score for each configuration. The best-performing configurations are tested on the validation dataset, and those maintaining high scores are retained. The final output is a list of top ensemble configurations based on the validation scores. A voting mechanism identifies consensus results from these ensembles, ensuring robust prediction accuracy. Together, these methods refine prompt combinations for improved outcomes.

4   Data Cleaning with LLMClean

In this section, we detail how LLMClean detects errors in tabular data by using the generated context model to create a set of OFD rules. The data-cleaning method targets two key data quality issues: missing values and functional dependency violations. First, it addresses missing values, often represented by placeholders like "N/A," "nan," "none," "null," or empty strings, which standard tools may not recognize. LLMClean translates these placeholders into a unified NaN, ensuring consistent identification of missing data. The algorithm then scans the dataset to locate and record each instance of missing data. Additionally, LLMClean identifies complex rule violations involving functional dependencies between fields. For example, it enforces rules ensuring that when the "SensingDevice" fields in two rows (from possibly the same table) are equivalent, the "Device" fields must also be identical. The algorithm segments the dataset by the determinant column and uses statistical methods to determine the modal (most frequent) dependent value within each group, flagging any deviations as anomalies. This approach efficiently isolates and identifies data inconsistencies, providing clear and actionable error indices for data practitioners. Consequently, LLMClean significantly enhances the data-cleaning workflow, leading to more accurate and reliable data analyses.

5   Performance Evaluation

In this section, we present an extensive evaluation of LLMClean in different scenarios. Through a series of carefully designed experiments, we aim to address the following key questions: (1) How effective is the proposed prompt ensemble technique at achieving its intended outcomes? (2) To what extent does fine-tuning the Llama model enhance the efficacy of our prompt ensemble technique? and (4) How does the performance of LLMClean, in terms of error detection and repair accuracy, compare to baseline methods? and By addressing these questions, we shed light on the effectiveness and potential advantages of LLMClean in the context of data cleaning. We first describe the setup of our evaluations, before discussing the results and the lessons learned throughout this study.

Experimental Setup. We conducted our experiments on an Ubuntu 20.04 LTS machine, equipped with 256 cores @ 2.45 GHz, 1 TB RAM, and four Nvidia A100 GPUs with 40GB VRAM each. However, the minimum requirement is at least one GPU with 40GB of memory. Several LLM models have been utilized in the evaluations. GPT4-turbo, the preview version, was selected for its enhanced performance and cost efficiency over GPT4. We also included GPT4, representing the series' fourth iteration, and GPT3.5, which was preferred in the development phase for its cost-effectiveness. These models provided a baseline for comparison. We tested various Llama2 configurations-70b, 13b, 7b-to leverage its open-source accessibility and parameter-driven versatility for local execution tailored to specific computational needs. To enable faster inference time, quantization has been applied to the weights of Llama2-70b from 16 bits to 4 bits. In the evaluations, we utilized three real-world datasets, namely *IoT*, *Hospital*, and *CONTEXT* datasets, all provided as reference data with intentional errors to assess the data cleaning method's efficacy. Additionally, we utilize the *LMKBC dataset* for evaluating the prompt ensembling method. In addition to the above datasets, we incorporated the LM-KBC dataset [3] as a benchmark for evaluating the efficacy of our proposed prompt ensembling algorithm.

5.1   Results

In this section, we begin with evaluating the prompt ensembling algorithm, before presenting the evaluation results of LLMClean in three scenarios, namely (S1) the *Context Change* scenario, (S2) the *Context Model* scenario, and (S3) the *Sensor Capabilities* scenario.

Prompt Ensembling. *Few-Shot Learning* In this experiment, we examined the impact of varying the number of examples in prompts on performance. We conducted experiments using randomly selected few-shot examples from the training dataset for each relation, ensuring a comprehensive range, including the largest and smallest answer sets, and considering empty sets where applicable. These examples were then integrated into either the best prompt (BP) or the best ensemble (BE) for evaluation. The results, illustrated in Fig. 4a, show a significant improvement in prediction accuracy, measured by the F1 score, with the addition of examples. The F1 score nearly doubles with just two examples, with further increases yielding more modest improvements. Additionally, the best ensemble consistently achieved a higher F1 score, outperforming the best prompt by an average of 11.2%.

Model Selection In this set of experiments, we explored the performance of the Llama2 model across various configurations, examining editions with 7 billion, 13 billion, and 70 billion parameters. In addition to these, we compared the outcomes of a non-fine-tuned model against those of a version that had undergone fine-tuning specifically for chat completion tasks. As depicted in Fig. 4b, our comparison focused on the differential impact of model size and the scenarios of using the best prompt and the best ensemble. The results from this figure indicate a clear trend: as the number of parameters in the Llama2 model increases, so does its performance. Notably, the 70 billion parameter variant, when combined with the best ensemble method, yields the highest F1 score. Another interesting insight from the figure is the fact that the 7 billion parameter model, when utilized in conjunction with the best ensemble, outperforms the 13 billion parameter model that employs only the best prompt. This observation suggests that the integration of the best ensemble technique can significantly elevate a model's effectiveness, to the extent that a less complex model can surpass a more complex one that does not utilize this optimization

Figure 4c depicts a comparative analysis between fine-tuned (FT) and non-fine-tuned (Non-FT) versions of the Llama2 model, each assessed using both the best prompt and the best ensemble. The depicted results highlight a significant finding, where the Non-FT model leveraging the best ensemble approach outperforms all other configurations in terms of F1 score. In a detailed breakdown of the performance metrics, the Non-FT model with the best ensemble

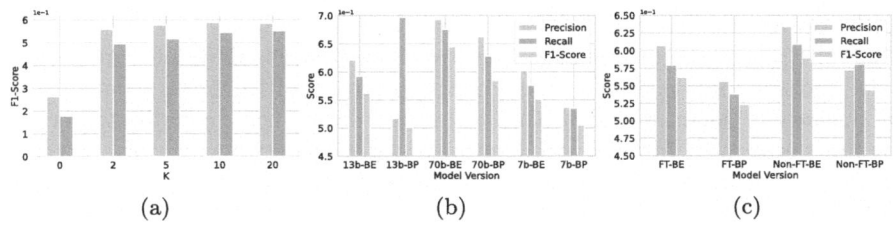

Fig. 4. LLM Results. (a) Impact of few-shot learning, where BE denotes the best ensemble and BP denotes the best prompt. (b) Comparisons of the Llama2 model with different parameter sizes, each with the best prompt or the best ensemble. (c) Comparisons of the fine-tuned (FT) and the non-fine-tuned (Non-FT) Llama2 model, each with the best prompt (BP) or the best ensemble (BE)

(Non-FT-BE) surpasses the FT model employing the best prompt (FT-BP) by a margin of 11.2%. It also outperforms the FT model paired with the best ensemble (FT-BE) by 4.6% and shows a 7.67% improvement over the Non-FT model that uses the best prompt. For development, the 13 billion parameter Non-FT model was selected. This choice was motivated by the model's computational efficiency, which provides a pragmatic balance between performance and resource expenditure. However, for the final iteration of our results, we capitalized on the superior capacity of the 70 billion parameter Non-FT version. The selection of this model was predicated on its enhanced capability to encode and process complex patterns, thereby optimizing the outcome of our data-cleaning task.

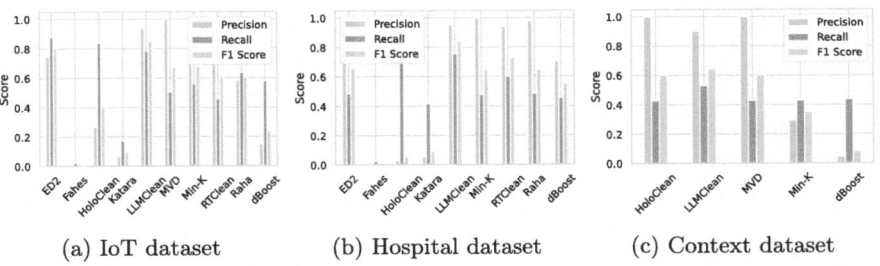

Fig. 5. Accuracy of error detection comparing LLMClean to the baselines

Figure 5 depicts the accuracy of LLMClean and the compared baseline tools–in terms of the detection precision, recall, and F-Score–while detecting errors in three real-world datasets. For the IoT datasets, Fig. 5a shows that LLMClean demonstrates superior performance in terms of the F1-score when compared to various baseline tools. It notably surpasses HoloClean, ED2, Pandas' Missing Value Detector (MVD), and Raha with substantial margins of improvement–53%, 5.4%, 21.3%, and 28.7% respectively. Further underlining its efficiency,

LLMClean identified 868 cells as containing errors, significantly lower than the 1222, 1131, and 3316 instances flagged by ED2, Raha, and HoloClean, respectively. These figures not only highlight the precision of LLMClean but also its effectiveness in accurately detecting erroneous data within a dataset. For the Hospital dataset, LLMClean demonstrates superior performance over conventional baseline tools when it comes to the detection F1-score. To illustrate, LLMClean surpasses ED2, RAHA, and dBoost by substantial margins of 22.7%, 23.2%, and 34.7%, respectively.

A closer look at the results reveals that LLMClean identified 404 cells as erroneous, which represents a significant increase in detection over the 253, 247, and 328 cells flagged by RAHA, ED2, and dBoost, respectively. Interestingly, even when operating under the same number of FD rules as HoloClean, LLMClean vastly overshadows its performance. HoloClean detected 13,044 cells, a number which, due to its magnitude, severely diminished its F1-score to less than 1%, highlighting the precision and efficiency of LLMClean. For the CONTEXT dataset, ML-based tools, such as RAHA and ED2, encountered operational challenges, specifically, they were unable to complete execution due to the dataset's extensive size. Yet again, LLMClean stands out, outstripping all baseline tools with an F1-score that is 6.2% higher in comparison to both MVD and HoloClean. This consistent outperformance across different datasets underscores the robustness of LLMClean.

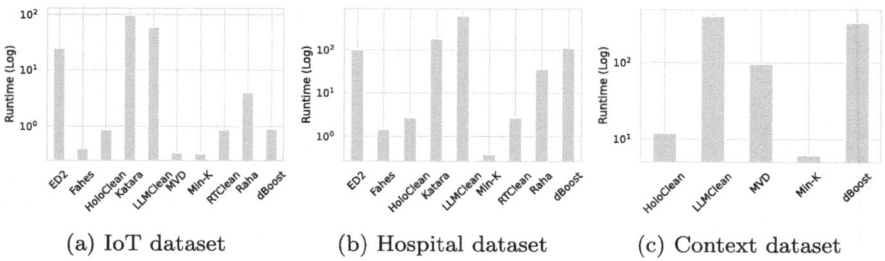

(a) IoT dataset   (b) Hospital dataset   (c) Context dataset

Fig. 6. Runtime of error detection comparing LLMClean to the baselines

Figure 6 presents the runtime analysis of LLMClean in comparison with baseline tools. From the figure, it is evident that LLMClean's runtime is marginally higher than that of the most competitive baselines. The reason for this could be attributed to LLMClean's thorough approach, which involves checking and validating numerous combinations of column pairs. Additionally, the runtime for LLMClean is largely influenced by the quantity of generated OFD rules. For instance, in the case of the IoT dataset, LLMClean takes approximately 0.98 min to apply two OFD rules. However, when applying 21 OFD rules to the Hospital dataset, the runtime extends to approximately 10.56 min. A similar pattern is observed with the CONTEXT dataset, where LLMClean spends 6.8 min to examine the data thoroughly. This contrasts with the 5.6 min taken by dBoost and a swift 1.6 min by the MVD detector. These runtime variances highlight the complexity and depth of analysis conducted by LLMClean, particularly concerning the number of OFD rules it enforces.

Error Repair Results. While LLMClean is primarily deployed for error detection within datasets, assessing the effectiveness of its detection in conjunction with subsequent repair processes is essential. This section provides an analysis of the combined performance of error detection and repair, utilizing LLMClean and various baseline tools alongside three SOTA repair tools: Baran, standard statistical imputation, and ML-based imputation. For statistical imputation, we apply mean-value imputation for numerical attributes and mode-value imputation for categorical ones. The ML-based imputation employs a K-Nearest Neighbors (KNN) regressor for numerical attributes and MissForest for categorical attributes. An examination of the results for the IoT and Hospital datasets, in Fig. 7, reveals noteworthy outcomes[1]. Specifically, Fig. 7a indicates that LLM-Clean consistently achieves the lowest RMSE of 0.22 across numerical attributes, independent of the repair mechanism employed. This performance is comparable to the results of other tools, such as HoloClean, MVD, and Nadeef.

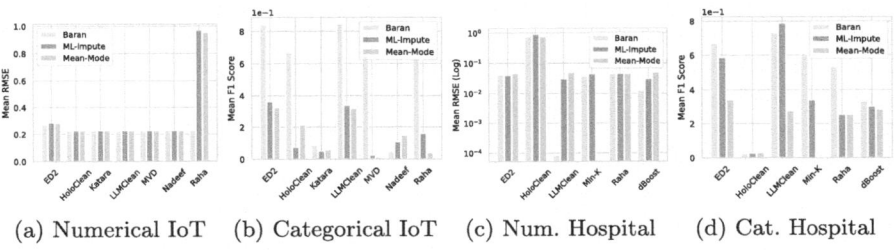

(a) Numerical IoT   (b) Categorical IoT   (c) Num. Hospital   (d) Cat. Hospital

Fig. 7. Accuracy of error repair comparing LLMClean to the baselines

For the categorical attributes of the IoT dataset, as shown in Fig. 7b, the combination of LLMClean and Baran attains the highest F1-score at 85%. This marginally surpasses the 84% F1-score observed with ED2 and Baran, and significantly outperforms the 67% F1-score seen with MVD and Baran. For the numerical attributes of the Hospital dataset, as depicted in Fig. 7c, the pairing of LLMClean with Baran again yields the most favorable RMSE value (8.44E-05). It is important to note that the vertical axis of this figure is in a logarithmic scale to appropriately represent the notably small values achieved by this combination. Figure 7d demonstrates that for the Hospital dataset's categorical attributes, LLMClean used with ML-based imputation (ML-Impute) secures the highest F1-score of 78.7%. This result outdoes the 73% F1-score obtained with LLMClean and Baran, as well as the 66.7% achieved by combining ED2 with Baran.

---

[1] The CONTEXT dataset analysis is omitted here for conciseness.

6   Conclusion and Future Work

This paper introduces a novel method that leverages LLM models to autonomously extract context models from datasets without requiring additional dataset information, addressing the labor-intensive nature of manual creation. These context models are essential for data cleaning and preparing data for AI applications, with OFD dependencies supporting this process. LLMClean streamlines context model creation for integration into data-cleaning workflows, focusing on two dataset types: IoT datasets with sensor data and non-IoT data in relational databases. Evaluations show that LLMClean effectively cleanses both types of datasets. Future research could enhance table-to-knowledge graph conversions by incorporating knowledge graph embeddings, improving accuracy and semantic richness.

References

1. Abdelaal, M., Hammacher, C., Schoening, H.: REIN: a comprehensive benchmark framework for data cleaning methods in ML pipelines. In: EDBT (2023)
2. Del Gaudio, D., Schubert, T., Abdelaal, M.: RTClean: context-aware tabular data cleaning using real-time OFDs. In: CoMoRea 2023. IEEE (2023)
3. Singhania, S., Nguyen, T.P., Razniewski, S.: LM-KBC: knowledge base construction from pre-trained language models, semantic web challenge. In: CEUR-WS (2023)

DOING 2024: 5th Workshop on Intelligent Data - From Data to Knowledge

Capturing Analytical Intents from Text

Gerard Pons[(✉)] , Miona Dimic, and Besim Bilalli

Universitat Politècnica de Catalunya, Barcelona, Spain
`{gerard.pons.recasens,miona.dimic,besim.bilalli}@upc.edu`

Abstract. The ability to extract valuable information from data is crucial for organizations and individuals who want to remain competitive in a constantly evolving data-driven environment. However, some of them lack the skills required to appropriately leverage the existing data analytics tools and methods. This problem is aggravated when the users are domain-experts but completely unfamiliar with data analytics terminology, as existing assistant tools, such as AutoML or Intelligent Discovery Assistants, require them to state their analytical intent (i.e., the type of data analysis they want to perform). To address this problem, we propose to capture the underlying analytical intent from textual problem descriptions by leveraging Large Language Models (LLMs). To this end, we propose a hierarchical categorization of analytical intents, along with a data collection methodology to obtain analytical problem descriptions for all of them in order to validate different approaches that aim to extract such intents from text. Next, we compare the performance of state-of-the-art approaches with LLMs, and then study the performance of different LLMs based on their characteristics and the impact of the source of validation data. Finally, we develop a prototype to showcase how our method could interact with existing AutoML systems.

Keywords: Intelligent User Assistance · Data Science · Natural Language Processing

1 Introduction

In an era characterized by a rapid growth of the volume and complexity of data, the ability to effectively analyze and interpret such information empowers companies, organizations and researchers to stay ahead in the competitive landscape. However, there exist individuals who may not possess the data analytics or programming skills needed to appropriately harness the knowledge extraction capabilities enabled by existing methods. To mitigate this issue, some systems have been lately developed to minimize the skill requirements for data analytics. Among them, Intelligent Discovery Assistants (IDAs) are tools that guide the end-users during the generation of analytical workflows, by recommending actions in each step and allowing them to easily interact with the system. Alternatively, AutoML tools have been developed to automate different steps of the

© The Author(s), under exclusive license to Springer Nature Switzerland AG 2025
J. Tekli et al. (Eds.): ADBIS 2024, CCIS 2186, pp. 81–94, 2025.
https://doi.org/10.1007/978-3-031-70421-5_8

creation of data analytics pipelines by directly generating optimized workflows, thus removing the need for end-users to understand the underlying processes.

However, these systems do not completely remove the entry barrier, as usually the purpose or type of the analysis (e.g., classification, regression, clustering, etc.), which we refer to as analytical intent, still needs to be specified. This initial step becomes challenging when users do not know how to translate an analytical intent or a business question into data analytics terms. Nevertheless, they do know how to express the purpose of their analysis in text, even if it is a high level domain-focused description. Hence, in this work we explore the possibility of capturing the underlying analytical intents from textual problem descriptions, so that an analytical workflow can be generated without any further compulsory user interaction.

Therefore, our approach consists of assigning to a document d composed by a sequence of words $\{x_1, x_2, \ldots, x_n\}$ one of k different target labels. This corresponds to a text classification task, which in multiple domains has traditionally been solved with standard machine learning models (e.g., Random Forests, K-Nearest Neighbors, etc.), by first extracting hand-crafted features from the text which are then fed to the classifiers to make a prediction. However, to deal with the drawbacks of having to manually design the features, more recent approaches leverage neural architectures and create models that generate embeddings from the textual input which are then transformed into the corresponding label. Even if these models are built on top of already powerful pre-trained architectures, such as BERT [3] or RoBERTa [7], they require specific fine-tuning to adapt them to the specific domain and to adjust the parameters of the newly added classification components. Lately, new advances in Large Language Models (LLMs) have demonstrated their ability to solve Natural Language Processing (NLP) tasks [13], even in a zero-shot manner. With this, the need for training specific models is replaced by the creation of concrete prompts that can solve the problem at hand. In this work, we study to what extent these LLMs are capable of correctly solving the classification of analytical intents, or if it is still worth to spend time and effort designing and training classification-specific models.

The generated text classifier is envisioned as an initial module that can be integrated to different IDA or AutoML tools, the latter exemplified in Fig. 1. There, a user starts by stating, with a textual input, the objective or description of their analysis. Then, the analytical intent is extracted from this text by the classifier and the answer is used as an input for the tool, so that the search space is appropriately restricted. To this end, in this work we also create a hierarchical representation of analytical intents, which are fine-grained enough to be restrictive for the automation tools but also general enough so that they can be detected from high-level problem descriptions. Finally, the AutoML tool generates an analytical workflow, which is presented to the user.

Therefore, the contributions of this work are:

- A hierarchical categorization of analytical intents, covering the data analytics algorithms going from descriptive analytics to clustering or time-series forecasting.

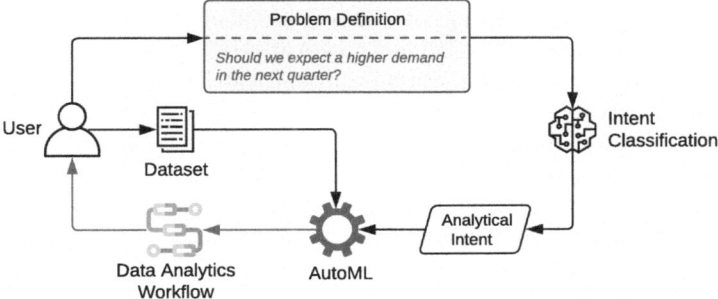

Fig. 1. Example of automatic workflow generation from a problem description.

– A preliminary study on the performance of different text classification methods in the analytical intent domain.
– A study on the performance of different LLMs and the impact of different data sources.
– A prototype that shows the feasibility of our approach and its potential integration with existing tools.

2 Related Work

In this section we first review how users are typically assisted in data analytics environments and then introduce the different methodologies currently used for text classification.

2.1 Assisting Users in Data Analytics Workflow Creation

Assisting users in the design of data analytics workflows is crucial for democratizing the access to analytical tools and algorithms. This is especially relevant for those users who are domain experts but do not possess sufficient data analytics or programming skills. Some of the initially proposed methods for advising users in the different steps of the generation of analytics workflows are IDAs [10], and they can be classified based on the methods they employ. For instance, Expert systems use a repository of hand-crafted rules by analytical experts, which depending on the circumstances, suggest different types of algorithms. Meta-learning approaches base the recommendations on the relationships between the datasets' characteristics and the algorithms' performance, hence moving away from the manual interaction of experts. Case-based reasoning systems contain an expert-supervised repository of successful cases, and base the recommendations on retrieving the most promising workflows obtained from similar datasets. Finally, planning-based systems build the workflow by taking into account the inputs, outputs and preconditions of the algorithms that compose it.

More recently, the automation of different steps of data analytics workflows creation has been proposed; this is known as AutoML [5]. Here, the user involvement is further reduced, as the system automatically optimizes a workflow by exploring the search space of algorithms and parameters. This is beneficial for all kinds of users, as it streamlines some of the most tedious tasks for data analytics experts while allowing novice users to effectively obtain optimized workflows.

However, the vast majority of these systems still require user interaction to select the analytical intent, which can be a source of doubt or even error for domain expert users who do not know how to translate their intentions using data analytics terminology. In [4], this issue is solved by deciding the intent based on the distribution of the dataset's target column. This is, if the target column type is numerical, the analytical intent is set to regression, and if it is a string, the intent is set to classification. This approach limits the scope to two prediction analytical intents, does not take into account users who may like to perform classification over a numerical column by first preprocessing it (e.g., data binning) and it is not extensible to analytical intents which do not require a target (e.g., clustering). These problems could be mitigated by asking the users to describe the problem they want to solve, even in domain-related terminology, and analyzing the resulting text to extract the analytical intent. This conversion from text to intent has been explored in [14], where features are extracted from the text (i.e., embeddings, keywords and topics) and are used as an input for a classifier. However, this work only considers prediction, clustering and frequent pattern mining as analytical intents, which are not extensive enough to encompass all the analytical purposes a user may request (e.g., descriptive analytics) and also not granular enough (i.e., prediction could be decomposed in classification and regression) to be used as an input for different IDAs or AutoML systems. To this end, in this work we explore how to further represent and detect the analytical intents, in different levels of granularity, in order to effectively reduce the entry barrier for the interaction with data analytics assistant tools for non-expert users.

2.2   Text Classification

Text classification is essential for extracting knowledge from unstructured textual data, enabling tasks such as text organization for improving information retrieval or document filtering for performing spam detection or feedback analysis. Initially, text classification has been based on the creation of hand-crafted rules to extract features from the texts, which are then used by machine learning classifiers (e.g., SVM, RF, KNN, etc.). For instance, some features take into account the appearance of certain keywords, syntactic patterns in the structure of the sentence or regular expression matching, hence a deep understanding of the domain is required for their creation.

In the past decades, neural architectures have gained relevance in the NLP community, and have also been explored to mitigate the problems originated by hand-crafting features. With these architectures, such as Recurrent Neural

Networks, Convolutional Neural Networks or LLMs, a low dimensional feature-vector is generated, which is then fed to a classifier. Specifically, Transformer-based pre-trained Language Models, such as BERT or RoBERTa, have been used effectively to build text classifiers. These models have deep architectures and are pre-trained on huge amounts of text data to learn contextual representations. Usually, the pre-training is continued with domain-specific data and a task-specific layer (i.e., a classifier with the desired labels) is added on top. Finally, the whole architecture is fine-tuned to perform the classification task with labeled data.

Recently, new advances in LLMs such as GPT-3 [2], GPT-4 [1] or LLaMA-2 [11], have positioned them as a promising approach for solving NLP tasks [13], even without the need of training or fine-tuning them for the task at hand (i.e., zero and few shot approaches). This is, these generative LLMs can be prompted to directly produce an answer, allowing, in the case of text classification, a very flexible adaptation to different target classes or domains. In this work, we study how these LLMs can be used for intent classification.

3   Intent Classification

In this section, we motivate the usage of LLMs for intent classification by comparing their performance with other methods for the intent classification task introduced in [14]. Then, we present an expanded categorization of intents and the proposed methodology to gather the data needed for the validation of the classification methods.

3.1   On the Use of LLMs for Capturing Analytical Intents

There exist different text classification techniques, with various levels of modeling effort, that can be employed for the classification of analytical intents. To explore how they affect the classification, we have conducted a preliminary analysis over the validation set provided in [14], which considers the *clustering* (CL), *prediction* (PR), and *frequent pattern mining* (FPM) intents. This dataset consists of 60 analytical problem descriptions for these analytical intents (i.e., 20 of each). The methods studied, presented in decreasing order of modeling effort, are:

- The best performing classifier proposed in [14] is assessed, corresponding to a high modeling effort approach. It uses FastText embeddings for word vectors, which then go through an ensemble method consisting of an LSTM, a one-to-one GRU and an SVM. To train this model, 22.000 text instances gathered from scientific databases and online journals are used.
- Then, the effort is decreased by leveraging pre-trained LLMs, in our case BERT and RoBERTa. To this end, a linear classification module has been built on top of the architectures, using the pooled output as the classifier input. Then, the full architecture has been fine-tuned with the same training data as in [14]. The architecture and fine-tuning details can be found in the GitHub repository.[1]

---

[1] https://github.com/mionaD-upc/Capturing-Analytical-Intents-From-Text.

Table 1. Preliminary Intent Classification Evaluation

Classifer	Precision			Recall			Accuracy
	CL	FPM	PR	CL	FPM	PR	
TbIAS [14]	1.00	0.79	0.86	0.70	0.95	0.95	0.87
BERT	1.00	0.79	0.90	0.75	0.95	0.90	0.87
RoBERTa	1.00	0.95	0.87	0.85	0.95	1.00	0.93
GPT-4	1.00	1.00	0.91	0.95	0.95	1.00	0.97
Mistral-7b-instruct	1.00	0.87	1.00	0.95	1.00	0.90	0.95

– Finally, and as the approach with the least modeling effort, the use of generative LLMs has been used to solve the task via prompting. Therefore, generic prompt templates have been designed to automatically predict the intent class from the textual input. Here, LLMs with various characteristics (i.e., number of parameters, open source) have been studied.

The results of the experiments are shown in Table 1, where the evaluations are reported in terms of *precision* and *recall* for individual classes, along with an assessment of the *overall accuracy*. It can be observed that a simple classification module over the BERT-like architectures achieves a better or similar performance than the best model in [14], without the effort of building different classifiers and ensembling them. Additionally, the performance is significantly improved by LLM prompting, even when using the small open-source models.

This preliminary study, based on data from [14], shows that, the use of LLMs for extracting analytical intents is not only the best performing strategy, but also the one that requires less modeling effort. Motivated by this, in the rest of this work we further explore the capabilities of LLMs using newly acquired validation data. To this end, in order to do it in a systematic manner, we propose the method depicted in Fig. 2. That is, we start by proposing a hierarchical categorization of analytical intents with the aim of being complete and thus categorizing intents that have not been considered in the state-of-the-art. Next, we propose a generic data generation process in order to generate validation

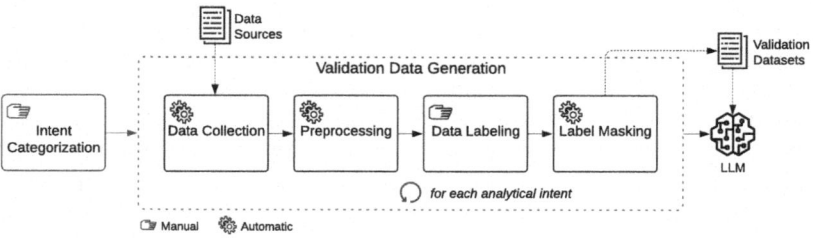

Fig. 2. A systematic method for the validation of analytical intent text classifiers.

data for each category. Finally, we consider using different LLMs to perform predictions over the validation data generated for each analytical intent.

3.2   Intent Categorization

The intents defined in [14] are not sufficient to cover all the users' analytical needs. Therefore, we expand their categorization with a hierarchical structure, having a top layer with more general intents based on the work presented in [12], and a second level with more concrete data analytics tasks. The classifier will have as labels the tasks defined on the second level, thus resulting in concrete enough predictions to narrow down the selection of fit-for-purpose data analytics methods. Therefore, the intents (see Fig. 3) can be organized into five different groups:

- **Describe:** the objective is to provide a descriptive view of the instances or variables of a dataset. Therefore, *Data Profiling* (DP) and *Clustering* (CL) techniques can be found in this group.
- **Assess:** the intent aims to study the presence of discrepancies, correlations or associations between instances or variables. Under this intent we can find data analytics methods corresponding to *Correlation Analysis* (CA), *Anomaly Detection* (AD) and *Association Rule Mining* (AR).
- **Explain:** this intent encompasses the techniques aimed at exploring the casualties towards the outcome, which corresponds to *Causal Inference* (CI).
- **Predict:** the goal is to forecast a future event or to predict the likelihood of a particular outcome. Therefore, algorithms addressing *Classification* (CS), *Regression* (RE) and *Forecasting* (FC) tasks correspond to this intent.
- **Suggest:** the aim of this approach is to give recommendations to guide future actions, thus it consists of inspecting the results from the other categories.

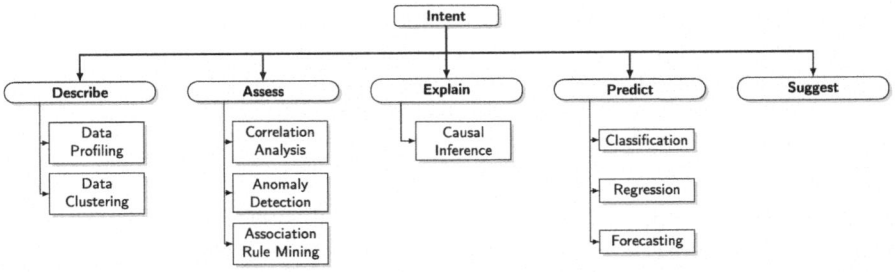

Fig. 3. Intent Categorization

Additionally, the proposed intents can be tied to distinct types of data analytics [9]. *Describe* aligns with summarizing historical data, which falls under descriptive analytics. *Assess* and *Explain* can be linked to diagnostic analytics, focused on identifying patterns and causal relationships toward the outcome. *Predict* corresponds to predictive analytics, forecasting future outcomes. Lastly, *Suggest* is associated with prescriptive analytics, aimed at finding optimal strategies.

3.3   Validation Data Collection

We propose a four-step pipeline to obtain the validation data, consisting of Data Collection, Preprocessing, Label Verification and Label Masking. This generic processes can be instantiated over different data sources provided that there exist a connector to crawl the data, and it is not specific for intent classification.

In this domain, the availability of datasets containing analytical problem descriptions across a wide spectrum of classes is notably scarce. Typically, analytical problems are documented through research papers, code repositories, or web articles, written in structured sentences that may not accurately reflect natural language usage. They are frequently presented as theoretical discussions on applications or as instructional step-by-step guides toward achieving analytical objectives.

Specifically, question-and-answer forums serve as excellent representatives of real user-defined problems, particularly focused on specific domains. Therefore, for the sake of the proposed experiments, the validation data has been crawled from the collaborative platforms *Kaggle*[2] and *Cross Validated*,[3] both oriented towards Data Science, generating the KG-QA and CV-QA datasets respectively. Additionally, another dataset, KG-Comp, has been generated from the competitions that different organizations propose to *Kaggle* users. To this end, two data collectors, *Kaggle-DataCollector* and *CrossValidated-DataCollector* have been implemented,[4] serving as semi-automated data retrieval tools.

Data Collection. The data has been collected using the search query functionalities enabled by both platforms (see Table 2). In the case of *Kaggle*, searches are conducted using keywords (i.e., exact matches of the word), while *CrossValidated* enables search by tags (i.e., the categories assigned to each post). These keywords and tags are derived from the intent categorization, under the criteria that a source is considered relevant for the intent if at least 50 search results are available for it.

Preprocessing. Competitions and question-and-answer forums are valuable sources of user-defined analytical problems, yet they are often accompanied by noise such as code snippets and images. Therefore, preprocessing is necessary to extract the relevant textual data for further analysis.

The preprocessing stage involves extracting the content from the first message of the user who posted the question for *Kaggle QA* and *Cross Validated*, as well

---

[2] https://www.kaggle.com/.

[3] https://stats.stackexchange.com/.

[4] https://github.com/mionaD-upc/Capturing-Analytical-Intents-From-Text.

Table 2. Search queries for the three validation datasets.

Class	KG–QA	KG-Comp	CV-QA
DP	–	–	[data-visualization] **or** [descriptive-statistics]
CL	clustering	clustering	[clustering]
CA	correlation	–	[correlation]
AD	anomaly detection	anomaly detection	[anomaly-detection]
AR	–	–	[association-rules]
CI	–	–	[causality]
RE	regression	regression	[regression] **and** [predictive-models]
CS	classification	classification	[classification] **and** [predictive-models]
FC	forecasting	forecasting	[forecasting] **and** [predictive-models]

as the *Kaggle Competition* descriptions. Only textual data is retained, with any additional elements removed. Furthermore, only results written in English are kept for further processing.

Label Verification. Each retrieved text is associated with the search query that produced it. However, this association does not guarantee accurate labeling. This noise could be assumed for the creation of training datasets, but it is not acceptable in a validation context. Therefore, a manual label verification feature has been implemented where an evaluator (i.e., a data analytics expert) is presented with the file content and asked whether the label is correctly assigned. If not, the file is excluded from the dataset.

With the purpose of ensuring a consistent label verification, the following rules are applied to flag as incorrect and thus discard a labeled document:

– If mentions of specific algorithms are present.
– If the content includes theoretical explanations.
– If textual data resembling tables/code are present.
– If descriptions of analytical workflows are provided.
– If the majority of the content consists of advertisements.

Table 3 shows the number of files that have been retrieved and how many have been disregarded in the preprocessing or in the label verification stages. The validation dataset size was targeted to 20 documents per class in order to be consistent with the dataset in [14]. This was achieved for the KG-QA and KG-Comp, but for CV-QA only 15 documents per class could be retrieved due to the high levels of noise in the documents (see Table 3).

Label Masking. Validation data was retrieved by keyword/tag and thus in many cases contains the search query. This could pose a problem as the intent completely or closely matches the search query, thus potentially introducing bias in the classification.

Table 3. Number of retrieved (Ret), disregarded in the preprocessing (D-Pre) or in the verification (D-Ver) files.

Class	KG-QA			KG-Comp			CV-QA		
	Ret	D-Pre	D-Ver	Ret	D-Pre.	D-Ver	Ret	D-Pre	D-Ver
Data profiling (DP)	–	–	–	–	–	–	250	1	234
Clustering (CL)	200	3	177	84	26	38	250	0	235
Correlation analysis (CA)	200	7	173	–	–	–	500	0	485
Anomaly detection (AD)	94	1	73	79	25	34	200	1	184
Association rules (AR)	–	–	–	–	–	–	105	0	90
Regression (RE)	200	5	175	200	24	156	392	0	377
Classification (CS)	200	4	176	200	15	165	312	0	297
Forecasting (FC)	200	1	179	200	40	140	200	0	185
Causal inference (CI)	–	–	–	–	–	–	500	0	485

To solve this issue, first the words sharing the same stem as the class name (e.g., 'clust' for clustering) were systematically identified and substituted with appropriate synonyms sourced from *WordNet* [8], taking into consideration linguistic attributes such as part-of-speech (POS) tag and plurality. However, certain class names such as 'anomaly' and 'correlation' posed challenges as WordNet failed to provide sufficiently distinct synonyms. To address this limitation, a mapping strategy leveraging synonyms sourced from a *Thesaurus synonym dictionary* was employed. For instance, 'correlation' was mapped with 'interdependence', and 'anomaly' was associated with 'aberration'. In instances involving the classes regression, classification, and forecasting, where the aforementioned substitution method was inapplicable, variations of the term 'prediction' (e.g., predicts, predict) were utilized. Additionally, occurrences of terms such as 'classifier' or 'regressor' were systematically replaced with 'model' to introduce lexical ambiguity.

4   Experiments

The objective of the following experiments is two-fold. First, we want to study the ability of LLMs to correctly classify analytical intents and explore the differences between currently available LLMs. Moreover, we also study the impact of the source of validation data to the classification performance.

4.1   Experimental Setting

In these experiments, the performance of LLMs on various datasets is explored:

Validation Datasets. The three datasets (i.e., KG-QA, KG-Comp and CV-QA) contain the intents presented in Table 3. These datasets have been used to study how the performance differs with different types of documents.

LLMs. The LLMs under consideration differ in their sizes and in their availability as open-source. For open-source LLMs, we have considered models from Mistral and MetaAI. From Mistral, a mixture of experts with 8 models of 22 or 7 billion parameters each has been studied (Mixtral-8 × 22b and Mixtral-8 × 7b), along with a smaller 7 billion parameter model (Mistral-7b). From MetaAI, the Llama 3 models have been explored: the current largest available, Llama3-70b, and the smallest one, Llama3-8b, with 70 and 8 billion parameters respectively. As proprietary models, the LLMs from OpenAI, specifically GPT3.5-turbo and GPT-4, have been considered, the latter being much larger and with enhanced reasoning capabilities.

4.2   Results

The results can be seen in Table 4. The best results are obtained with the largest model available, GPT-4, and are followed by the open-source model with the highest number of active parameters, Llama3-70b. Interestingly, the latter outperforms GPT-3.5, which has an estimated number of parameters of 175 billion. For each family of models, the smallest ones are expected to have the worst performance. This is not the case for Mixtral-8 × 22b, which has been seen to produce answers with multiple classes or without giving a conclusive result. The results of this experiment are worse in terms of accuracy compared to the preliminary analysis. This observation could be attributed to the lower number of intent classes and the curation of the validation dataset of [14], which contain descriptions written or re-written by the authors. Comparing the datasets, it can be observed that on average KG-QA is the one that obtains a better performance, and KG-Comp and CV-QA obtain similar results. The difference between KG-QA and CV-QA can be explained by the difference in the number of classes, which leads to confusion on some specific intents (see Table 5). The drop in performance in KG-Comp, which is the dataset with fewer classes, could be attributed to questions being more specific than the listed problems.

Table 4. Accuracy on expanded validation sets.

Classifier	KG-QA	KG-Comp	CV-QA	Average
GPT-4	**0.93**	0.86	**0.88**	**0.89**
GPT-3.5 Turbo	0.91	0.78	0.79	0.82
Mixtral-8 × 22b-instruct	0.68	0.66	0.75	0.69
Mixtral-8 × 7b-instruct	0.8	**0.89**	0.73	0.80
Mistral-7b-instruct	0.78	0.76	0.61	0.71
Llama3-70b	**0.93**	0.82	0.85	0.86
Llama3-8b	0.74	0.65	0.61	0.66

Table 5. Precision and recall for each intent class with GPT-4.

Dataset	Precision									Recall								
	DP	CL	CA	AD	AR	CS	RE	FC	CI	DP	CL	CA	AD	AR	CS	RE	FC	CI
KG-QA	–	1.00	1.00	0.95	–	0.85	0.94	0.82	–	–	1.00	1.00	1.00	–	0.85	0.80	0.90	–
KG-Comp	–	1.00	–	1.00	–	0.70	1.00	0.77	–	–	0.75	–	0.85	–	0.95	0.75	1.00	–
CV-QA	1.00	0.93	0.88	0.83	1.00	0.83	0.83	0.79	0.93	0.53	0.93	0.93	1.00	0.93	1.00	0.67	1.00	0.93

5   End-to-End Prototype

The generated text classifier is envisioned as a module that can be added to different AutoML tools and IDA systems, with the purpose of removing the need for manually selecting the intent of the analysis. In order to exemplify the integration, a prototype[5] has been created by leveraging an AutoML tool, concretely Hyperopt [6], which focuses on model and hyperparameter selection for classification and regression tasks. Therefore, as the system expects the user to manually choose the type of the analysis, we replace this selection stage by asking for the problem's description or objective, and use the classifier to select the appropriate intent. This intent is then fed as an input to the AutoML tool, which produces an optimized workflow. Then, a final visualization module displays the results and the generated pipeline. In Fig. 4 an example of the interaction is shown, where from the user's textual problem description a final optimized workflow is generated.

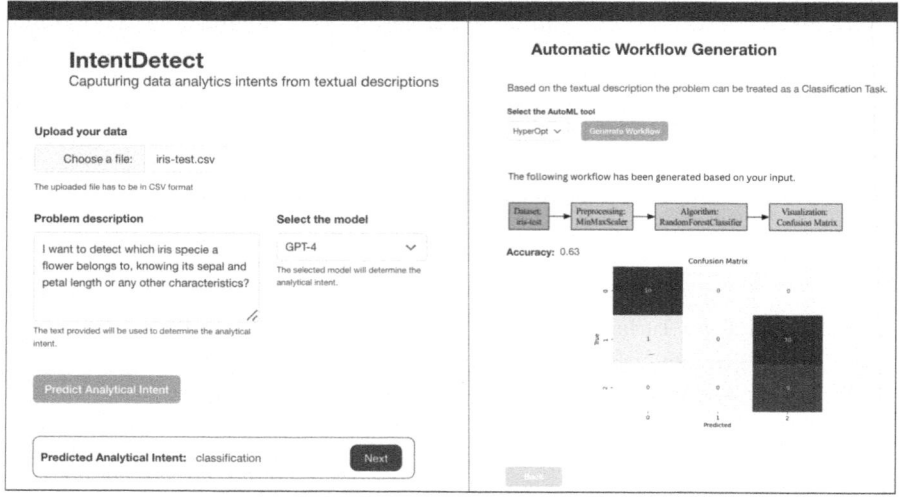

Fig. 4. Example of the integration of the text classifier with an AutoML tool.

---

[5] https://github.com/mionaD-upc/Capturing-Analytical-Intents-From-Text.

6 Conclusions

In this work we have focused on the translation of textual descriptions of data analytics problems to their underlying intent. For this purpose, we have defined an intent categorization and explored how to solve the classification problem with LLMs. To this end, we have studied the performance of LLMs with different characteristics and how it is affected by changes in the source of validation data. It has been observed that the number of parameters still has a large impact on the results, but the gap is being progressively closed by the appearance of new open-source LLMs. Furthermore, the sources from which the data is collected affect the performance, due to the style of the question or problem definition. Finally, we have integrated the different classifiers to an existing AutoML tool, to showcase how it could be used to automatically generate analytics workflows from textual descriptions.

Acknowledgments. Gerard Pons is supported by the EU's Horizon Programme call, under Grant Agreement No. 101093164 (ExtremeXP), and Besim Bilalli is partially supported by the DOGO4ML project, funded by the Spanish Ministerio de Ciencia i Innovación under the funding scheme PID2020-117191RB-I00/AEI/10.13039/501100011033.

Disclosure of Interests. The authors have no competing interests to declare that are relevant to the content of this article.

References

1. Achiam, J., et al.: GPT-4 technical report (2023). arXiv preprint arXiv:2303.08774
2. Brown, T., et al.: Language models are few-shot learners. Adv. Neural. Inf. Process. Syst. **33**, 1877–1901 (2020)
3. Devlin, J., Chang, M.W., Lee, K., Toutanova, K.: BERT: pre-training of deep bidirectional transformers for language understanding. In: Burstein, J., Doran, C., Solorio, T. (eds.) Proceedings of the 2019 Conference of the North American Chapter of the Association for Computational Linguistics: Human Language Technologies, Volume 1 (Long and Short Papers). pp. 4171–4186. Association for Computational Linguistics, Minneapolis, Minnesota (2019). https://doi.org/10.18653/v1/N19-1423, https://aclanthology.org/N19-1423
4. Helali, M., Mansour, E., Abdelaziz, I., Dolby, J., Srinivas, K.: A scalable AutoML approach based on graph neural networks. Proc. VLDB Endowment **15**(11), 2428–2436 (2022). https://doi.org/10.14778/3551793.3551804
5. Karmaker, S.K., Hassan, M.M., Smith, M.J., Xu, L., Zhai, C., Veeramachaneni, K.: AutoML to date and beyond: challenges and opportunities. ACM Comput. Surv. (CSUR) **54**(8), 1–36 (2021)
6. Komer, B., Bergstra, J., Eliasmith, C.: Hyperopt-Sklearn. In: Hutter, F., Kotthoff, L., Vanschoren, J. (eds.) Automated Machine Learning. The Springer Series on Challenges in Machine Learning. Springer, Cham (2019). https://doi.org/10.1007/978-3-030-05318-5_5
7. Liu, Y., et al.: RoBERTa: A robustly optimized BERT pretraining approach (2019). arXiv preprint arXiv:1907.11692

8. Miller, G.A.: WordNet: a lexical database for English. Commun. ACM **38**(11), 39–41 (1995)
9. Sarker, I.H.: Data science and analytics: an overview from data-driven smart computing, decision-making and applications perspective. SN Comput. Sci. **2**(5), 377 (2021)
10. Serban, F., Vanschoren, J., Kietz, J.U., Bernstein, A.: A survey of intelligent assistants for data analysis. ACM Comput. Surv. (CSUR) **45**(3), 1–35 (2013)
11. Touvron, H., et al.: Llama 2: Open foundation and fine-tuned chat models (2023). arXiv preprint arXiv:2307.09288
12. Vassiliadis, P., Marcel, P., Rizzi, S.: Beyond roll-up's and drill-down's: an intentional analytics model to reinvent OLAP. Inf. Syst. **85**, 68–91 (2019). https://doi.org/10.1016/j.is.2019.03.011
13. Yang, J., et al.: Harnessing the power of LLMs in practice: a survey on ChatGPT and beyond. ACM Trans. Knowl. Discov. Data **18**(6), 1–32 (2023)
14. Zschech, P., Horn, R., Höschele, D., Janiesch, C., Heinrich, K.: Intelligent user assistance for automated data mining method selection. Bus. Inf. Syst. Eng. **62**, 227–247 (2020)

The Effect of Text Normalization on Mining Portuguese Man-of-War Instagram Posts

Heloisa F. Rocha[1]🆔, Carlos A. Prolo[2]🆔, Aurora R. Pozo[1]🆔,
and Carmem S. Hara[1](✉)🆔

[1] Universidade Federal do Paraná, Curitiba, PR, Brazil
{heloisarocha,aurora.pozo,carmemhara}@ufpr.br
[2] Universidade Federal do Rio Grande do Norte, Natal, RN, Brazil
prolo@dimap.ufrn.br

Abstract. Text extracted from social media have distinct characteristics, such as: use of hashtags, emojis, neologisms, mix of languages and informal writing. These features increase the challenge of determining which normalization techniques should be applied in the preprocessing step in order to maximize the performance of classifiers. In this work we evaluate the impact of applying these techniques on the result of models trained to classify Instagram posts as legitimate occurrences of the Portuguese man-of-war. We performed experiments with different techniques and individually evaluated their performance in comparison to models trained with raw data. The results showed that some normalization can be interesting to apply when training the language representation model BERT, while none of them significantly outperformed the classical models trained with raw text.

Keywords: natural language processing · social networks

1 Introduction

The Portuguese man-of-war (PMW - cnidarian *Physalia physalis*) is an animal that is usually mistaken as a jellyfish and has beautiful and eye-catching colors. However, it poses risk to the population, as it may cause severe burns if touched. Data on their occurrences on the Brazilian coast are not always available in traditional sources. On the other hand, previous studies [10] show that social media can be an effective source of information for monitoring animal sightings.

In this work, we are interested in determining whether posts extracted from Instagram are legitimate occurrences of PMW on the Brazilian coast. The posts were extracted based on a set of hashtags, such as #caravelaportuguesa, and

This work was partially supported by CNPq - Edital Universal 407644/2021-0 and CAPES-PrInt-UFPR.

J. Tekli et al. (Eds.): ADBIS 2024, CCIS 2186, pp. 95–103, 2025.
https://doi.org/10.1007/978-3-031-70421-5_9

#physaliaphysalis [3]. However, some of the collected posts refer to other subjects, such as a type of Portuguese ship ("caravela") and tattoos. Thus, there is a need to filter out such posts. A previous work has investigated the problem based solely on their image [4]. Here, we are considering the text associated with the post. More specifically, we are interested in applying text mining techniques to classify Instagram posts as legitimate occurrences of PMW based solely on their text. Future work includes the combination of image and text classifiers.

Text extracted from social media have distinct characteristics, such as: use of hashtags, emojis, neologisms, mix of languages and informal writing. These features increase the challenge of determining which normalization techniques should be applied during the preprocessing step in order to maximize the performance of the classifiers. Furthermore, each treatment consumes time and processing energy. Thus, it is important to determine which ones may in fact affect the classification result. This is especially important when the classification model is continuously used in real applications, on new incoming text.

Although there are studies in other languages and domains [7], in this paper we focus on preprocessing techniques applied to text in Portuguese extracted from Instagram. We came across several natural language processing (NLP) works applying different techniques, usually in sequence, as in the works of [9] and [6]. We also found few works that proposed to compare the impact of preprocessing applied to machine learning models [2,11,14,15]. However, there are no general guidelines to follow in order to determine which techniques to apply. In [8], the book authors state that before almost any NLP task, the text must be normalized, but they leave it up to the developer to choose which treatments to apply.

But what is the impact of removing numbers, mentions, urls, emojis, hashtags on the performance of the model? Is it better to tokenize these elements instead of removing them? Should we lemmatize or stemmize? Stopwords should be removed? Could spelling correction improve the performance of the models? The purpose of this paper is to evaluate the impact of normalization techniques commonly used in NLP works on the result of models trained to classify Instagram posts about PMW. We performed experiments with different techniques and individually evaluated their performance in comparison to models trained with raw data. We considered both a classic machine learning model, Logistic Regression (LR), as well as a state-of-the-art deep learning method for NLP tasks: multilingual BERT.

The rest of the paper is organized as follows. Section 2 presents some related works that used data in Portuguese and evaluated the impact of normalization techniques on the performance of the models. The dataset used in the experiments is described in Sect. 3, as well as the text normalization processes considered. An experimental study is presented in Sect. 4. Section 5 concludes the paper presenting some future work.

2   Related Work

In this section, we present some works that used text extracted from social media in Portuguese to train machine learning models and evaluated the performance of the models according to the normalization applied in the preprocessing step. [14] used a dataset of evaluations extracted from Google Play, and considered combinations of preprocessing techniques to determine their impact in sentiment analysis tasks. They created scenarios where treatments are gradually applied. Results showed that the preprocessing was insignificant in the performance of the models. On the other hand, results shown by [15] on data extracted from Twitter and Google Play, showed superior performance of models trained with data that received some treatment as opposed to raw data. Sentiment analysis was also the application considered by [11] to evaluate combinations of preprocessing treatments. Their results showed a variation in the performance of the models depending on the combinations applied.

Another work [2] used a dataset of messages collected from whatsapp groups to train classifiers to detect fake news. They created scenarios with combinations of treatments and vectorization. The experiments showed that some classifiers achieved the best results regardless of the scenario used.

Although there are works that used data in Portuguese and evaluated the impact of some normalizations on the performance of the models, to the best of our knowledge, there are no studies that uses text in Portuguese and *individually* evaluated the effect of treatments commonly seen in NLP works, such as preprocessing of hashtags and Instagram Fonts[1], removal versus tokenization of numbers, mentions, urls and emojis. Furthermore, we did not find works comparing the impact of different normalization in BERT models [5], which is considered the state-of-the-art in some NLP tasks.

3   Dataset and Text Normalization

In this work we used a dataset composed of 6,204 posts extracted from Instagram based on a set of hashtags search. The posts were classified as ACCEPTED (positive class) or REJECTED (negative class) for legitimate occurrences of PMW on the Brazilian coast. Details on the dataset construction are described in [13].

Some of the dataset characteristics are the following:

- **Presence of hashtags:** 6,025 (97%) posts have one or more hashtags in the caption, while 458 (7%) posts have only hashtags in their caption.
- **Presence of emojis:** 2,705 (43%) posts have one or more emojis in the caption, while 7 (<1%) posts have only emojis in the caption.

---

[1] Some Instagram users have been posting content generated by Instagram Fonts Generator. They are websites that create "pseudo-alphabets" by taking advantage from Unicode symbols that look like the normal Latin alphabet, but have some differences, such as being in bold or italic (https://igfonts.io/).

- **Mention of other users in the caption:** 1,250 (20%) posts have one or more mentions of other users in the caption.
- **Presence of URLs:** 375 (6%) posts have one or more URLs in the caption.
- **Presence of numbers:** 2,486 (40%) posts have one or more numbers in the caption; these numbers can be contact numbers, prices, dates and others.
- **Presence of Instagram Fonts:** 73 (1%) posts have Instagram Fonts. To identify the use of Instagram Fonts we used Unicodedata[2]

The transformations considered for text normalization are described below. They have been chosen either because they consider the distinct features of our dataset or because they are commonly applied in NLP works.

- **Hashtags**: To determine whether hashtag-related treatments improve the performance of the models, we considered 3 different ways to normalize the text: removal of all hashtags, replacement of hashtags by a `hashtag` token; and keeping only hashtags in the text.
- **Emojis**: We considered 3 ways to normalize the text in the post caption: removal of emojis; replacement of emoji by an `emoji` token; and translation of emojis to their meaning in Portuguese, using the Emoji library[3].
- **User Mentions, Numbers and URLs**: Replacement and removal of user mentions, numbers and URLs are common normalization processes found in the literature. Even if these elements do not have a large number of occurrences in our data, we wanted to determine if treatments related to them can improve the performance of the models. Thus we considered the following transformations: removal of user mentions; replacement of user mentions by a `user` token; removal of URLs; replacement of URLs by a `url` token; removal of numbers; and replacement of numbers by a `number` token.
- **Lemmatization and Stemming**: Preprocessing aimed at vocabulary reduction is common in NLP works. To understand the effectiveness of using lemmatization and stemming to train the models for our problem, we carried out experiments with these two treatments.
- **Stopwords**: Another treatment commonly found in the literature is the removal of stopwords. Although [8] already pointed out in their work that in many applications removing stopwords does not improve performance, we would like to determine if the same applies in our context. Thus this transformation has also been considered.
- **Instagram Fonts**: The problem with the use of Instagram Fonts is that different symbols generate different tokens. As an example, consider the words "physalia" and "𝔭𝔥𝔶𝔰𝔞𝔩𝔦𝔞". They create two distinct tokens by the TfidfVectorizer method, but both have the same semantics, while the BERT tokenizer turns the second one into a `UNK` token. To normalize the caption, we considered 2 methods: conversion of the text to ASCII; and conversion of the text with Unicodedata.

---

[2] Unicodedata is a Python 3 module that can be used to normalization tasks. https://docs.python.org/3/library/unicodedata.html.

[3] https://pypi.org/project/emoji/.

– **Spelling Correction**: While some posts correctly follow the grammatical rules, others contain errors, and use slang and/or abbreviations. By applying spelling correction, it is possible to decrease these linguistic variations and consequently reduce the dimensionality of the feature vector. To find out if there is any impact of applying spelling correction on models training for our problem, we considered the following transformation: normalization with Enelvo[4].

4   Experimental Study

We conducted an experimental study to determine the effect of the normalization transformations described in the previous section on the performance of models trained for our problem.

First we filtered the dataset as follows. We deleted posts that have the following characteristics: empty text (12 posts); repeated posts (164 posts); with video only (643 posts); and location outside Brazil (1,991 posts). Moreover, we kept only posts identified as Portuguese language or emoji only in the caption because emojis can be translated to Portuguese. After applying the filters, 2,610 posts remained in the dataset, distributed between 648 positive and 1,962 negative labels.

We trained a classic machine learning model available in the Scikit-learn library [12]: Logistic Regression (LR). LR has been chosen because it is a strong baseline for text classification tasks. We also considered a state-of-the-art deep learning method for NLP tasks: multilingual BERT, which is available in the Tensorflow library[5]. For training with LR we kept its default parameters with the exception of the parameter `class_weight`, which was set to '`balanced`'. As vectorization methods, TF-IDF and BERT were used. For TF-IDF, we kept the default values of the method. By default, this method converts data to lowercase, ignores tokens less than 2 characters long, emojis, punctuation and signs, with the exception of underline. At the end, the method normalizes the resulting TF-IDF vectors by the Euclidean norm.

BERT was fine-tuned with our data. We used as hyperparameters: 50 epochs, batch size of 16 and AdamW[6] optimizer starting with a learning rate of 3e-5. Furthermore, we use the same class weighting calculation as LR to train BERT.

We divided the dataset into 70% (train) and 30% (test), then the train dataset was divided into 70% (train) and 30% (validation) and applied cross-validation with 5 folds.

Some of the treatments applied in our experiments resulted in some empty samples. In that case, the sample in question was removed from the experiment. Considering emojis, as TF-IDF ignores them, in the case of a caption composed only of emojis, the vectorization process with this method results in a vector composed only of zeros. In other words, in the experiment with raw text this

---

[4] Enelvo is a normalizer for user-generated content in Portuguese [1].

[5] https://www.tensorflow.org/.

[6] https://www.tensorflow.org/api_docs/python/tf/keras/optimizers/AdamW.

Table 1. Classifiers F1-score with standard deviation

Treatment	TF-IDF + LR	BERT + LR	BERT
raw text	0.861 (0.020)	0.717 (0.013)	0.858 (0.012)
hashtags			
remove hashtag	0.754 (0.030)	0.586 (0.017)	0.715 (0.012)
token hashtag	0.746 (0.027)	0.591 (0.027)	0.720 (0.020)
hashtag only	0.813 (0.029)	0.714 (0.039)	0.820 (0.036)
emojis			
remove emoji	0.848 (0.022)	0.716 (0.023)	0.859 (0.027)
token emoji	0.855 (0.015)	0.708 (0.020)	0.856 (0.022)
translate emoji	0.854 (0.015)	0.702 (0.008)	0.858 (0.017)
user mentions			
remove user	0.860 (0.020)	0.701 (0.015)	0.865 (0.016)
token user	0.861 (0.017)	0.710 (0.015)	0.867 (0.027)
numbers			
remove number	0.860 (0.017)	0.715 (0.016)	0.875 (0.018)
token number	0.865 (0.019)	0.714 (0.012)	0.873 (0.022)
urls			
remove url	0.862 (0.021)	0.714 (0.017)	0.861 (0.018)
token url	0.862 (0.021)	0.714 (0.015)	0.857 (0.017)
lemmatization/stemming			
lemmatization	0.858 (0.013)	0.716 (0.006)	0.859 (0.011)
stemming	0.858 (0.011)	0.697 (0.019)	0.866 (0.016)
stopwords			
remove stopwords	0.857 (0.013)	0.726 (0.025)	0.846 (0.007)
instagram fonts			
ascii	0.841 (0.021)	0.706 (0.023)	0.859 (0.021)
unicodedata	0.858 (0.019)	0.714 (0.023)	0.864 (0.016)
spelling correction			
enelvo	0.860 (0.021)	0.709 (0.009)	0.860 (0.029)

zeroed sample was kept, while in the experiment with emoji removal this sample was excluded.

To compare the results obtained with models trained with normalized text we trained a model with raw text. The results obtained with all experiments are presented in Table 1.

Consider first the treatments related to hashtags. Although they did not outperform the model trained with raw data, models trained only with hashtags performed better than models trained with other treatments. In addition, the absence of hashtags significantly worsened the performance of the models, which demonstrates that hashtags have a good discriminative power for our research problem and should be maintained.

Emoji-related treatments worsened the performance of LR models, with the exception of emoji removal, which showed no significant difference in the result of the BERT+LR model compared to the model trained with raw text. Likewise, BERT models showed no difference in performance when trained with normalized text.

For LR models, removal or tokenization of users, numbers and urls did not presented significant difference in the results compared to the model trained with raw data, with the exception of user removal which worsened the performance of the BERT+LR model. In contrast, the BERT model benefited from these treatments, outperforming the model trained with raw data, with the exception of url tokenization, which did not show a significant difference. Even if they did not improve the performance in the case of training with LR, these treatments can be interesting for reducing the size of the vocabulary.

Although lemmatization and stemming did not improve the performance of LR models, BERT model trained with stemmed data outperformed the model trained with raw data by 1%.

The removal of stopwords was the only treatment that outperformed the model trained with raw data for the BERT+LR model. For the other models, this treatment worsened the performance of the models.

Conversion to ASCII, although tested for the purpose of normalizing Instagram Fonts, changes all the text and did not benefit LR models. The text treated with Unicodedata, despite not showing a significant difference in the LR models, performed better than the raw data in the BERT model.

The difference in performance of models trained with Enelvo-normalized data compared to models trained with raw data was not significant. Among all the preprocesses tried, this one took the longest to run.

In general, in experiments with TF-IDF+LR, it is noted that none of the models trained with normalized data outperformed significantly the model trained with raw data. In the experiments with BERT+LR, only the model trained with text without stopwords outperformed the model trained with raw data by 1%. In the experiments with BERT, 8 models trained with normalized data outperformed the model trained with raw data by up to 2%.

5 Conclusion

This work presented a comparison in the performance of models trained with raw data and normalized text with different treatments, such as removal and tokenization of hashtag, emojis, user mentions, numbers; lemmatization; and spelling correction. The results showed that, for our application of classifying PMW posts, some treatments such as: removing or tokenizing user mentions, removing or tokenizing numbers, removing URLs, stemming and normalization of Instagram Fonts, can be interesting to apply when training BERT models.

Despite being customarily used in the preprocessing step for training classical models, none of the treatments experimented in our work significantly outperformed the classical models trained with raw text when considered individually

As future work, we intend to experiment with the combination of some normalization techniques, in particular with BERT models. Developing general guidelines of recommended preprocessing techniques based on text characteristics is an interesting and useful line of investigation.

References

1. Bertaglia, T.F.C., Nunes, M.D.G.V.: Exploring word embeddings for unsupervised textual user-generated content normalization. In: Proceedings of the 2nd Workshop on Noisy User-generated Text (WNUT), pp. 112–120 (2016)
2. Cabral., L., Monteiro., J.M., Franco da Silva., J.W., Mattos., C.L., Mourão.. P.J.C.: FakeWhastApp.BR: NLP and machine learning techniques for misinformation detection in Brazilian Portuguese whatsapp messages. In: Proceedings of the 23rd International Conference on Enterprise Information Systems, pp. 63–74 (2021). https://doi.org/10.5220/0010446800630074
3. Camargo, L., Rocha, H., Nascimento, L., Hara, C.: Coleta de dados do instagram sobre ocorrências de caravelas-portuguesas na costa brasileira. In: Anais da XVIII Escola Regional de Banco de Dados, pp. 51–59. SBC, Porto Alegre, RS, Brasil (2023). https://doi.org/10.5753/erbd.2023.229499
4. Carneiro, A., Nascimento, L., Noernberg, M., Hara, C., Pozo, A.: Social media image classification for jellyfish monitoring. Aquat. Ecol. **58**, 3–15 (2024)
5. Devlin, J., Chang, M.W., Lee, K., Toutanova, K.: BERT: pre-training of deep bidirectional transformers for language understanding. In: Proceedings of the 2019 Conference of the North American Chapter of the Association for Computational Linguistics: Human Language Technologies, Volume 1 (Long and Short Papers), Minneapolis, Minnesota, pp. 4171–4186. Association for Computational Linguistics (2019). https://doi.org/10.18653/v1/N19-1423
6. Diniz, E.J., et al.: Boamente: a natural language processing-based digital phenotyping tool for smart monitoring of suicidal ideation. Healthcare **10**(4) (2022). https://doi.org/10.3390/healthcare10040698
7. Groppe, J., Schlichting, R., Groppe, S., Möller, R.: Deep learning-based classification of customer communications of a German utility company. In: International Semantic Intelligence Conference, pp. 205–222 (2022). https://doi.org/10.1007/978-981-19-7126-6_16
8. Jurafsky, D., Martin, J.H.: Speech and Language Processing, 2nd edn. Prentice Hall, Hoboken (2008)
9. Mota, A., Franco, W., Mattos, C.: Detecção de desinformação sobre covid-19 no twitter. In: Anais do XIII Simpósio Brasileiro de Tecnologia da Informação e da Linguagem Humana, pp. 172–181. SBC, Porto Alegre, RS, Brasil (2021). https://doi.org/10.5753/stil.2021.17796
10. do Nascimento, L.S., Hara, C.S., Júnior, M.N., Noernberg, M.: Instagram como fonte de dados alternativa no monitoramento da #caravelaportuguesa (physalia phisalis, cnidaria). In: Livro de Memórias do IV SUSTENTARE e VII WIPIS: Workshop internancional de Sustentabilidade, Indicadores e Gestão de Recursos Hídricos (2022). https://doi.org/10.29327/sustentare_wipis_2022.584935

11. de Oliveira, D.N., Merschmann, L.H.D.C.: Joint evaluation of preprocessing tasks with classifiers for sentiment analysis in Brazilian Portuguese language. Multimedia Tools Appl. **80**(10), 15391–15412 (2021). https://doi.org/10.1007/s11042-020-10323-8

12. Pedregosa, F., et al.: Scikit-learn: machine learning in python. J. Mach. Learn. Res. **12**, 2825–2830 (2011)

13. Rocha, H.F., Nascimento, L.S., Camargo, L., Noernberg, M., Hara, C.S.: Labeling Portuguese man-of-war posts collected from instagram. In: Abelló, A., et al. (eds.) ADBIS 2023. CCIS, vol. 1850, pp. 369–381. Springer, Cham (2023). https://doi.org/10.1007/978-3-031-42941-5_32

14. dos Santos, F.L., Ladeira, M.: The role of text pre-processing in opinion mining on a social media language dataset. In: 2014 Brazilian Conference on Intelligent Systems, pp. 50–54 (2014). https://doi.org/10.1109/BRACIS.2014.20

15. Stiilpen Junior, M., Merschmann, L.H.C.: A methodology to handle social media posts in Brazilian Portuguese for text mining applications. In: Proceedings of the 22nd Brazilian Symposium on Multimedia and the Web, pp. 239–246 (2016). https://doi.org/10.1145/2976796.2976845

A Preliminary Investigation: Strategies for Incorporating Logical Rules Into Knowledge Graph Embeddings

Jacques Chabin🆔, Mirian Halfeld-Ferrari🆔, and Lingchen Wang$^{(\boxtimes)}$🆔

Université d'Orléans, INSA CVL, LIFO UR 4022, Orléans, France
{jacques.chabin,mirian}@univ-orleans.fr,
lingchen.wang@etu.univ-orleans.fr

Abstract. This paper presents our preliminary state-of-the-art studies and bibliographic research on the fusion of knowledge derived from logical rules into machine learning (ML) approaches for knowledge graph completion. It is intended to serve as an initial guide for researchers and practitioners interested in this emerging field of study.

Keywords: knowledge graph · logic rules · embedding · link prediction

1 Introduction

Logical rules have been used to complete or ensure the consistency of databases. A rule such as $c_1 : Follows(X_{stud}, X_{course}) \wedge OfferedBy(X_{course}, X_{univ}) \rightarrow Enrolled(X_{stud}, X_{univ})$ is a way of completing the database information by declaring that any student following a course offered by a given university is enrolled at that university, or, in a more restrictive way, to impose an update policy that enforces compliance with such a rule. For example, if $Follows(Lea, CS)$ is inserted into a database containing $OfferedBy(CS, UOrleans)$, then $Enrolled(Lea, UOrleans)$ should also be present in the database.

Due to our increasingly interconnected world and the need to analyse these connections, graph databases are experiencing rapid growth. In such a model, nodes are used to represent entities (for example, in our initial example, nodes can represent entities such as $Students$, $Courses$, or $Universities$), while edges represent relationships between them (for example, edges can represent relationships such as $Follows$ and $OfferedBy$). In this evolving landscape, our focus is particularly drawn to the adaptation of intentional knowledge, as exemplified by a rule like the aforementioned c_1, within data graphs. It is a complex question that leads to many different solutions. On one hand, some approaches extend the application of logical rules [5], as constraints, to graphs, and adjust data modeling to address efficiency concerns [3,4]. Concurrently, alternative approaches introduce novel constraints tailored for graphs, albeit typically less expressive than traditional logical rules [1,11]. On the other hand, an emerging trend involves machine-learning (ML) based methods for completing knowledge graphs. While

J. Tekli et al. (Eds.): ADBIS 2024, CCIS 2186, pp. 104–116, 2025.
https://doi.org/10.1007/978-3-031-70421-5_10

these techniques can predict new links within graphs, they often overlook the constraints or dependencies articulated by traditional logical rules commonly associated with databases.

We focus on the integration of ML methods into graph database management. This paper outlines our preliminary state of the art studies on the fusion of knowledge derived from logical rules into ML approaches for knowledge graph completion. To achieve this goal, we need to perform graph embeddings, which transform graphs into vectors. In order to exploit the knowledge encoded in logical rules, we are particularly interested in methods that provide embeddings designed to incorporate these rules.

Our purpose is to step into the world of multidisciplinary research that has now become a reality, particularly in the fields of Database (DB) and Artificial Intelligence (AI). Both DB and AI focus on the act of exploring, reasoning, correlating facts and discovering implicit knowledge from data. The challenge today is to break down the boundaries between the disciplines and seamlessly integrate database principles that bring robust and efficient solutions to AI, and vice versa, AI to the DB that opens data to more sophisticated and analytical queries. The integration of AI into query processes represents a noticeable shift within the DB domain. This shift moves away from certainties to embracing possibilities, and from general deduction to relying on sampling and instantiation to drive results.

Paper Organization. Section 2 reviews notions of logic, database chasing, and graph embedding, while Sect. 3 presents an overview of the state of the art in injecting logic rules into knowledge graph embeddings. Some concluding remarks are dressed in Sect. 4.

2 Background

2.1 Graph Models

When we look at data graphs, we have different models to consider. Two common models in graph databases are LPG (Labelled-Property Graphs, also known as Property Graphs) and RDF (Resource Description Framework) graphs. The latter is often called a knowledge graph. Both of these models are based on the notion of node- and edge-labelled directed graphs, which allow complex information to be represented by multiple graphs. Knowledge graphs consist of nodes and arcs. Property graph enriches nodes or edges with attributes or properties. Each property is a tuple (key, value), enriching the graph data. Graphs are also generally categorized into two types: homogeneous and heterogeneous. A homogeneous graph is a graph with a single type of node and a single type of edge. A heterogeneous graph is a graph which contains either multiple types of nodes (heterogeneous nodes) or multiple types of edges (heterogeneous edges).

2.2 Logical Rules

Data is often accompanied by additional mechanisms such as constraints, knowledge rules, data dependencies and more, which support knowledge generation or ensure the integrity, accuracy and reliability of stored data. While there are various methods for representing these mechanisms, First-Order Logic (FOL) is one of the most commonly used because it provides a precise, declarative, high-level language that is unambiguous.

We assume a standard FOL alphabet composed of three pairwise disjoint sets: CONST, a set of constants, VAR, a set of variables and PRED, a set of predicates, each predicate is associated with a positive integer called its arity. A *term* is a constant or a variable and an atomic formula, or an atom, is a formula of the form $P(t_1, \ldots, t_n)$ where P is a predicate of arity n and $t_1, \ldots, t_n$ are terms. Every atom in which no variables occur is a *ground atom*, also called a *fact*. In this paper we deal with rules that correspond to Horn clauses. In other words, a rule c is a rule of the form $P_1(x, z_1) \land P_2(z_1, z_2) \land \ldots \land P_n(z_{n-1}, y) \rightarrow P(x, y)$. The atom $P(x, y)$ is the head of c (denoted by $head(c)$) and the set of atoms $\{P_1(x, z_1), P_2(z_1, z_2), \ldots, P_n(z_{n-1}, y)\}$ is the body of c (denoted by $body(c)$). The set $atoms(c)$ contains all the atoms of c.

A *homomorphism* h from a set of atoms A_1 to a set of atoms A_2 is a mapping from the terms of A_1 to the terms of A_2 such that: (i) if $t \in$ CONST, then $h(t) = t$, and (ii) if $P(t_1, ..., t_n)$ is in A_1, then $P(h(t_1), ..., h(t_n))$ is in A_2. The set A_1 is *isomorphic* to the set A_2 if there exists a homomorphism h_1 from A_1 to A_2 which admits an inverse homomorphism (from A_2 to A_1). We usually use the expression *instantiating* or *grounding* to indicate a homomorphism that replaces all variables in a formula with constants.

Let $\mathbf{C}$ be a set of rules. For a given $c \in \mathbf{C}$, a set of facts M satisfies c ($M \models c$) if for every homomorphism h such that $h(body(c)) \subseteq M$, we have $h(head(c)) \in M$. Now, a database instance I can be seen as a set of facts. I is consistent concerning a set of rules $\mathbf{C}$ if and only if, for every rule $c \in \mathbf{C}$, I satisfies c.

Our rules express data dependencies. The *chase* algorithm serves as a fixed-point method for evaluating and enforcing the implication of these data dependencies within a database. This concept is presented in the following example.

Example 1. Consider a database instance

$$I_0 = \{Parent(Alice, Jac), Parent(Alice, Lea), Male(Jac), Female(Lea)\}$$

and the set of constraints $\mathbf{C}$ containing the following three rules:

$c_1 : Parent(x, y), Parent(x, z) \rightarrow Siblings(y, z)$
$c_2 : Siblings(x, y), Male(y) \rightarrow BrotherOf(y, x)$
$c_3 : Siblings(x, y), Female(y) \rightarrow SisterOf(y, x)$

I_0 does not satisfy c_1. To ensure that I satisfies c_1, the facts $Siblings(Jac, Lea)$ and $Siblings(Lea, Jac)$ must be incorporated into the instance. Consequently,

I_1 $=$ $\{Parent(Alice, Jac), Parent(Alice, Lea), Male(Jac), Female(Lea),$
$Siblings(Lea, Jac), Siblings(Lea, Jac)\}$ satisfies c_1.

Roughly speaking, given I_0 and c_1, we can consider c_1 is triggered, *i.e.* $body(c_1)$ is instantiated by replacing the variables x, y, z with the constants *Alice, Jac*, and *Lea*, respectively, to generate $Siblings(Jac, Lea)$. Another possible instantiation involves replacing the variables x, y, z with the constants *Alice, Lea*, and *Jac*, respectively, to generate the symmetric relationship $Siblings(Lea, Jac)$. Similarly, only when the database instance includes facts capable of instantiating the bodies of constraints c_2 and c_3, will the relationships *BrotherOf* and *SisterOf* be generated.

Starting with I_0 and **C** the *chase* algorithm iterates until reaching the fixed-point represented by the instance:

$$I = \{Parent(Alice, Jac), Parent(Alice, Lea), Male(Jac), Female(Lea),$$
$$Siblings(Lea, Jac), Siblings(Lea, Jac), BrotherOf(Jac, Lea), Sister(Lea, Jac)\}.$$

Clearly, $I \models \mathbf{C}$. We say that the database is consistent with respect to the set of constraints **C**.   □

2.3 Embedding

Feature embedding is a method that maps categorical variables, *i.e.*, those that take on a limited, and usually fixed, number of possible values, into a continuous vector space. This technique, often used in NLP (natural language processing) to map words and phrases to vectors of real numbers, are recently also used to deal with graphs. *Graph embeddings* are the transformation of graphs into continuous vectors. Graphs are mostly represented by $\mid V \mid \times \mid V \mid$ matrices and thus considered as sparse high-dimensional spaces. Given a graph $G(V, E)$, the task of node embedding consists in learning a projection ϕ such that each graph node can be encoded into a low-dimension space, *i.e.*, into a vector of dimension $L \ll \mid V \mid$. The aim of the transformation is to align the similarity observed in the latent embedded space closely with that in the original graph, while preserving the graph structure property. Essentially, most of the graph node embedding techniques learn low-dimension node representation by solving an optimization problem. The generated graph embeddings in the latent space can be conveniently used for various subsequent graph analysis tasks, e.g. node classification, link prediction, community detection, visualisation, etc.

According to [26] graph embedding approaches have three important components:

(1) *Preservation of the graph structure property.* Embeddings preserve the topological properties of the original graph. The aim is to introduce an embedding approach that will allow a connectivity algorithm to retrieve the set of neighbours of a node with high accuracy. This aspect is important for several downstream tasks, such as graph reconstruction, where two similar vectors symbolize nodes that are adjacent in the graph.

(2) *Node similarity measures in the original graphs.* Similarity between nodes in the original graphs can be characterized by different aspects such as: presence of edge, overlapped neighbourhood, reachability in k-hops (*i.e.*, the set of vertices that are reachable from a node by following a path with k or fewer edges), reachability through random walks (similarity is characterized by the probability of reaching a node v_j through a random walk of length l across the graph starting from the source node v_i) and similar node attributes.

(3) *Similarity measures in the embedded space.* According to the specific learned node embedding form, different metrics to measure the node similarity can be used in the embedded space. For example, one possible form for node embedding is through deterministic point vectors $\{\mathbf{v}_i \in \mathbb{R}^L \mid i = 1, 2, \ldots, \mid V \mid\}$ over which similarity measures such as dot product, euclidean distance and cosine similarity can be applied. However, for node embeddings as Guassian distribution in the latent space, there are other similarity measures.

Recently, [9] has divided KG embedding into two broad groups: *triple fact-based representation learning methods* and *description-based representation learning methods*. The former revolves around KGs as triples, where a fact like *Parent(Alice, Jac)* is represented as the triple (*Alice, Parent, Jac*) in the form (*subject entity, relationship, object entity*), abbreviated as (s, r, o). The latter includes additional contextual information, such as textual data associated with KG, to improve the performance of the embedding model.

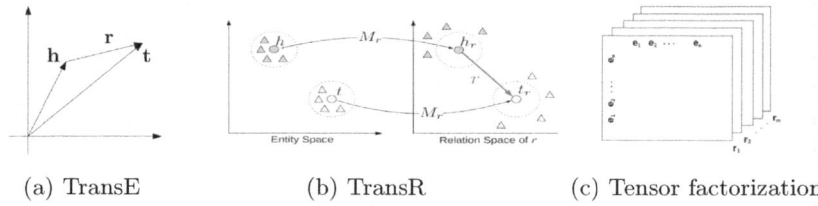

(a) TransE   (b) TransR   (c) Tensor factorization

Fig. 1. Translation-based models [8] and Tensor factorization [9]

Along the lines of [8,9], KG embeddings in the triple fact-based approaches can then be studied by considering the following classes:

(1) *Translation-based models.* These models use distance-based measures to generate the similarity score of a pair of entities and their relationships. The translation model aims to find a vector representation of entities with relation to the translation of the entities. For instance, TransE [2] embeds all entities and relations in the same space: if $\mathbf{v}_s$, $\mathbf{v}_r$ and $\mathbf{v}_o$ are vector embeddings for s, r and o, respectively, then, $\mathbf{v}_s + \mathbf{v}_r \approx \mathbf{v}_o$. For a true triple, the relation-specific translation of subject embedding $(\mathbf{v}_s + \mathbf{v}_r)$ is close to the object embedding $\mathbf{v}_o$ in embedding's vector space. On the other hand, TransR [17] embeds entities and relations in different spaces: for each relation r, a projection matrix $\mathbf{M}_r \in \mathbb{R}^{k \times d}$ (where k and d are dimensions) is set.

The role of $\mathbf{M}_r$ is to project entities from an entity space to a relation space. Thus $\mathbf{v}_{s_r} = \mathbf{v}_s \mathbf{M}_r$ and $\mathbf{v}_{o_r} = \mathbf{v}_o \mathbf{M}_r$. Figure 1a, 1b illustrate the difference between these two methods.

(2) *Tensor factorization-based models.* Methods using tensor factorization see KG in a three-way tensor $\mathcal{X} \in \mathbb{R}^{n \times n \times m}$ (Fig. 1c) where n is the number of entities (for both s or o in the triples) and m is the number of relationships. $\mathcal{X}_{i,j,k} = 1$ indicates that the triple *(ith-entity, kth-relation, jth-entity)* exists in the graph; otherwise $\mathcal{X}_{i,j,k} = 0$ indicates that it does not exist (it is false or unknown). The idea is then to use techniques of tensor decomposition, *i.e.*, to express our three-way tensor $\mathcal{X}$ as a sequence of elementary operations on other, simpler, tensors. Usually tensor decompositions generalize some matrix decompositions.

(3) *Neural Network-based models.* These methods use a multilayer neural structure to learn embeddings and estimate the plausibility of triples with nonlinear features, for example, ConvE [10] concatenes the embeddings of the head and the relation of the triple for a 2D convolutional layer, the resulting feature map tensor is vectorised and projected into a k-dimensional space and matched with all candidate object embeddings. Besides, R-GCN [22] extends traditional GCNs by incorporating multiple types of relations in graph data, enabling more expressive and accurate learning of node representations in complex relational domains.

The aforementioned techniques primarily rely on learning representations from existing triples in KGs, but their effectiveness is limited by the sparse nature of the data. To address these challenges, some recent methods aim to converge the benefits of triple-centred embedding strategies with first-order domain knowledge. This convergence is the focus of our investigation.

3 Injecting Logical Rules in Knowledge Graph Embedding

As discussed in Sect. 2, the data management community typically represents important properties of data by constraints or dependencies expressed as logical rules, leading to implementations which may be not very efficient. Graph embedding helps to complete the database, but ignoring the intentional knowledge conveyed by logical rules. This section attempts to categorize how to integrate the logical rule with embeddings.

Some work propose to put together logic rule reasoning and embedding. For example, [28] considers both the mechanisms to inject different types of rules into embeddings, and how to use logic and embeddings for query answering and rule mining. In our work, we are particularly interested in investigating how embedding methods and logical rule reasoning can be used together to obtain a complete and consistent knowledge graph.

We firstly identify two main classes of methods, methods based on Markov logic networks and PSL regularisations. However, most of them are not iterative, i.e. they capture only one step of logical inference, ignoring the iterative nature of logical inference. In order to obtain a complete knowledge graph we need to extend our study to iterative methods which try to maximise the benefits of combining the logic rules and embeddings. Table 1 gives an overview of the methods we have studied.

Table 1. Methods of embedding with constraints: MLN, PSL, Chase and NN are used to incorporate logic rules. Single iteration: grounding rules injected in the embedding. Multiple iterations: iterative steps of chase and embedding.

Method	MLN	PSL	Chase	NN	Single iteration	Multiple iterations
Chen et al. [6]			✓			
ExpressGNN [31]	✓					
IterE [29]			✓			✓
Iterlogic-E [15]		✓				✓
KALE [12]		✓			✓	
Li et al. [16]			✓			✓
LNN [20]				✓		
Melo [24]			✓		✓	
pGat [14]	✓					
pLogicNet [18]	✓					
Rocktäschel et al. [21]		✓			✓	
RUGE [13]		✓				✓
RulE [23]				✓		
SoLE [27]		✓	✓			✓
UniKER [7]			✓			✓
Wang et al. [25]		✓			✓	

3.1 MLN-Based Methods

Methods such as pLogicNet [18], pGat [14] and ExpressGNN [31] use *Markov Logic Network* (MLN) [19]. MLN combines first-order logic with probabilistic graphical models. Indeed, the idea of approaches based on MLN is to use probabilistic model to approximate the exact logical inference. The critical problem on knowledge graphs, *i.e.*, to predict the missing facts, is reformulated in a probabilistic way. Each triple (e_i, r_k, e_j) is associated with a binary indicator variable, we note $(e_i, r_k, e_j) = 1$ when (e_i, r_k, e_j) is true and $(e_i, r_k, e_j) = 0$ otherwise. Let $\mathbf{v}_O$ be the set of embedded observed facts, *i.e.*, $\mathbf{v}_O = \{\mathbf{v}_{(e_i,r_k,e_j)} \mid (e_i, r_k, e_j) \in O\}$. Given some true facts, those in $\mathbf{v}_O$, we aim to predict the remaining hidden triples H, *i.e.*, $\mathbf{v}_H = \{\mathbf{v}_{(e_i,r_k,e_j)} \mid (e_i, r_k, e_j) \in H\}$. A Markov network is designed to define the joint distribution of the observed and the hidden triples. Let F denote a set of first-order logic (FOL) formulas. Each logic rule f in F generates a collection of groundings by instantiating the logic rules with real entities in knowledge graphs. For instance, consider a logical formula like $BornIn(x, y) \rightarrow LiveIn(x, y)$, which holds true for $(Alice, Paris)$ but false for $(Bob, London)$. To incorporate uncertainty into logic rules, Markov logic networks assign a weight w_f to each rule f. Then, the joint distribution over all such instances is defined as: $p(\mathbf{v}_O, \mathbf{v}_H) = \frac{1}{Z} \exp(\sum_{f \in F} w_f . n_f(\mathbf{v}_O, \mathbf{v}_H))$, where, n_f represents the count of true groundings of the logic rule based on the values of $\mathbf{v}_O$ and $\mathbf{v}_H$, and Z is a normalization constant. With such a formulation, predicting the missing triples essentially becomes inferring the posterior distribution $p(\mathbf{v}_H \mid \mathbf{v}_O)$, i.e., the probability of $\mathbf{v}_H$ given the evidence $\mathbf{v}_O$.

The model can be trained by maximizing the log-likelihood of the observed indicator variables, i.e., $ln(p(\mathbf{v}_O))$. However, directly optimizing the objective is usually infeasible [18], thus, several methods propose approximation inference of MLN to alleviate the procedure. For instance, *pLogicNet* [18] optimizes the evidence lower bound (ELBO) of the log-likelihood function. Such a lower bound can be effectively optimized with the variational Expectation Maximization (EM) algorithm. EM is used to estimate the maximum likelihood of data given the model parameters in cases where the data has some latent variables. In order to do so, EM repeats the following two steps until convergence:

(i) E-step: Estimate the latent variables according to posterior distribution calculated with the model parameters;
(ii) M-step: Update the model parameters by maximizing the likelihood.

In *pLogicNet*, the E-step uses the logic rules to predict the missing triples, and treat them as extra training data while the M-step updates the weights of rules to maximize the joint probability of observed and hidden triples.

3.2 PSL-Based Regularization

Approaches called *PSL(Probabilistic Soft Logic)-based regularization in embedding loss* in [7] consider logical rules as supplementary regularization for embedding models. PSL serves as a modelling language tailored for learning and predicting within relational domains, employing soft truth values for predicates within the interval of $[0, 1]$. Embedding techniques within the category of PSL-based regularization incorporate a loss function composed of two primary components: the first component addresses the triples representing the dataset, while the second component considers the rules. In simpler terms, the overarching objective of PSL-based regularization methods can be expressed as

$$\mathcal{L}_{KGE} + \lambda \mathcal{L}_{PSL} \tag{1}$$

where $\mathcal{L}_{KGE}$ represents the loss of the base KGE model, and $\mathcal{L}_{PSL}$ corresponds to the satisfaction loss of the sampled ground rules. A problem with these methods is the need to instantiate universally quantified rules prior to model learning. As the total number of ground rules is impractical to handle, only a sample is typically used, resulting in a loss of logical information. In addition, many of these techniques perform a single injection, which is equivalent to only one step of the chase procedure.

KALE [12] is an example of methods in this category which enriches the TransE embedding method with logic rules. Firstly a set of ground formulas $\mathcal{F}$ is prepared: atomic formulas (positive and negative triples) and ground rules (positive and negative ones[1]). Secondly, as in TransE, given a triple (e_i, r_k, e_j), a relation embedding $\mathbf{v}_{r_k}$ is modelled as a translation $\mathbf{v}_{e_i} + \mathbf{v}_{r_k} \approx \mathbf{v}_{e_j}$ (as explained in Fig. 1a), then each (e_i, r_k, e_j) is assigned with a soft truth value on the basis of $||\mathbf{v}_{e_i} + \mathbf{v}_{r_k} - \mathbf{v}_{e_j}||_1$. That is, $score(\mathbf{v}_{e_i}, \mathbf{v}_r, \mathbf{v}_{e_j}) = 1 - \frac{3}{\sqrt{d}}||\mathbf{v}_{e_i} + \mathbf{v}_{r_k} - \mathbf{v}_{e_j}||_1$ where d is the dimension of the embedding space. Thirdly, the soft truth value of a rule is defined on the basis of the product t-norm fuzzy logic. For instance, given formulas f_1 and f_2, we have $score(f_1 \wedge f_2) = score(f_1).score(f_2)$ and $score(f_1 \vee f_2) = score(f_1) + score(f_2) - score(f_1).score(f_2)$. Finally, according to the framework outlined in Eq. 1, KALE establishes a loss function across formulas originating from $\mathcal{F}$, encompassing triples and ground rules. Embeddings are learned via maximizing the difference between the soft truth value of positive formulas, denoted as $score(f)^+$, and negative ones, denoted as $score(f)^-$. Alternative approaches, such as RUGE [13], Rocktäschel et al. [21], Wang et al. [25] propose similar methods.

3.3   Multi-iterative Completion

Some methods are concerned by iteration in reasoning. They focus on the combination of logical inference and KGE. IterE [30] proposes an iterative framework that combines embedding learning and rule learning to explore mutual benefits. In this paper, axioms from OWL are considered, and only those with good support during an iteration are used to derive more facts. The method can be summarised as follows (see Fig. 2a). First, *an embedding learning phase* learns embeddings of relations and entities by minimising a loss function computed over triples. The inputs are triples in the dataset and those inferred by rules during the iteration process. The next step concerns the *induction of a set of axioms* based on the relational embeddings from the embedding learning step and the assignment of a score to each axiom. In fact, in [30], the score of the grounded OWL axioms is calculated and those above a threshold are used to infer new triples. For example, suppose the symmetric axiom holds for *friendOf*, *i.e.* the score of this rule is 0.8 for a threshold of 0.5. In this case, if the dataset already has a triple $(Bob, friendOf, Lea)$, then the symmetric rule is applied to produce $(Lea, friendOf, Bob)$. This newly produced triple (inferred from grounding of quality axioms with high scores) will be injected in the next step of the KGE. Actually, the third step involves *injecting axioms*. It is concerned with the injection of new triples, in particular about sparse entities, to help improve their

---

[1] Assuming that $A(x, y), B(y, z) \rightarrow C(x, z)$, $A(bob, alice)$ as a true fact and $B(alice, tom)$ as a false fact, the ground rule $A(bob, alice), B(alice, tom) \rightarrow C(bob, tom)$ is an example of a false ground rule.

poor embeddings caused by insufficient training. After axiom injection, with an updated dataset, the process goes back to embedding learning again.

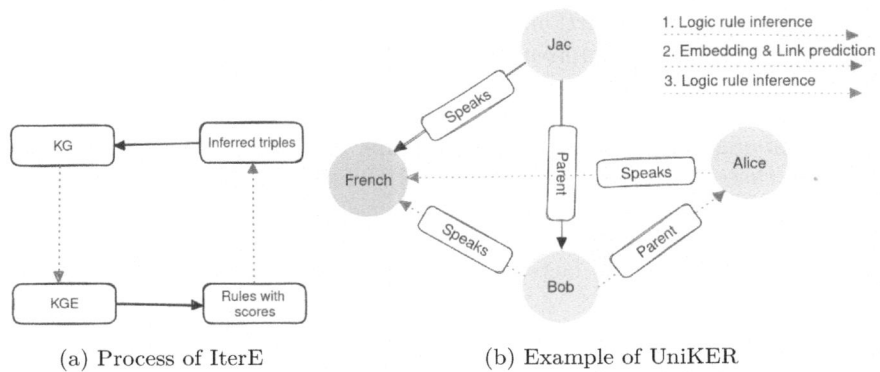

(a) Process of IterE   (b) Example of UniKER

Fig. 2. IterE and UniKER.

UniKER [7] (Unified framework for combining Knowledge Graph Embedding with Logical Rules) proposes an iterative mechanism to enable collaboration and mutual enrichment between logical inference and KGE. In contrast to methods which involve the computation of all ground predicates (true and false), UniKER proposes to initiate a chase step, triggered by the observed facts as illustrated in the following example.

Example 2. Let us consider rule $c : Parent(x, y) \land Speaks(x, z) \rightarrow Speaks(y, z)$ that establishes that children speak the same language of their parents. Suppose the KG in Fig. 2b, where, initially, we know that $Parent(Jac, Bob)$ and $Speaks(Jac, French)$. Thus, by applying the logical rule (a one-step chasing) it is possible to derive $Speaks(Bob, French)$. Methods such as IterE and UniKer inject this result in the next embedding step. Supposing that one of the predicted triples after the embedding is $Parent(Bob, Alice)$, rule c will then be used again to produce $Speaks(Alice, French)$. Figure 2b shows the derivations obtained in each step. Notice that, as mentioned before, some methods consider all possible grounding rules produced with c and the constants present in the graph. This is not the case of UniKER that considers only the instantiations corresponding to the triples (facts) in the graph (a way of reasoning that the authors refer to as *lazy inference*). ☐

As UniKER focus on definite Horn rules, the knowledge in rules is guaranteed to be exploited: derivations obtained in one step of the chase procedure are also true. However, the incompleteness of KG limits reasoning with rules. The idea is to take advantage of KGE models to add useful hidden triples to the data set and eliminate incorrect ones. Indeed, the application of a rule depends on the satisfaction of its body by positive ground atoms. UniKER defines "potentially useful" triples as those considered positive enough to trigger a rule. However, incorrect observations can be added to the inference process, leading to incorrect conclusions. In order to mitigate this, UniKER cleans up the KG by excluding triples with prediction scores below a certain threshold, since incorrect triples tend to result in lower prediction scores.

Similarly to the above approaches, the Soft Logical Rules enhanced Embedding (SoLE) method [27] is structured into two stages. Initially, during the grounding generation phase, it extracts soft rules with specific confidences utilizing a rule mining system. Subsequently, it employs a rule engine to execute one step of the chase over these soft rules and KG facts, thereby generating additional groundings. In the following stage, namely embedding learning, a joint training algorithm is employed. This algorithm learns KG embeddings by simultaneously incorporating KG facts and soft rules. Here, groundings are modelled using t-norm fuzzy logics.

3.4   Comments and Other Possible Directions

References for other methods are offered in Table 1. Methods such as [6], LNN [20] and RulE [23] propose to transform the logic rules as the input of Neural Network (NN) to obtain a probabilistic graph.

Overall, the MLN-based methods and the PSL-based regularization are both combinations of logic rules and embedding. MLN-based methods are more suitable for probabilistic graphical models with the Graph Autoencoder (GAE). PSL-based regularization optimizes the translation-based or factorization-based methods with the traditional chase and fuzzy logic. Both types of methods can be used in iterative approaches, usually starting with a rule mining step.

One of the great advantages of multiple loops is the ability to make full use of logical rules, especially when one rule's head is a part of another rule's body. On this basis, we will be able to apply more logic rules to data completion, not only the rules found by data mining, but also some industry and social common sense. In addition, with the development of the Large Language Model (LLM), we can envisage that embedding with the injection of logical rules could enrich the Retrieval-Augmented Generation (RAG).

4   Concluding Remarks

Integrating logical rules, chase reasoning, KG, and embeddings of KG for fusion between databases(DB) and artificial intelligence (AI) presents a formidable challenge. Much work is still necessary. The mostly rigid structures of databases, governed by logical rules, must seamlessly merge with the fuzzy nature of AI systems, represented by embeddings. Successful integration demands a delicate balance between preserving the integrity of existing data structures and leveraging the richness of semantic representations. The references listed in this paper provide an overview of various approaches to initiate work in this direction. However, research avenues such as constraints on LPG and their integration with AI prediction algorithms remain relatively unexplored, presenting more intricate challenges.

Acknowledgement. This work was done in the context of the regional project APR-IA DOING and the action DOING of the GDR MADICS.

References

1. Bonifati, A., Fletcher, G.H.L., Voigt, H., Yakovets, N.: Querying Graphs. Synthesis Lectures on Data Management. Morgan & Claypool Publishers, San Rafael (2018)
2. Bordes, A., Usunier, N., Garcia-Duran, A., Weston, J., Yakhnenko, O.: Translating embeddings for modeling multi-relational data. In: Advances in Neural Information Processing Systems, vol. 26 (2013)
3. Chabin, J., Eichler, C., Halfeld Ferrari, M., Hiot, N.: Graph rewriting rules for RDF database evolution: optimizing side-effect processing. Int. J. Web Inf. Syst. **17**(6) (2021)
4. Chabin, J., Halfeld Ferrari, M., Hiot, N., Laurent, D.: Managing linked nulls in property graphs: Tools to ensure consistency and reduce redundancy. In: Abelló, A., Vassiliadis, P., Romero, O., Wrembel, R. (eds.) ADBIS 2023. LNCS, vol. 13985, pp. 180–194. Springer, Cham (2023). https://doi.org/10.1007/978-3-031-42914-9_13
5. Chabin, J., Halfeld Ferrari, M., Laurent, D.: Consistent updating of databases with marked nulls. Knowl. Inf. Syst. **62**(4), 1571–1609 (2020)
6. Chen, B., Hao, Z., Cai, X., Cai, R., Wen, W., Zhu, J., Xie, G.: Embedding logic rules into recurrent neural networks. IEEE Access **7**, 14938–14946 (2019)
7. Cheng, K., Yang, Z., Zhang, M., Sun, Y.: Uniker: a unified framework for combining embedding and definite horn rule reasoning for knowledge graph inference. In: Proceedings of the 2021 Conference on Empirical Methods in Natural Language Processing, pp. 9753–9771 (2021)
8. Choudhary, S., Luthra, T., Mittal, A., Singh, R.: A survey of knowledge graph embedding and their applications. CoRR abs/2107.07842 (2021). https://arxiv.org/abs/2107.07842
9. Dai, Y., Wang, S., Xiong, N.N., Guo, W.: A survey on knowledge graph embedding: approaches, applications and benchmarks. Electronics **9**(5) (2020)
10. Dettmers, T., Minervini, P., Stenetorp, P., Riedel, S.: Convolutional 2D knowledge graph embeddings. In: Proceedings of the AAAI Conference on Artificial Intelligence, vol. 32 (2018)
11. Fan, W., Lu, P.: Dependencies for graphs. In: Proceedings of the 36th ACM SIGMOD-SIGACT-SIGAI Symposium on Principles of Database Systems, PODS, Chicago, USA, pp. 403–416 (2017)
12. Guo, S., Wang, Q., Wang, L., Wang, B., Guo, L.: Jointly embedding knowledge graphs and logical rules. In: Proceedings of the 2016 Conference on Empirical Methods in Natural Language Processing, pp. 192–202 (2016)
13. Guo, S., Wang, Q., Wang, L., Wang, B., Guo, L.: Knowledge graph embedding with iterative guidance from soft rules. In: Proceedings of the AAAI Conference on Artificial Intelligence (2018)
14. Harsha Vardhan, L.V., Jia, G., Kok, S.: Probabilistic logic graph attention networks for reasoning. In: Companion Proceedings of the Web Conference 2020, pp. 669–673 (2020)

15. Lan, Y., He, S., Liu, K., Zhao, J.: Knowledge reasoning via jointly modeling knowledge graphs and soft rules. Appl. Sci. **13**(19), 10660 (2023)
16. Li, G., Sun, Z., Qian, L., Guo, Q., Hu, W.: Rule-based data augmentation for knowledge graph embedding. AI Open **2**, 186–196 (2021)
17. Lin, Y., Liu, Z., Sun, M., Liu, Y., Zhu, X.: Learning entity and relation embeddings for knowledge graph completion. In: Proceedings of the AAAI Conference on Artificial Intelligence, vol. 29 (2015)
18. Qu, M., Tang, J.: Probabilistic logic neural networks for reasoning. In: Advances in Neural Information Processing Systems, vol. 32 (2019)
19. Richardson, M., Domingos, P.: Markov logic networks. Mach. Learn. **62**, 107–136 (2006)
20. Riegel, R., et al.: Logical neural networks. arXiv preprint arXiv:2006.13155 (2020)
21. Rocktäschel, T., Singh, S., Riedel, S.: Injecting logical background knowledge into embeddings for relation extraction. In: Proceedings of the 2015 Conference of the North American Chapter of the Association for Computational Linguistics: Human Language Technologies, pp. 1119–1129 (2015)
22. Schlichtkrull, M., Kipf, T.N., Bloem, P., van den Berg, R., Titov, I., Welling, M.: Modeling relational data with graph convolutional networks. In: Gangemi, A., et al. (eds.) ESWC 2018. LNCS, vol. 10843, pp. 593–607. Springer, Cham (2018). https://doi.org/10.1007/978-3-319-93417-4_38
23. Tang, X., Zhu, S.C., Liang, Y., Zhang, M.: Rule: neural-symbolic knowledge graph reasoning with rule embedding. arXiv preprint arXiv:2210.14905 (2022)
24. Wang, L., Lu, J., Sun, Y.: Knowledge graph representation learning model based on meta-information and logical rule enhancements. J. King Saud Univ.-Comput. Inf. Sci. **35**(4), 112–125 (2023)
25. Wang, P., Dou, D., Wu, F., de Silva, N., Jin, L.: Logic rules powered knowledge graph embedding. arxiv 2019. arXiv preprint arXiv:1903.03772 (2019)
26. Xu, M.: Understanding graph embedding methods and their applications. CoRR abs/2012.08019 (2020). https://arxiv.org/abs/2012.08019
27. Zhang, J., Li, J.: Enhanced knowledge graph embedding by jointly learning soft rules and facts. Algorithms **12**(12), 265 (2019)
28. Zhang, W., Chen, J., Li, J., Xu, Z., Pan, J.Z., Chen, H.: Knowledge graph reasoning with logics and embeddings: survey and perspective. arXiv preprint arXiv:2202.07412 (2022)
29. Zhang, W., et al.: Iteratively learning embeddings and rules for knowledge graph reasoning. In: The World Wide Web Conference, pp. 2366–2377 (2019)
30. Zhang, W., et al.: Iteratively learning embeddings and rules for knowledge graph reasoning. In: Liu, L., White, R.W., Mantrach, A., Silvestri, F., McAuley, J.J., Baeza-Yates, R., Zia, L. (eds.) The World Wide Web Conference, WWW 2019, San Francisco, CA, USA, 13–17 May 2019, pp. 2366–2377. ACM (2019)
31. Zhang, Y., et al.: Efficient probabilistic logic reasoning with graph neural networks. arXiv preprint arXiv:2001.11850 (2020)

Construction of Open Data Sources for Data Interoperability in Brazilian Health Information Systems

Márcia Jacobina Andrade Martins$^{(\boxtimes)}$ and Claudia Bauzer Medeiros ⓘ

Institute of Computing, University of Campinas (UNICAMP), Campinas, Brazil
m905106@dac.unicamp.br, cmbm@ic.unicamp.br

Abstract. Health Information Systems (HIS) rely on correlation of heterogeneous data sources, which are often closed and in proprietary formats. Moreover, studies conducted to process and link such data sources are based on corpora and databases in the English language. This hinders an adaptation to non-English speaking countries, not only because of the language but also due to the lack of authoritative integrated bases in these countries - such as Brazil, in which our research is being conducted. To help solve these issues, we have designed and implemented a set of modules to create consolidated curated open data sources, in Portuguese, thereby helping health data interoperability for HIS in Brazil. The resulting database, called HealDB, created from public authoritative sources, provides data on diseases, symptoms, medications and their indications, and drug-drug and drug-food interactions. All data are in Portuguese, and consonant with medications, indications and treatments officially approved by the Brazilian Ministry of Health. We have moreover created an ontology, still in a prototypical stage, to allow linking different attributes and sources, and support connection to additional data, thereby facilitating HIS interoperability.

Keywords: Open Science · NLP · Health data management · Portuguese-language health data

1 Introduction

The primary objective of Health Information Systems (HIS) is to provide high-quality and efficient patient care, based on managing healthcare data. There is a wide variety of such systems, which may include both patient-geared functionality (e.g., management of electronic health records, intensive care monitoring) and healthcare institution management (e.g., billing, accounting).

Some HIS infrastructures are dedicated to healthcare professionals (such as those based on handling Electronic Health Records - EHR). Others also cater to patients (e.g., Patient Portals).

J. Tekli et al. (Eds.): ADBIS 2024, CCIS 2186, pp. 117–129, 2025.
https://doi.org/10.1007/978-3-031-70421-5_11

Regardless of their functionality, and in spite of extensive research efforts to improve HIS performance, flexibility and interoperability, such systems still face many challenges due to, among others, data quality, heterogeneity and different use authorizations for drugs across countries. Moreover, one often disregarded aspect are the effects of drug interactions. The work of [3], for instance, esti-mates that there were 85,811 hospitalizations and deaths in Brazil related to drug intoxication between 2009 to 2018 with 97% linked to prescribed drugs. In Europe and the USA, adverse drug reactions reach up to 300,000 deaths per year [9]. This highlights the importance of organizing information regarding drug interactions to prevent such hospitalizations and deaths.

Given these challenges, this research aims to work in two directions. The first is to design and create consolidated and curated open data sources, in Por-tuguese, to help the construction of generic HIS and health data interoperability. The second is to design and create an ontology that will allow connecting these sources to others – this is still ongoing work, and will not be detailed here. Both directions are geared towards helping construction of HIS for Brazil, where solu-tions are often closed and suffer from poor interoperability. Moreover, several such solutions do not operate well with official government sources, requiring additional effort from healthcare practitioners.

The choice of creating data sources in Portuguese, based on Brazilian official health-related data, is motivated by the fact that most open curated data sources used in HIS research are in English. This hampers construction of HIS using Brazilian officially recognized medications, and complicates research on national scenarios, in particular for plant-based medications - many of which are specific to a geographic region. As presented throughout this text, creating curated data sources in Portuguese is not just a matter of automatic translation of the contents of the English sources, but also ensuring their compatibility with official sources. An yet bigger challenge is the varying quality of open official data.

We have implemented an integrated database, called HealDB [15], which sup-ports two main kinds of functionality. The first involves processing and curating data from drug leaflets and medications to produce consolidated data on Medi-cations X Indications (including drug components, symptoms and others). Our work here has revealed shortcomings of Brazilian official open health-related data, highlighting the need for considerable data curation before HIS utiliza-tion. The second focuses on enhancing HealDB with drug-drug and drug-food interactions. Here, we needed to match drugs, perform cross-translations between Portuguese and English bases, and identify interactions of official Brazilian med-ication data. Our translation efforts, discussed here, showed the shortcomings of NLP-based tools when it comes to drug and drug component-related names.

Throughout the work, we encountered additional challenges such as the dif-ficulty in locating structured sources of open health data in Brazil, implying the necessity to construct them. Furthermore, when using Portuguese-related data sources, since many consensual vocabularies and data structures are mostly available in English, there was the need for a specialist to review translation of medical and chemical terms.

Our main contributions are thus

– creation of curated open data sources to be used by HIS, involving medications and treatments approved in Brazil. These sources were consolidated into a relational database, called HealDB, containing medications, their indications, diseases, drugs, their ingredients and interactions, among others.
– discussion of our solution for extracting, curating and translating health data to Portuguese, pointing out some quality concerns and how to circumvent them. Curation and translation involved, among others, experimenting with distinct kinds of NLP tools.

We have also designed and developed an ontology to help link these sources to other datasets. This is still in a prototypical stage, and will just be briefly mentioned here.

The Medications x Indications part of our research has already been published in a conference paper [16], which is summarized in this paper for completeness. Thus, most of this text is centered on work on drug-drug and drug-food interactions.

The rest of this article is organized as follows: Sect. 2 discusses related work. Sections 3 and 4 discuss the HealDB database, respectively presenting the Medications x Indications and the Drug Interactions data design and creation. Finally, Sect. 5 concludes the article, providing insights into ongoing work.

2   Related Work

Related work involves algorithms and systems that curate, integrate and link public open data for healthcare, sometimes constructing ontologies. Data sources of interest involve information on medications, symptoms, diseases, drug interactions, and other available public data sources to support health practitioners, health-related research, and the creation of integrated datasets. While the term "integration" traditionally considers connecting data sources using their data types and values, linkage and semantic integration involve work on ontologies. Our work belongs to the latter field, and thus we will not discuss more traditional integration approaches.

2.1   Semantic Integration and Interoperability in Healthcare

Healthcare interoperability means the capacity of heterogeneous health information systems to share, exchange, and reveal their health data. Its benefits are significant, including improved patient care, reduced duplication of clinical tests and procedures, and expanding the availability of health data for scientific investigation. However, due to the great variety of data sources used, their volume, and disparity of needs and uses, interoperability continues to present many challenges, which often include scarcity of quality data and even language barriers - e.g., when patients move from one country to another their clinical records and medications cannot be easily ported across HIS.

Several aspects can be explored to achieve semantic interoperability for healthcare systems. Most involve constructing or improving quality of ontologies – e.g., [22] or [20], sometimes with help of NLP models. Our ontology uses our curated data, and relies on the relationships that we elicited from distinct data sources.

2.2   Data Linkage in Healthcare

Research on data linkage in healthcare often focuses on analyzing drug leaflets, electronic health records, or extracting data from corpora, ontologies, and curated and integrated data sources - all of which in English. An exception in the Brazilian context is the work of [19], which processes Brazilian leaflets available at the ANVISA[1] site, extracting their main elements. Unlike our work, however, it does not link such elements to other official data sources nor does it construct any associated ontology or integrated databases.

The extraction of information to support linkage follows two main approaches. Some studies focus on algorithmic extraction, e.g., using NLP (see [1]), while others undertake a broader analysis through systematic reviews of public knowledge bases, as shown in [6].

Our work on Medications x Indications focuses on drug leaflets and additional external open data sources – we do not consider Electronic Health Records, since we concentrate on public open data sources. Research on extracting information from drug leaflets concentrates on identifying key elements (e.g., compounds, indications), often with the help of the *Structured Drug Labels (SPL)*[2] standard, using NLP. Alternatively, other papers propose APIs to query an SPL-encoded drug leaflet base – e.g., [5]. SPL is an XML-based standard adopted by the USA *Food and Drug Administration (FDA)* for managing products. The main result of this kind of research is the creation of structured files containing, for instance, pairs <medication, indication> [8]. Our work also uses NLP to extract information from official public drug leaflets; however, the fact that they are in PDF format without any standard encoding complicated the extraction.

This discussion exemplifies the variety of well-structured, annotated health-related corpora in English, which facilitate health data processing. Brazilian drug leaflets, for instance, are only available in PDF and need to be retrieved one by one via a query interface of the ANVISA system, complicating the construction of extraction processes and data linkage structures.

2.3   Drug and Food Interactions

A drug interaction occurs when the effect of a drug is altered in some way due to the presence of another drug, food, or environmental exposure [17]. The effectiveness of the drug diminishes, and it may cause side effects or enhance

---

[1] Brazilian Health Regulatory Agency.

[2] https://www.fda.gov/industry/fda-data-standards-advisory-board/structured-product-labeling-resources.

the action of a particular drug. According to the FDA [4], there are three main categories of drug interactions: (i) drug-drug interactions, when two or more drugs react with each other; (ii) drug-food/beverage interactions, resulting from the drug's interaction with food or drink; (iii) drug-condition interactions, when a medical condition makes a drug potentially hazardous. Our work deals with the first two.

Table 1. Related Work - Drug and food interactions

Articles	Sources			Machine Learning			DDI Analysis			Result
	Publication	Corpora for NLP	Public Dataset	Transformer	Neural Network	Other	Identification	Classification	Prediction	Structured Dataset
[9]		X		R-BioBERT	BLSTM		X	X		
[14]		X		BioBERT		X	X			X
[12]			X		KGNN				X	
[13]		X			CNN		X	X		
[2]	X	X	X				X			X
[10]	X					X	X			
[7]			X			X	X			
Our Work			X	GPT			X			X

Table 1 presents our categorization of a few papers that represent different kinds of research on identifying Drug-Drug Interactions (DDI). Data sources can be publications (full texts or abstracts), corpora for NLP such as DDI Corpus, and public datasets such as Drugbank, SemMedDB, or Twosides. Most research employs machine learning techniques for DDI - with methods based on transformers, neural networks, or other supervised algorithms. The method to identify DDI varies – with some papers incorporating DDI classification approaches. Labels *advice, mechanism, effect*, and *int* (refers to an interaction without any further information) represent the types of interactions between two drugs, while type *false* indicates the absence of interaction. Some studies generate structured datasets for consultation, while others do not, varying in their approaches to data dissemination and accessibility.

Most of the studies listed use machine learning (ML) and NLP to identify or extract DDI from different sources. Our work uses NLP to extract information from drug leaflets and to perform translations of active ingredients, to construct DDI tuples from DrugBank. In more detail, [14] and [9] explore different techniques for Drug-Drug Interaction (DDI) extraction, utilizing transformers associated with ML classifiers and neural networks, respectively. While [14] trains a Drug Entity Recognition Model using BioBERT [9,11] employs Relation-BioBERT and recurrent neural networks to detect and classify DDIs. BioBERT is a specialized version of BERT for biomedical text mining.

Several articles adopt neural network approaches to identify or extract DDIs. such as [13] or [12]. While the first applies Convolutional Neural Networks (CNNs) to extract DDIs from biomedical texts, the second combines Knowledge Graphs (KG) to neural networks to predict such interactions.

Other ML-based methods include [10] and [7]. [10] employed a Support Vector Machine (SVM)-based approach for Drug Named Entity Recognition (NER) and

DDI Extraction. Bjorne [7] used several kinds of classifiers applied to PubMed abstracts and annotated sentences.

Similar to our proposal, [2] uses public data sources to create potential DDIs into a unified dataset, comprising 5 clinically-oriented information sources, 4 NLP corpora, and 5 public datasets including DrugBank. Our work also explores the DrugBank public data source; we employ a transformer model (GPT-3.5 and GPT-4.0) to help translations, aligning with studies that provide or intend to provide a structured and public dataset.

3   Creating an Open Database for Medications and Their Indications

The extraction of Medications x Indications was published in a conference paper [16], being summarized here for completeness. Our inputs are data extracted from drug leaflets and medications in Portuguese (ANVISA[3]), ICD codes and disease symptoms (from BIREME MESH[4]). The results are a MySQL database (HealDB) integrating Medications, Diseases and Symptoms, and a small prototypical ontology that mirrors this database.

This work can be divided into three major processes: (i) leaflet preprocessing, including web crawling of leaflets from the ANVISA website and subsequent processing and curation of these leaflets; (ii) disease and symptom extraction using NLP algorithms from the "Indication" text field of the leaflets, resulting in the linkage between the medication and associated diseases and symptoms; and (iii) creation and generation of the medication ontology, using data from additional curated repositories.

Leaflet Preprocessing. We initially downloaded, preprocessed and deduplicated 8472 leaflets, resulting in 7476 leaflets. Preprocessing included many curation steps – e.g., eliminating image files, or non-leaflet ones.

Disease and Symptom Extraction from the Indications Field of the Leaflets. We extracted diseases and symptoms from the drug leaflets using the SpaCy NLP library. Since many diseases (and symptoms) are described in various forms, we had to use two similarity measures, levenshtein and jarowinkler, to better identify diseases/symptoms from the leaflets' indication section "What this medication is intended for"[5]. Unfortunately, the outcome was not as anticipated. Of the 7,476 leaflets, we found diseases and symptoms in only 4,250 (57% of the total), confirming that the adopted strategy is not the most appropriate. Hence, we are still working on this part to improve disease and symptom recognition.

Creation and Automatic Generation of Our Medication Ontology. Once the database was built, we used it to create our prototypical Protégé Ontology,

---

[3] Brazilian Health Regulatory Agency.

[4] https://decs.bvsalud.org/en/.

[5] "Para que este medicamento é indicado", in Portuguese.

whose properties reflect the relationships between the elements of the various database tables. Here, we intend to construct the turtle code to instantiate the Ontology using data stored in the database.

4   Constructing Drug and Food Interactions

We next designed and developed a framework to identify drug-drug as well as drug-food interactions for the medications available in our HealDB database. Since there are many medications with similar effects, interactions between drugs are described in terms of their active ingredients, rather than the medication names themselves.

To create our interactions, we used the DrugBank XML File, accessible on the DrugBank website[6], which contains approximately 15 thousand drugs along with various properties such as drug and food interactions. We restricted ourselves to the active ingredients of medications officially approved in Brazil. The process involved identifying the interactions relevant to our medications, translating them from English to Portuguese, and subsequently storing them in HealDB. The decision to utilize DrugBank [21] is based on the reliability of that data source. At the moment, we are extending our Medication Ontology with this information.

4.1   Architecture and Workflow

Figure 1 describes the architecture and workflow for constructing the Drug Interactions modules. As seen in the Fig. 1, its main results are the extension of the HealDB database with curated drug and food interactions (in Portuguese).

The first hurdle was that our data was all in Portuguese, extracted from Brazilian sources, while DrugBank data is in English. We first tried translating DrugBank data into Portuguese, but there were too many mismatches, which hampered discovering the appropriate interactions. Thus, we took a different approach. First, we translated our curated Medication data into English, to check interactions against DrugBank. Next, once we had built all interactions, we translated them into Portuguese, which ensured better curation results, since we already had the original Portuguese terms. It must be stressed that we have kept the initial Portuguese-to-English translation in the database, to ensure research auditing and provenance verification.

Let us now discuss the steps in detail. The initial step, labeled "Configure and create tables to store DrugBank data" (A), involved importing the external DrugBank XML file (1) and storing this data in a relational database (2). This external data source comprises around 15 thousand drugs (all in English) with hundreds of properties, including those that concern drug and food interactions.

In parallel, process B translated the active ingredients found in the Portuguese-language HealDB database (3) (Medications X Indications) into English and subsequently creates new files in HealDB (4) with these translations. This translation process uses the GPT-4 model.

---

[6] https://go.drugbank.com.

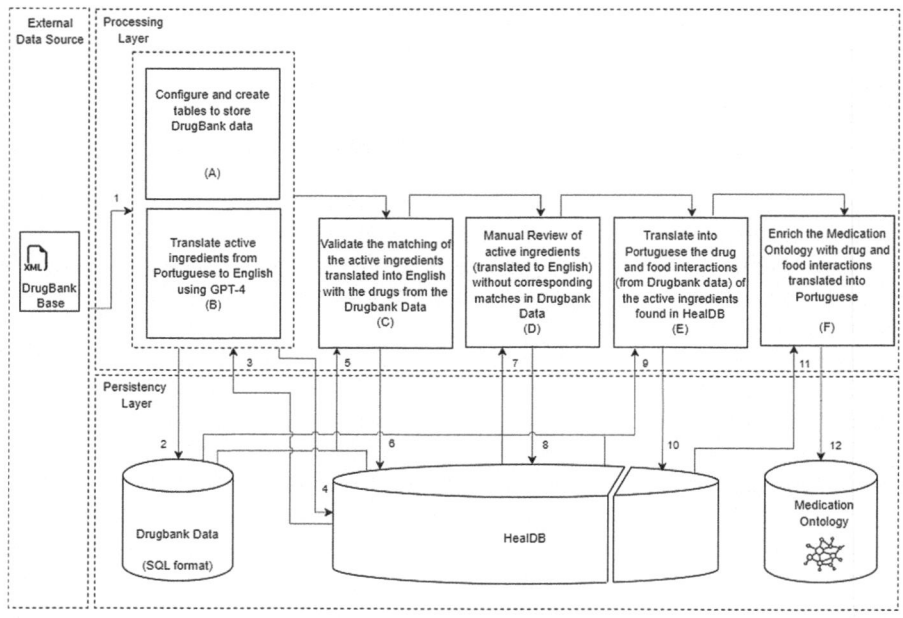

Fig. 1. Architecture and Workflow - Extracting, translating and storing interactions

Step C validates the English translations of active ingredients from HealDB against the drugs listed in the DrugBank data (5) and updates auxiliary translation tables in HealDB, flagging the correct translations (6).

The next step, D, involved a manual review of the active ingredients with no match in the Drugbank data (7). The outcomes of this manual review were stored in auxiliary translation tables in HealDB (8).

Now that all active ingredients are correctly translated into English, the next step, E, aims to identify drug and food interactions (all in English) from the DrugBank data related to the active ingredients HealDB (9). These interactions were then translated into Portuguese using GPT-3.5 and stored in the same repository. This process expanded HealDB with Portuguese-language drug and food interactions, associating them with the active ingredients previously stored in HealDB (10).

The final step (F) uses the curated Portuguese-translated data from HealDB (11) related to drug and food interaction to extend the Medication Ontology (12) – this is ongoing work.

4.2   Implementation Details and Results

Our implementation used Python and a MySQL Database to store the data. All the steps listed below were automated, except for a manual review of the translation of the active ingredients from Portuguese to English.

To enrich HealDB with drug interaction information, the initial idea was to use Drugbank's online interaction checker. However, this strategy would be time-consuming and could be potentially blocked by its website due to the large amount of active ingredient combinations – around 2 million pairs. After evaluating the Drugbank site contents, we decided to directly import data from a subset of its XML files into our database, creating an internal Drugbank repository – again, ensuring auditability. These files contain drug attributes, indications, synonyms, product ingredients, and food and drug interactions

The Drugbank repository contained 2.8 million drug interactions in English. To integrate this data into HealDB, we identified the correspondence between drugs (from Drugbank data) and active ingredients (from the HealDB repository, translated into English). Subsequently, we translated and stored this subset of interactions.

We first tried to use the Google Translate API for the task of translating active ingredients from Portuguese to English. However, Google soon blocked our access due to the volume of terms to translate. Thus, we decided to compare two different NLP models for the translation task - an API from OpenAI employing the GPT-4 model, and the SeamlessM4T model from META.

The GPT-4 model successfully matched 1,436 active ingredients (60% of the total) from the Drugbank repository. In contrast, the SeamlessM4T model only managed to match 143 active ingredients (6% of the total). Thus, we decided to use the GPT-4 model. The performance of the META model was disappointing. Several translation errors occurred, such as translating "Insulina Glulisina" to "Insuline and glutamine", or "Óxido Cúprico" to "Oxygenated copper". Translations were validated when there was a correspondence in Drugbank. For example, 'sulfate of sodium' and 'sodium sulfate' are both correct, but the DrugBank match is 'sodium sulfate', which was thus deemed the most appropriate translation. Table 2 presents some translation examples, with the translations we considered best highlighted in bold. The first two rows exemplify cases in which neither tool provided an adequate translation, and thus we needed to perform manual translation. The last two rows exemplify cases in which both models succeeded, and the rest are examples where the GPT-4 model was accurate. The last column shows cases in which manual translation was required, because neither model provided a satisfactory result.

Although the GPT-4 translations were more accurate, they still required manual reviews for the cases that did not match the Drugbank entries. Out of the 2,398 active ingredients analyzed, 1,436 were accurately matched to Drugbank, while 962 required manual review – which took three weeks, leading to a total of 2,032 active ingredients correctly translated. However, 367 active ingredients, primarily from herbal and homeopathic medicines, remained unchecked.

Once we had our active ingredients translated into English, we were able to identify which Drugbank interactions applied to our data.

We identified 775,212 distinct drug interactions and 1,781 food interactions. The next step was to automatically translate these drug and food interactions from English to Portuguese using OpenAI's API, with the simpler but effective

GPT-3.5 model. We also optimized the translation process by handling text repetition, substituting drug names with placeholders like 'drug1' and 'drug2'. This reduced the number of sentences for translation. Afterward, we restored the original drug names. These interactions were then stored in HealDB.

Table 2. Translations of Active Ingredients: Portuguese to English

Active Ingredients (in portuguese)	SeamlessM4T Model	GPT-4 Model	Required Manual Translation
Riociguate	Ryougouat	Riociguate	*Riociguat*
Policresuleno	Polysaccharide Resin	Polychlorinated	*Policresulen*
Insulina Glulisina	Insulin and Glutamine	*Insulin Glulisine*	
Oxido Cuprico	Oxygenated Copper	*Cupric Oxide*	
Acido Acetilsalicilico	*Acetylsalicylic Acid*	*Acetylsalicylic Acid*	
Ácido Ascórbico	*Ascorbic Acid*	*Ascorbic Acid*	

Table 3. Number of Records in MyHealthDB and DrugBank

Repository	Tables	Number of Records
HealDB *(Portuguese)*	medications	10,458
	drug leaflets	7,476
	active ingredients	2,398
	diseases	14,771
	symptoms	432
	medication indications	10,622
	drug interactions	775,212
	food interactions	1,781
DRUGBANK *(English)*	drugs	15,235
	food interactions	2,456
	drug interactions	2,866,522
	synonyms	34,680
	product ingredients	2,693

Table 3 provides an overview of the number of curated records of HealDB, and some of the corresponding data in DrugBank. We point out that in spite of all of our automated NLP efforts, the translation of 962 active ingredients (out of the final HealDB records – see line 3 of the table) had to be checked manually with expert help. In the HealDB Database, 'medications' refer to all medications approved by ANVISA, limited to active ingredients approved in Brazil. Conversely, 'drugs' in DrugBank includes active ingredients and compounds at various stages of clinical trials worldwide, such as approved, experimental, investigational, and withdrawn stages. As a result, DrugBank's broader scope includes a wider range of active ingredients from various countries and stages, document-ing approximately 15,235 active ingredients and 2.8 million drug interactions, compared to 2,398 active ingredients and 775,212 interactions in HealDB.

5   Conclusions and Ongoing Work

This paper presented our ongoing work to design and construct curated open databases to allow interoperability in Brazilian HIS, linking heterogeneous data sources for drugs officially approved in Brazil. This fills a much-needed gap, since there is a lack of reliable open data in Portuguese, consolidating information across several official data sites. Since most related research is based on open English sources, our results can also serve for research on, e.g., clinical practices, linking to Brazilian EHR. All the data we use as input is already open and publicly available, thereby avoiding issues such as privacy violation.

Our open database, HealDB, contains more than 20 tables, the largest of which with over 770,000 records on drug-drug and drug-food interactions. Some tables are for internal purposes only, providing auxiliary information to ensure research auditing and replicability. The discussion of our implementation efforts, and of how we overcame varying data quality issues is part of the contribution – this kind of discussion is not often found in papers that discuss creation of reliable data sources.

Throughout the work we faced several challenges, including the identifica-tion of relevant data sources in Portuguese, and the non-uniform quality of such sources. Data curation, checking the results of running different NLP transla-tions, manual verification, and the construction of the ontology also proved to be major blocks for the research.

Ongoing work on "Medications X Indications" involves additional curation and extraction of more key concepts from drug leaflets, as well as running other NLP models. Our ongoing strategy involves utilizing Named Entity Recognition (NER) with a pretrained BioBERT model, known for its efficacy in biomedical text analysis. Studies by [9] and [14] underscore its effectiveness. BioBERTpt [18], pretrained on Portuguese text, handles the complexities of biomedical text. We are now annotating diseases and symptoms from approximately 200 leaflets using an open-source tool and training an appropriate NER model using BioBERTpt, to be used on the full set of leaflets.

Another ongoing task involves transforming our prototypical Medication ontology into a full-fledged ontology, including extending it to cover drug inter-actions. Moreover, we aim to automate ontology generation by developing Turtle code to instantiate the ontology using data stored in HealDB.

Acknowledgements. Work partially funded by projects FAPESP 2013/08293-7 and CNPq #308018/2021-4. We thank Dr. Nestor Andrade Piva for his help in checking our work on the drug leaflets, and members of the LIS research laboratory for suggestions.

References

1. Jagannatha, A., Liu, F., Liu, W., Yu, H.: Overview of the first natural language processing challenge for extracting medication, indication, and adverse drug events from electronic health record notes (made 1.0). Drug Saf. **42**(1), 99–111 (2019)
2. Ayvaz, S., et al.: Toward a complete dataset of drug-drug interaction information from publicly available sources. J. Biomed. Inform. **55**, 206–217 (2015). https://doi.org/10.1016/j.jbi.2015.04.006
3. Duarte, F.G., de Paula, M.N., Vianna, N.A., de Almeida, M.C., Junior, D.D.M.: Deaths and hospitalizations resulting from poisoning by prescription and over-the-counter drugs in brazil. Revista de Saúde Pública **55**, 81 (2021). https://doi.org/10.11606/s1518-8787.2021055003551
4. FDA: Drug interactions: what you should know (2013). https://www.fda.gov/drugs/resources-drugs/drug-interactions-what-you-should-know. Accessed Sept 2023
5. Flynn, A., et al.: An experiment to convert structured product labels into computable prescribing information. In: 2021 IEEE 9th International Conference on Healthcare Informatics (ICHI), pp. 296–300 (2021)
6. Salmasian, H., Tran, T.H., Chase, H.S., Friedman, C.: Medication-indication knowledge bases: a systematic review and critical appraisal. J. Am. Med. Inform. Assoc. **22**(6), 1261–1270 (2015)
7. Björne, J., Kaewphan, S., Salakoski, T.: Uturku: drug named entity recognition and drug-drug interaction extraction using SVM classification and domain knowledge. In: Second Joint Conference on Lexical and Computational Semantics (* SEM), Volume 2: Proceedings of the Seventh International Workshop on Semantic Evaluation (SemEval 2013), pp. 651–659 (2013)
8. Fung, K.W., Jao, C.S., Demner-Fushman, D.: Extracting drug indication information from structured product labels using natural language processing. J. Am. Med. Inform. Assoc. **20**(3), 482–488 (2013)
9. KafiKang, M., Hendawi, A.: Drug-drug interaction extraction from biomedical text using relation BioBERT with BLSTM. Mach. Learn. Knowl. Extr. **5**(2), 669–683 (2023). https://doi.org/10.3390/make5020036
10. Kolchinsky, A., Lourenço, A., Wu, H., Li, L., Rocha, L.M.: Extraction of pharmacokinetic evidence of drug-drug interactions from the literature. PLoS ONE **10**(5), e0122199 (2015). https://doi.org/10.1371/journal.pone.0122199
11. Lee, J., et al.: BioBERT: a pre-trained biomedical language representation model for biomedical text mining. Bioinformatics **36**(4), 1234–1240 (2019). https://doi.org/10.1093/bioinformatics/btz682
12. Lin, X., Quan, Z., Wang, Z., Ma, T., Zeng, X.: KGNN: knowledge graph neural network for drug-drug interaction prediction. In: IJCAI, vol. 380, pp. 2739–2745 (2020)
13. Liu, S., Tang, B., Chen, Q., Wang, X.: Drug-drug interaction extraction via convolutional neural networks. Comput. Math. Methods Med. **2016**, 6918381 (2016). https://doi.org/10.1155/2016/6918381
14. Machado, J., Rodrigues, C., Sousa, R., Gomes, L.M.: Drug-drug interaction extraction-based system: an natural language processing approach. Expert Syst. e13303 (2023). https://doi.org/10.1111/exsy.13303
15. Martins, M.J.A., Medeiros, C.B.: Medications, symptoms and drug leaflets extracted from public Brazilian sources (2023). https://doi.org/10.25824/redu/JUHFWF. Repositório de Dados de Pesquisa da Unicamp, DRAFT VERSION

16. Martins, M.J.A., Medeiros, C.B.: Linking heterogeneous health data sources in brazil centered on drug leaflet processing. In: Proceedings of XXXVIII Brazilian Database Symposium, pp. 366–371. SBC - Brazilian Computer Society (2023). https://doi.org/10.5753/sbbd.2023.233356
17. Robertson, S., Penzak, S.: Chapter 15 - drug interactions. In: Principles of Clinical Pharmacology, 2nd edn, pp. 229–247. Academic Press (2007). https://doi.org/10.1016/B978-012369417-1/50055-9
18. Schneider, E., et al.: BioBERTpt - a Portuguese neural language model for clinical named entity recognition. In: Proceedings of the 3rd Clinical Natural Language Processing Workshop, pp. 65–72. Association for Computational Linguistics, Online (2020). https://doi.org/10.18653/v1/2020.clinicalnlp-1.7. https://aclanthology.org/2020.clinicalnlp-1.7
19. Silva, J.V.F.: Facil Bula: Sistema que Estrutura o Bulario Eletronico da ANVISA. Master's thesis, UTFPR (2016)
20. Silveira, R., Cavalcanti, M.: Método para rotular ligações semânticas na web de dados. In: Anais do XXXV Simpósio Brasileiro de Bancos de Dados, pp. 49–60 (2020). https://doi.org/10.5753/sbbd.2020.13624
21. Wishart, D.S., et al.: Drugbank: a comprehensive resource for in silico drug discovery and exploration. Nucleic Acids Res. **34**(Database Issue), 668–672 (2006). https://doi.org/10.1093/nar/gkj067
22. Zhang, G.Q., Bodenreider, O.: Using SPARQL to test for lattices: application to quality assurance in biomedical ontologies. In: The Semantic Web – ISWC 2010, pp. 273–288 (2010)

Brazilian Political Study with Topics Analysis and Complex Networks

Tiago Toledo Junior[(⊠)] [iD], Diego Raphael Amancio [iD],
and Roseli Aparecida F. Romero [iD]

Institute of Mathematics and Computer Science, University of São Paulo, São Carlos,
SP, Brazil
tiago.toledo@usp.br, {diego,rafrance}@icmc.usp.br

Abstract. The Brazilian Chamber of Deputies is a complex system
composed of conflicting interests, ideological points of view, and social
interactions between deputies. The present work expands the social analysis of this system, usually analyzed with complex networks, to also consider the ideological aspect of the house by conducting a topic analysis to
generate thematic complex networks that are analyzed to uncover how
the theme of a voting proposition affects the behavior of the deputies.
We show that the parties do not maximize modularity inside the thematic networks, and therefore the social aspect of the network presents
a more prominent group structure. Also, we show that the topic being
voted on highly influences how parties from the same political spectrum
behave and that a set of 10 topics seems to indicate the overall themes
that are voted on in this House. The source code for this analysis can
be found in Github (Available in https://github.com/TNanukem/topic_
networks_congress. Accessed in July 2024).

Keywords: Complex Networks · Topics Analysis · Political Analysis

1 Introduction

Composed of the Senate and the Chamber of Deputies, the Brazilian Congress
is one of the most complex legislative systems in the world because of its high
number of members, parties, and interactions among them. This complexity
makes the system suitable to be analyzed with complex networks [4].

Several works have analyzed the Brazilian Congress using complex networks
[4,5,9,10,13]. However, one crucial aspect of the legislative voting process, often
not analyzed by these works, is the content of the propositions set up to vote.

Although textual analyses were made to analyze speeches from the standing
committees [15], the administrative decrees of the executive power [14], and even
for the propositions itself [1], the integration of textual models with complex
networks analysis for political analysis is still a new avenue to be walked on.

In the present work, we propose an automatic pipeline for analyzing the
dynamics of the Brazilian Congress using complex networks and topic analysis,

J. Tekli et al. (Eds.): ADBIS 2024, CCIS 2186, pp. 130–141, 2025.
https://doi.org/10.1007/978-3-031-70421-5_12

aiming to uncover how parties and congress members behave according to each type of proposition presented in the plenary. We will show that the content of the propositions highly affects how parties vote and how disciplined the members of that party are.

We will also show that, for a voting network constructed exclusively on a single topic of propositions, the parties are not the structure group that maximizes modularity and that the number of effective groups, by voting behavior, is much smaller than the number of parties.

2   Related Work

2.1   Topic Analysis in Brazilian Politics

The work by [14] analyzed the decrees from the Brazilian federal executive power for the two starting decades of the 2000s. They used a Latent Dirichlet Allocation (LDA) algorithm to uncover 25 topics from the set of decrees constituted by abstracts and full decrees. These topics were then exchanged for a predefined classification, shown to better represent the documents, which were transformed into a complex network to verify relationships between topics.

The Brazilian Chamber of Deputies has standing committees, which are permanent groups responsible for discussing themes before they go to the plenary to vote. The work by [15] analyzed the textual transcriptions of the discussions for these committees from 1995 to 2021. They have used an LDA algorithm, uncovering 21 topics. They showed that each committee, as expected, has its group of main topics of discussion. However, these topics present some temporal variation.

The voting propositions were analyzed with an LDA algorithm, using data from 1995 to 2014 [1] to analyze the saliency of each topic for each party using the government, the opposition, and the state of each deputy. They define salience as the importance of the topic to some entity and uncovered 7 topics, from Taxes to Regulation. It is noticeable the lack of education and health-related topics. They concluded that most variation in topic behavior is given by individual incentives to legislators, and not exactly by the party alignment.

2.2   Complex Networks for Political Analysis

The usage of complex networks to uncover how parties and party members behave over time is extensive. The work in [4] showed that the parties are not the group structure that maximizes modularity and that there was an increase in the isolation of the president's party preceding the impeachment process that happened in Brazil in 2015–2016. For that, they transformed the voting data into yearly complex networks, applied a backbone extraction methodology, and conducted a group analysis followed by an isolation and fragmentation analysis.

It can be found in [10], the fact that the Brazilian Chamber of Deputies has a higher variability in ideological behavior when compared to the US Congress and that the parties in Brazil can be reduced to a smaller set of ideological communities. To uncover those insights they used temporal embeddings in complex networks constructed from voting data.

It was also shown, for the Canadian House of Commons, by [6], that party cohesion is variable per party and is influenced by the position of the party in the current governmental structure, depending on whether it is from the government or the opposition.

The work proposed in [9] proposed a backbone extraction methodology for co-interaction networks based on two principles: topological analysis and contextual analysis. The first one uses topological metrics to define the most suitable network, whereas the latter tries to verify the predictive capacity of the resulting network in predicting the edges of the original network as an identification proxy of the underlying phenomenon of interest.

3   Methodology

In this section, are described the methodology used to achieve the topic networks and how their analysis was conducted. In Fig. 1 is presented the visual representation of the steps of the methodology.

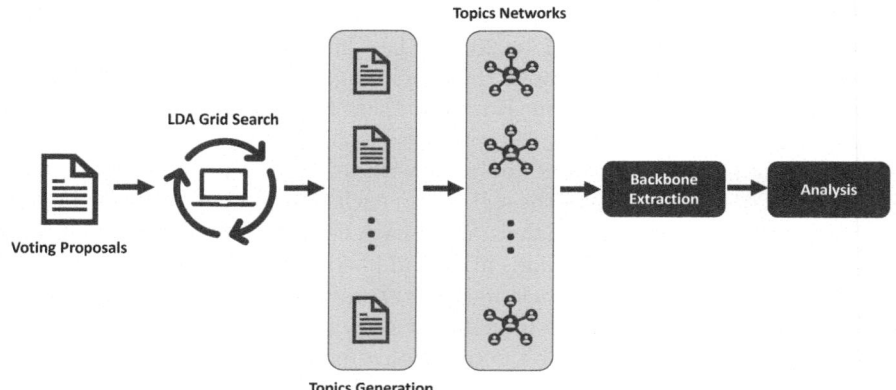

Fig. 1. Used methodology to generate and analyze the topics networks

3.1   Topics Analysis

The first step is to conduct a topic analysis of all the propositions for the Chamber of Deputies. For that, an LDA algorithm [2] was used, following a grid search over the following parameters (Table 1):

It is important to notice that the available token types to be used in the analysis were the full description of the proposition, the keywords defined by the

Table 1. LDA Grid

Variable	Values
Tokens	Tokens, Bigrams, Trigrams, Fourgrams
Alpha	0.01, 0.51, Symmetric, Asymmetric
Number of Topics	2, 3, 5, 7, 9, 10, 15
Token Type	keys, description, both

house on that proposition, and a concatenation of both. The implementation provided by the Gensim Python package [18] was used.

Once this grid is run, the following metrics are used to evaluate the resulting topics:

- **Coherence** - A measure using pointwise mutual information and cosine similarity between the top words of the topics.
- **U-Mass** - Based on counting of concurrence of documents.
- **UCI** - Measures the pointwise mutual information between all pairs of words of the top words.
- **Perplexity** - A measure of the degree of surprise of a model in new data. It is measured as the normalized log-likelihood of the test set.

After a manual inspection of the behavior of each metric, the method that maximizes the coherence was selected.

3.2 Network Construction

To construct the networks, all voting sessions ranging from 2002 to 2022 for a given topic are selected and used to generate the network. Therefore, one network for each topic, over all years, has been created.

Given the set of votes from each deputy, each deputy is defined as a node of the network. To create the edges between the deputies, the method described in Eq. 1 proposed by [4] is adopted.

$$w_{ij} = \frac{1}{N} \sum_{k=1}^{N} v_k^{(i)} v_k^{(j)} = \frac{\mathbf{v}^{(i)} \cdot \mathbf{v}^{(j)}}{N} \qquad (1)$$

where given N votes in a period, $\mathbf{v}^{(i)}$ is the N-dimensional vector of the congress member i in which each position of the vector is 1, for a yes vote, -1 for a no vote and 0 for an abstention and $\mathbf{v}^{(i)} \cdot \mathbf{v}^{(j)}$ is the dot product between the two congress members vectors.

3.3 Backbone Extraction

Then, each generated network is exposed to a backbone extraction pipeline in which a set of different algorithms was tested against the data. The following algorithms are used:

- **Disparity Filter** - As proposed in [16].
- **Locally Adaptive Network Sparsification Filter (LANS)** - As proposed in [11].
- **High Salience Skeleton Filter** - As proposed in [12].
- **Doubly Stochastic Filter** - As proposed in [17].
- **Noise Corrected Filter** - As proposed in [8].

For the pipeline definition, a topological analysis and then a contextual analysis, as proposed by [9] are applied. For the topological analysis, the node fraction, the LCC size, and the modularity of the resulting networks were used.

For the contextual analysis, the party for each node of the edge and the prevalence of each of those parties are used as features for the machine learning algorithm, XGBoost [7], aiming to predict the edge weight of the testing set.

In order to avoid a preference for smaller networks, that naturally yield higher modularities, all methods were tested over a range of different resulting densities. Finally, the selection method consists of:

- Only methods that yielded networks with LCC size greater than 70% were kept
- Methods that could not be trained were removed
- The higher modularity method, removing ties by the smaller MSE, was selected

3.4 Community Detection

For the community detection, the Louvain algorithm [3] is applied on the network for each topic after the process of the backbone extraction is finished.

Once the communities are generated, they are evaluated into the following dimensions: the number of communities and their modularity, and the composing political spectrum for each community.

3.5 Network Characterization

To characterize the resulting networks, three metrics are used as defined by [4]. To use them, an edge weight conversion, to transform it from a similarity to a dissimilarity index was applied using Eq. 2.

$$\Delta(w_{ij}) = 2(1 - w_{ij})^{\frac{1}{2}} \tag{2}$$

The coalition between two groups (parties or communities) is defined as in Eq. 3.

$$d(A, B) = \frac{1}{|A \times B|} \sum_{(a,b) \in A \times B} l(a, b) \tag{3}$$

where $l(A, B)$ is the shortest path length between two groups A and B.

Then, the isolation of a group was calculated using Eq. 4 and the Fragmentation of a group according to Eq. 5.

$$I(A) = \frac{\sum_{X \neq A} |X| d(A, X)}{\sum_{X \neq A} |X|} \tag{4}$$

$$F(A) = d(A, A) \tag{5}$$

4 Dataset and Network Assessment

4.1 Dataset

Two main datasets were gathered using the Chamber of Deputies Open Data API[1]: the voting data for each deputy and the propositions data for each voting session held.

The resulting dataset was comprised of 4094 voting sessions, from 2002 to 2022 with 1330 associated propositions. The first was then used to generate the edges of the networks and the latter was used to generate the topics.

4.2 Networks Statistics

The descriptive statistics for the topic networks can be found in Table 2, in which Topic stands for the detected topic by the LDA, N is the number of nodes of the network, M is the number of edges, Components is the number of connected components of the network, and C.C. is the Clustering Coefficient.

Table 2. Network Statistics for the Topics Networks

Topic	N	M	Components	C.C.	Transitivity	Assortativity	Degree
0	1595	2943	1	0.09	0.00	−0.74	3.69
1	1614	2866	1	0.13	0.00	−0.66	3.55
2	1726	3171	1	0.09	0.00	−0.89	3.67
3	1549	2826	1	0.11	0.01	−0.71	3.65
4	1614	2883	1	0.07	0.00	−0.79	3.57
5	1488	2612	1	0.14	0.01	−0.85	3.51
6	1334	2448	1	0.17	0.01	−0.67	3.67
7	1580	3194	1	0.11	0.01	−0.88	4.04
8	1732	3192	1	0.07	0.00	−0.71	3.69
9	1647	2936	1	0.12	0.00	−0.97	3.57

It is worth noting that the generated networks consistently have a low clustering coefficient, low transitivity, and high negative assortativity. The first two

---

[1] Available in https://dadosabertos.camara.leg.br/. Accessed in July 2023.

can be explained by the fact that the networks are not segmented by year, and therefore, several nodes do not coexist in time. The last shows a high degree of heterophily in the network, which probably represents the presence of hubs, possibly deputies with several mandates.

5   Results

5.1   Topics

The LDA analysis resulted in 10 topics exposed in Table 3 with their respective top-5 words. The words are stemmed and in Portuguese.

Table 3. Found Topics for the Chamber of Deputies Propositions

Topic	Top-5 Words	Brief Description
0	penal, codig, deputy, crim, administr	Law and Security
1	soc, previdenc, regim, segur, contribuica	Social Security
2	impost, tribute, rend, tributari, lucr	Taxes
3	fisc, financ, unia, publ, feder	Fiscalization and Financing
4	rural, are, imovel, urban, fund	Ruralism, Agronomy and Environment
5	eleitor, part, poli, propaganda, eleico	Elections
6	trabalh, empreg, pandm, serv, rural	Work
7	eletr, energ, contribuica, set, brasil	Energy
8	carg, transport, brasil, carr, agenc	Transportation
9	fund, educaca, financ, constituic, ensin	Education

The selected grid for this model was the usage of single tokens, with both token types, 10 topics with an alpha of 0.51. It achieved a coherence value of 0.497654. It is noticeable the lack of a topic related to health issues.

5.2   Backbone Extraction

The resulting backbone extraction algorithms selected by the pipeline, and their respective fraction, are shown in Table 4.

It can be seen that each topic was favored by a different algorithm and that all methods required a sizable fraction of edges, always with at least half of the original fraction of the network, probably because the networks span several years, which could make them very sparse with a strong removal of edges.

5.3   Number of Communities and Political Parties

The number of found communities, for every network, is smaller than the number of parties whereas they also present a higher modularity, showing that they have a better representation of the coalitions inside the network. The number of communities can be seen in Fig. 2 and the resulting modularity in Fig. 3.

Table 4. Resulting Backbone Extraction Methods and Fractions for the Topics Networks

Topic	Method	Fraction of Edges
0	LANS	0.6
1	LANS	0.6
2	High Salience Skeleton	0.6
3	High Salience Skeleton	0.5
4	LANS	0.5
5	LANS	0.5
6	High Salience Skeleton	0.6
7	LANS	0.5
8	Doubly Stochastic Filter	0.8
9	High Salience Skeleton	0.6

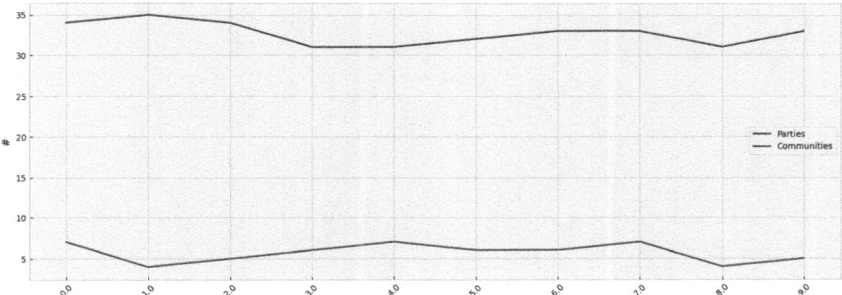

Fig. 2. Number of parties and communities for each topic

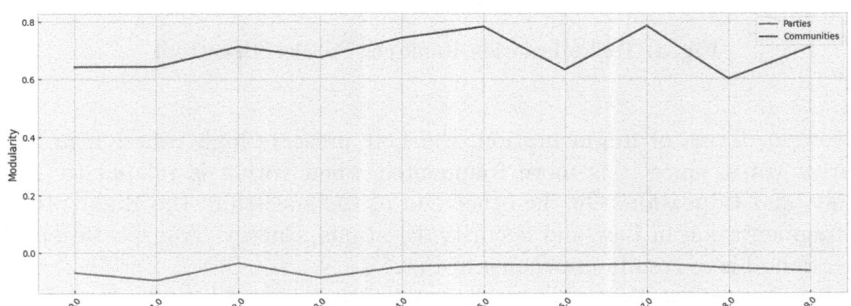

Fig. 3. Modularity for parties and communities for each topic

5.4 Isolation and Fragmentation Analysis

The political parties were ranked according to their isolation and fragmentation scores. The ranks were assigned in an ascending order so the lower the rank, the

lower the isolation or the fragmentation of that party. The spectrum rank was
calculated as the average of the rank of the parties of that spectrum.

The Left political spectrum is highly isolated for the following topics: Social
Security (topic 1), Taxes (topic 2), Fiscalization and Financing (topic 3), Work
(topic 6), Energy (topic 7), Transportation (topic 8), and Education (topic 9).
All topics that require government expenditure and financing, usually caring
topics for the Left. The isolation for Social Security is also high for the Right,
which shows that both extremes of the spectrum behave isolated from the rest
of the house in these themes when compared to parties more to the center. The
resulting isolation for each spectrum in each topic can be found in Fig. 4.

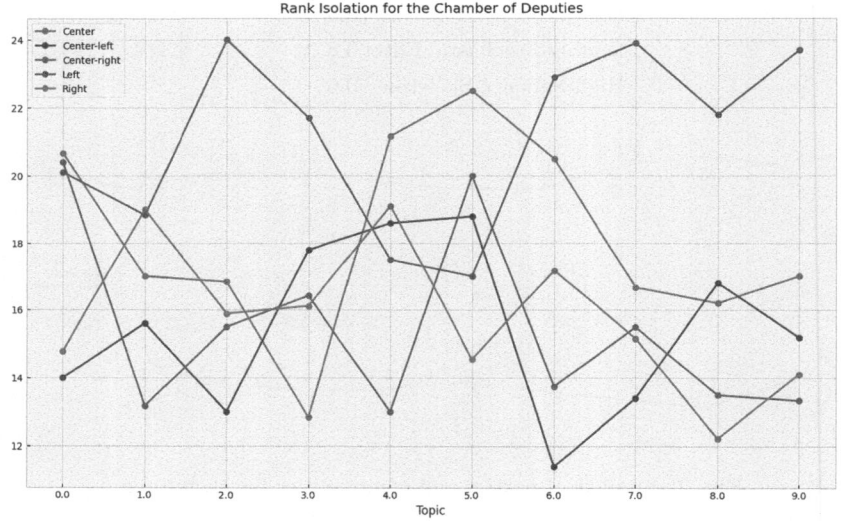

Fig. 4. Topics Isolation Rank per Political Spectrum

Now, in terms of fragmentation, the Left presents high cohesion in Social
Security votes, since it is more fragmented when voting is related to Taxes,
Energy, and Education. On the other end of the spectrum, the Right presents
low fragmentation in Law and Security, Elections, Energy, Transportation, and
Education. These results are shown in Fig. 5.

This shows that the topic of the proposition is highly influential for the
behavior of parties, as both extremes of the spectrum become less fragmented
and more isolated for topics that are core to their line of thought. The Left
appears isolated more often, while the Right is more in line with parties to the
center. This can also indicate a less extreme Right or a more right-aligned Center
spectrum.

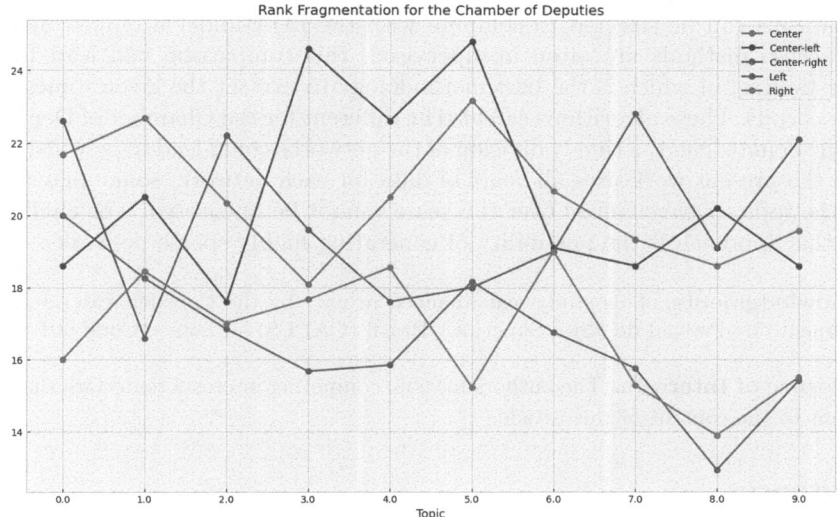

Fig. 5. Topics Fragmentation Rank per Political Spectrum

6   Conclusion and Future Works

In this paper, it was shown that the text of the voting propositions in the Brazilian Chamber of Deputies could be used to refine relevant voting topics with the usage of an LDA algorithm. All of the found topics are shown to be happening with frequency by common news media sources in Brazil.

The results obtained showed firstly that complex networks can be used to analyze topics of voting in the Brazilian Chamber of Deputies and that in this type of network, the parties are not the group structure with higher modularity. Further, they also showed that the best backbone algorithm for each topic network is variable and depends on the topic. Finally, the topic in question highly affects the behavior of parties inside a given political spectrum. Both the Left and the Right, the most extreme sides of the spectrum, were often more isolated from the rest of the parties, and the fragmentation of their parties became smaller for more important themes for the line of thought of the side of the spectrum.

Since the entire methodology depends on only two types of data: the voting for each member of the political house and the text for the proposition being voted, we believe that this could easily be applied to other political scenarios in different democratic countries. We notice, however, that the conclusions could be highly different in these scenarios since they are specific to each country. More than that, they could change given the temporal window being analyzed since these are dynamical systems.

Future works include analyzing the Senate behavior since the methodology presented here could easily be replicated in that house. Other topic extraction mechanisms can be selected to compete with the LDA under a unified metric. Since these methods are often unsupervised, this comparison can lead to an understanding of which is the best methodology to extract the given topics of a house corpus. These algorithms can also be different for the Chamber of Deputies and the Senate. Next, a timely division of the networks could lead to new insights. Since the present work uses all years of data for each network, some nuances of how the topics have changed over the years cannot be uncovered. The challenge with this approach is the possibility of generating highly sparse networks.

Acknowledgments. This study was financed in part by the Coordenação de Aperfeiçoamento de Pessoal de Nível Superior - Brasil (CAPES) - Finance Code 001.

Disclosure of Interests. The authors have no competing interests to declare that are relevant to the content of this article.

References

1. Batista, M.: Quais polÍticas importam? usando ênfases na agenda legislativa para mensurar saliência. Revista Brasileira de Ciências Sociais **35** (2020). https://doi.org/10.1590/3510411/2020
2. Blei, D., Ng, A., Jordan, M.: Latent dirichlet allocation, vol. 3, pp. 601–608 (2001)
3. Blondel, V., Guillaume, J.L., Lambiotte, R., Lefebvre, E.: Fast unfolding of communities in large networks. J. Stat. Mech. Theory Exp. **2008** (2008). https://doi.org/10.1088/1742-5468/2008/10/P10008
4. Brito, A., Silva, F., Amancio, D.: A complex network approach to political analysis: application to the Brazilian chamber of deputies. PLOS ONE **15**, e0229928 (2020). https://doi.org/10.1371/journal.pone.0229928
5. Bursztyn, V., Nunes, M., Figueiredo, D.: How congressmen connect: analyzing voting and donation networks in the Brazilian congress (2016). https://doi.org/10.5753/brasnam.2016.6451
6. Chartash, D., Caruana, N.J., Dickinson, M., Stephenson, L.B.: When the team's jersey is what matters: network analysis of party cohesion and structure in the Canadian house of commons. Party Polit. **26**(5), 555–569 (2020). https://doi.org/10.1177/1354068818795196
7. Chen, T., Guestrin, C.: Xgboost: a scalable tree boosting system, pp. 785–794 (2016). https://doi.org/10.1145/2939672.2939785
8. Coscia, M., Neffke, F.: Network backboning with noisy data, pp. 425–436 (2017). https://doi.org/10.1109/ICDE.2017.100
9. Ferreira, C.: Modeling and analyzing collective behavior captured by many-to-many networks. Ph.D. thesis, Universidade Federal de Minas Gerais (UFMG) (2022)
10. Ferreira, C., Murai, F., Matos, B., Almeida, J.: Modeling dynamic ideological behavior in political networks, pp. 1–14 (2019)
11. Foti, N., Hughes, J., Rockmore, D.: Nonparametric sparsification of complex multiscale networks. PLOS ONE **6**, e16431 (2011). https://doi.org/10.1371/journal.pone.0016431
12. Grady, D., Thiemann, C., Brockmann, D.: Robust classification of salient links in complex networks. Nat. Commun. **3**, 864 (2012). https://doi.org/10.1038/ncomms1847

13. Levorato, M., Frota, Y.: Brazilian congress structural balance analysis (2016)
14. Ribeiro, A., Rudda, O., Oliveira, L., Inacio, M.: The executive branch decisions in Brazil: a study of administrative decrees through machine learning and network analysis. PLOS ONE **17**, e0271741 (2022). https://doi.org/10.1371/journal.pone.0271741
15. dos Santos, M., Andrade, N., Morais, F.: Topic modeling of committee discussions in the Brazilian chamber of deputies. In: Anais do IX Symposium on Knowledge Discovery, Mining and Learning, pp. 49–56. SBC, Porto Alegre, RS, Brasil (2021). https://doi.org/10.5753/kdmile.2021.17460. https://sol.sbc.org.br/index.php/kdmile/article/view/17460
16. Serrano, M., Boguñá, M., Vespignani, A.: Extracting the multiscale backbone of complex weighted networks. Proc. Natl. Acad. Sci. USA **106**, 6483–6488 (2009). https://doi.org/10.1073/pnas.0808904106
17. Slater, P.: A two-stage algorithm for extracting the multiscale backbone of complex weighted networks. Proc. Natl. Acad. Sci. USA **106**, E66; author reply E67 (2009). https://doi.org/10.1073/pnas.0904725106
18. Řehůřek, R., Sojka, P.: Software framework for topic modelling with large corpora, pp. 45–50 (2010). https://doi.org/10.13140/2.1.2393.1847

K-GALS 2024: 3rd Workshop on Knowledge Graphs Analysis on a Large Scale

Transforming Text Into Knowledge with Graphs: Report of the GDR MADICS DOING Action

Mirian Halfeld-Ferrari[1(✉)], Anne-Lyse Minard[2], and Genoveva Vargas-Solar[3]

[1] Univ. Orléans, INSA Centre Val de Loire, LIFO UR 4022, 45067 Orléans, France
`mirian@univ-orleans.fr`
[2] Univ. Orléans, LLL UMR 7270, Orléans, France
`anne-lyse.minard@univ-orleans.fr`
[3] CNRS, INSA Lyon, Univ. Claude Bernad Lyon 1, LIRIS, UMR5205, 69622 Villeurbanne, France
`genoveva.vargas-solar@cnrs.fr`

Abstract. This paper provides an overview on graph databases for the retrieval and the integration of knowledge originating from textual data, attempting to bring together different bricks that are usually addressed separately. It explores concepts and insights that result from the scientific activities promoted by the GDR MADICS DOING Action (Intelligent Data: turning information into knowledge). The action promoted scientific discussion on the challenges, current findings, and open issues in converting textual data into information and, ultimately, knowledge. This topic has been investigated within a multidisciplinary context, involving specialists in Databases (DB), Natural Language Processing (NLP), Artificial Intelligence (AI), and professionals in various application domains.

Keywords: Text processing · graph data models · graph analytics · data science queries on graphs

1 Overview

This paper explores concepts and insights related to graph-based modelling of textual content that result from scientific activities promoted in the coordination action Intelligent Data: turning information into knowledge (DOING)[1]. The action promoted scientific discussion on the challenges, current findings, and open issues in converting textual data into information and, ultimately, knowledge. Leveraging expertise from scientists in natural language processing (NLP), databases (DB), and artificial intelligence (AI), the study and reflection focuses

---

[1] DOING is a coordination action funded by the network MADICS of the French Council of Scientific Research - CNRS http://www.madics.fr/actions/doing/. Created as a regional initiative in 2019, DOING extended its scope to a national level within the GDR MADICS in 2020 before attaining official status as an action.

J. Tekli et al. (Eds.): ADBIS 2024, CCIS 2186, pp. 145–159, 2025.
https://doi.org/10.1007/978-3-031-70421-5_13

on two main research directions: extracting valuable insights from textual data to enrich graph databases and developing intelligent and user-friendly techniques for effectively maintaining, querying, and conducting analytical studies while ensuring quality standards. The application domains of the study encompass medical and environmental fields.

The working method adopted by the coordination action[2] has addressed diverse perspectives about the problem and associated challenges through webinars, study days, featuring discussions, panels, and roundtable sessions to synthesize viewpoints from diverse scientific communities and methods for achieving the transformation of data into knowledge. Challenges and insights have emerged in aligning communication among scientists from diverse domains, primarily from vocabulary and conceptual understanding discrepancies. Fostering fruitful interaction and facilitating collaborations between French colleagues and international partners, including those engaged in multidisciplinary research projects in Brazil, DOING has produced a multi-perspective understanding of extracting, structuring, and querying textual content using graphs for knowledge extraction.

The research contribution of this paper is to make an overview on graph databases for the retrieval and the integration of knowledge originating from textual data, attempting to bring together different bricks that are usually treated separately. In this paper, we summarise the concepts, remarks and scientific vision about adopting graph databases as the primary component for organizing information extracted from texts and developing an intelligent querying framework. This framework allows users to execute analytic queries on data within the graph, targeting diverse user requirements. Therefore, the remainder of the paper is organized as follows: Sect. 2 provides an overview of the fundamental concepts of graph databases. Section 3 delves into textual data structuring, while Sect. 4 enumerates the components and functions of an intelligent query framework. Section 5 concludes the paper by underscoring the significance of multidisciplinarity (within the computer science domain) in this study.

2   Graph Databases: The Structure

A graph database management system (GDBMS) employs graphs as its fundamental data model. A graph structures data as a "network" of entities and relations, for example, in the Semantic Web, social networks, connected businesses, digital networks, knowledge networks, and Internet networks.

GDBMSs are engineered to prioritize the relationships between data that are equally significant as the data itself. These relationships enable stored data to be directly linked together, often facilitating rapid retrieval. Consequently,

---

[2] The impact of the DOING coordination action goes beyond national boundaries, inspiring international initiatives: (i) A regional project, APR-IA, supported by the Centre Val de Loire region (2021–2024). (ii) The international workshop DOING@ADBIS, marking its 5th edition this year, underscores the far-reaching influence of DOING.

graph query languages are formulated to articulate paths on a graph, highlighting the interconnected nature of the stored information. A native GDBMS is purposefully designed to handle graph workloads across the entire computing stack.

2.1 Graph Models and Query Languages

RDF and Property Graphs. Graph databases adopt two models: LPG (labelled-property graphs or just property graphs) and RDF (Resource Description Framework) graphs. Property graphs prioritize analytics and querying, whereas RDF graphs underscore the importance of data integration. Both models are centered around the concept of node- and edge-labelled graphs, allowing for the encoding of intricate information in multi-graphs. Unlike RDF graphs, which consist of nodes and arcs, a property graph augments nodes or edges with attributes or properties. Each property is represented as a pair (key, value), enhancing graph data. In the study and discussion promoted by the DOING community, LPG is favored for its alignment with classical database concepts. It utilizes properties to precisely define nodes and relations, making primary relationships among entities visible. This approach prevents dense (RDF) graphs, where relations serve as links between entities and as a means to express entity properties. Neo4J stands out as the most popular property graph database system, but other noteworthy options exist, such as Memgraph and TigerGraph.

Query Languages. Various query languages have emerged for querying graphs, such as SPARQL for RDF graphs, Cypher for property graphs in Neo4J, and Gremlin for property graphs in Apache TinkerPop. Despite differences in style, purpose, and implementation, these languages share a core focus on two fundamental operations: *graph pattern matching* and *graph navigation*.

Graph pattern matching involves identifying matches of a graph pattern within a graph database. In basic graph patterns, variables can represent node labels or relation types. Matches are defined as homomorphisms from the pattern to the graph instance. Various semantics exist for determining these matches. Navigational queries offer versatile querying methods that enable the exploration of the database's *topology*.

A path query specifies conditions that paths on a graph must meet. Paths can be defined using regular expressions denoted by L. Evaluation of a path query over a graph involves finding all paths whose labels satisfy L. Handling potentially infinite answers to regular path queries requires adopting specific semantics to ensure finiteness.

Designing a Graph Database. There is no methodology for designing a graph database for a dataset. Instead, there are general guidelines that can help model a representative graph. Unlike relational databases, the design of a graph database significantly influences query performance. Queries not optimized for the database model can result in poor performance. Graph databases excel in path traversal queries, as nodes typically store information about their neighboring nodes. This characteristic makes them attractive for exploring data relationships and employing data analytics techniques such as predicting node connections. Therefore, it is crucial to consider the types of queries the application will handle and design the graph database accordingly[3] [27].

2.2 Related Work in the DOING Community

Graph databases and their query languages have been a recurring focus in the seminars organized within the DOING coordination action. George Fletcher's research group at Eindhoven University of Technology, has emphasized the expressive power of graph query languages. They are particularly interested in characterizing the ability of these languages to constrain and shape concrete graph instances based solely on their structural properties[4]. Meanwhile, the research group led by Domagoj Vrogoc at Pontificia Universidad Católica de Chile[5] focuses on the theoretical aspects of queries that facilitate traversing paths of arbitrary lengths. Additionally, the LIGM-CNRS[6] group delves into theoretical and practical aspects of query languages for property graphs, providing a foundational basis for languages like Neo4J.

3 Structuring Textual Data

Databases store structured information, enabling efficient and practical querying. When working with unstructured textual data for answering users' queries, structuring the data allows leveraging the full range of database functionalities, improving accessibility. This context shows the intersection of two disciplines operating at different levels of abstraction: databases (DB) and natural language processing (NLP). The database perspective relies on high-level abstraction modelling that abstractly represents the original text, emphasizing the specification of concepts and their interrelationships. The objective is to encapsulate the types of essential information to be preserved. Conversely, the NLP community pursues a lower level of abstraction that directly refers to textual data without modeling concepts that generalize data.

The *database approach* (DB) contrasts with the *textual graph approach* that uses graphs to represent the text content directly *i.e.*, without attempting to deal with the abstractions a general schema would represent. The view promoted in the DOING action brings together the DB and NLP communities to integrate

---

[3] https://cambridge-intelligence.com/graph-data-modeling-101/.

[4] DOING Webinar: Language-aware indexing for conjunctive path queries, George Fletcher, Eindhoven University of Technology, Netherlands, Mars, 2021.

[5] DOING Webinar: Evaluating navigational queries over graphs, Domagoj Vrogoc, Pontificia Universidad Católica de Chile, Chili, July 2021.

[6] MADICS Symposium, DOING workshop. On July 2022: Aperçu général des langages de requêtes pour graphes propriétés by Victor Marsault. On May 2023: A Researcher's Digest of GQL by Liat Peterfreund.

their perspectives to explore and promote the interests of their complementarity. Figure 1 illustrates the *database approach* with a graph database that can be the result of structuring, as a property graph, information obtained from the given text (an extract of PubMed).

A female patient in the age group 55–60 years presented to us with blurring of vision in both eyes. On slit-lamp examination, numerous circular to oval fleck-like discrete blue opacities at the level of deep corneal stroma and Descemet's membrane was observed.

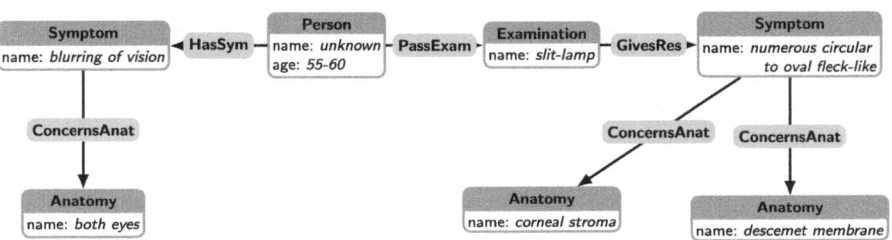

Fig. 1. Example of a Neo4J graph database instance obtained from the given text (edge properties are not shown).

The key idea is that the schema comprises notions that aggregate or specialize concepts detected in the text. For instance, the database schema may store *generalized* information concerning a *symptom* (*temperature of* 39 °C), *i.e.*, aggregating the variable *temperature* and its value 39 °C in a unique entity. The schema may store *specialized* information concerning a *symptom* (e.g., *discrete blue opacities*) and the *anatomy* (e.g., *the cornea*). The database approach focuses on how to instantiate the classes composing the database schema: on how to populate the graph specifying instances of nodes and edges types defined by a graph database schema. A structured instance built according to the schema is the database on top of which queries and analytics are performed.

An essential challenge of NLP is grouping initial NLP abstractions into more generalized ones. For instance, in the phrase 500 mg *of paracetamol*, entities value (500) and unit (*mg*) can be grouped into the concept of dosage, which, along with the entity drug, can be further grouped into the concept of treatment. For instance, this latter concept can be represented as a node and attributes in an LPG model.

The study and discussion promoted in DOING are guided by the hypothesis that data from texts are instances of a database whose schema is (or is not) predefined. Accordingly, structuring textual data implies using NLP tools and techniques and relying on a higher level of abstraction by grouping and occasionally factorizing the extracted information. This hypothesis assumes that information may only sometimes be uniformly represented, leading to the acceptance of missing data (often seen as anomalies that can harm the database consistency and query answer completeness).

3.1   Information Extraction

NLP techniques can accurately extract structured semantic information from unstructured texts and insights to categorise and organise them. These techniques can help uncover insights contained within text in a scalable and efficient manner by identifying and analysing explicit concepts and relationships. Information extraction includes two tasks: entity recognition and relation extraction.

Entity Recognition (NER). Consists in detecting and classifying entities in the text. The most current detected entities are the so-called named entities, which traditionally include person names, location names, organisation names and numerical expressions. In the medical domain, the entities of interest can be signs or symptoms, pathology, drugs, clinical tests, etc. [24] In the example text of Fig. 1, an entity recognition system can be used to detect the anatomy information, the symptoms, the examination and the patient's age. For the most common entity classes in English or French, some existing annotated corpora can be used to train supervised machine learning algorithms. For example, for the medical domain, we can cite the CAS corpus [15] (French) or the E3C corpus [20] (five languages, including English and French). In other specialised domains (e.g. geological domain), it is sometimes necessary to use techniques other than supervised machine learning, such as dictionary-based extraction or rules, because no annotated corpora are available. A BERT-based language model can be used to apply many machine-learning algorithms, such as CRF, Bi-LSTM CRF, or transformers. Once entities have been detected and classified, an entity-linking task can be performed. It consists of a disambiguation task by linking entities to a unique identifier. These identifiers can come from a knowledge base (e.g., DBpedia), a thesaurus (e.g., UMLS for the medical domain), or other knowledge representations. It is an important step to reduce variation and ambiguity while structuring textual information into a graph database.

Relation Extraction (RE). Consists in detecting and classifying the relationships between entities. Different types of relations can be of interest, mainly temporal, causal, and other semantic relations between specific entities. In the example text of Fig. 1, a RE system can be used to extract the temporal relation between the symptom "blurring of vision" and the examination or the semantic relation "ConcernsAnat" that links a symptom and a specific anatomy element. Various machine learning algorithms, such as SVM, CNN, and RNN, are used for the RE task. Bi-LSTM is used for long-distance relations. Unfortunately, few annotated corpora exist for relation extraction, especially for French, which makes the task more difficult to resolve. Syntactic structures, such as constituency trees [23] or dependency trees [28], are often exploited to detect the relations in supervised or rule-based systems.

The RE task often focuses on binary relations, but extracting n-ary relations or grouping entities is also needed. For example, in the text shown in Fig. 1, all information about one patient must be gathered: his gender, his age, his origin, etc. In medical documents, it is also often the case for the medication for which we have various information: substance, dosage, frequency, method for administration, etc. However, this task is rarely treated in the literature, and few annotated corpora are available.

The described information extraction tasks and techniques can structure textual information into graph databases. The main difficulty is the need for large quantities of annotated data to train supervised learning systems, particularly for relation extraction. The following lines present different perspectives for textual structuring depending on how an annotation schema is implied.

3.2 Exploring Approaches for Textual Structuring

Text structuring can be done through top-down and bottom-up methodologies. In the top-down approach, a schema is predetermined, tackling the problem as a query of the text to extract or identify pertinent "relevant" entities. Conversely, the bottom-up approach involves extracting terms, entities, and relationships from the text, followed by their classification and grouping, often based on a similarity distance metric. Hybrid methodologies, which combine these two perspectives, have demonstrated potential for enhanced efficiency.

Top-Down Methods. can rely on information extractors and ontologies. In the theoretical database community, extractors are viewed as functions [25], referred to as *spanners*, designed to extract a relation of spans from a document d. In this context, d is a string over a finite alphabet, and a span x is an interval of positions within d that represents a substring d_x. Regular spanners are the most studied. However, they are not capable of handling various complex scenarios found in natural language, such as nested structures. Significant formal advancements have been made in this field, offering insights into the expressive power (such as [11,26]) and complexity of these extractors (such as [1,14]).

Bottom-Up Methods. implement an Open Information Extraction task (OpenIE) to extract triples from raw texts without using pre-defined relations or an annotation schema [3]. OpenIE systems rely on linguistic features obtained with PoS tagging, chunking or dependency parsing. Other bottom-up methods are applied for ontology learning, defining a sequence of tasks [10,29]. (1) First, identify the natural language terms associated with the target domain. (2) Then, identifying synonyms to avoid redundant concepts, since two or more natural language terms can represent the same *concept*. The notion of "concept" is controversial but is usually associated with a knowledge abstraction that can be found with some regularity in a text from a given domain. Finding concepts in a text means finding regularities that the same abstraction can generalize. Most research in this area uses unsupervised machine learning techniques like clustering [6]. However, some works use supervised learning methods to improve results [13]. When dealing with ontologies, hierarchical relationships (the *is_a* relationship) between pairs of concepts are established after determining the concepts. This step leads to a taxonomy of concepts. (3) The following task concerns finding non-hierarchical relations. Usually, it combines statistical analysis and linguistic analysis as in [5]. (4) In ontology learning, the last step focuses on discovering inference rules from text, such as *If X is author of Y, then X wrote Y*. Constructing a database schema from text follows the same steps except for finding taxonomies.

Modelling a database under a bottom-up approach by structuring textual data implies following learning steps (i.e., similar to ontologies). For instance, in [16], syntax trees enriched with entities and relationships found in a pre-processing step are analysed, and similar sub-trees are grouped until reaching a tree format that respects specified rules. This strategy can be considered a hybrid method but connected to the ontology learning ideas.

While there is a notion that a database schema and an ontology may be equivalent, it is essential to recognize their distinct purposes. A database schema describes a set of instances tailored for efficient storage and querying, for example, structuring data in graph databases profoundly influencing query efficiency. Conversely, while constraints in an ontology primarily convey machine-readable semantics to facilitate automated reasoning, those in databases mainly serve to ensure data integrity. In contrast, an ontology, which may serve as a schema, serves broader objectives such as interoperability, reasoning, and knowledge representation.

3.3   Related Work in the DOING Community

The study of the DOING action discussed domain-specific, domain adaptation and domain-independent systems. The research conducted at the LISN by Aurélie Névéol[7] and by Perceval Wajsbürt at HP-HP[8] show concrete applications of domain specific systems for information extraction in the medical domain.

Another relevant topic DOING addresses is modelling extracted information and large-scale textual structuring. Davide Buscaldi at the LIPADE[9] has developed research on knowledge graph generation from scientific literature [9]. The SCICERO system described in [9] extracts entities and relations, then merges entities and relations, leading to a set of triples used to build the knowledge graph.

4   Towards Intelligent Querying on Graph Databases

Graph querying techniques can be grouped into two families. The first encompasses traditional querying approaches commonly adopted by database and information retrieval engines [12,30]. These techniques allow the expression of

---

[7] DOING Webinar: Natural language processing for epidemiology & public health, Aurélie Névéol, CNRS, LISN, France, 5th July, 2021.

[8] MADICS Symposium, DOING workshop: Tools for processing clinical reports in health data warehouses, Perceval Wajsbürt, Assistance de Paris - Hôpitaux de Paris (AP-HP), France, 25th June, 2023.

[9] DOING Webinar: Building Scientific Knowledge Graphs from Scholarly Data, Davide Buscaldi, LIPN, Université Sorbonne Paris Nord, France, 17th May, 2021.

graph traversal operations. In this scenario, results are characterized by their completeness concerning the graph database being queried [4]. The second family, in contrast, concerns the analysis of graphs with modelling, prediction objectives and recommendations. These data (i.e., graph) samples are processed with artificial intelligence algorithms (i.e., models), suggesting that outcomes are based on partial or representative data sets rather than exhaustive analyses. This distinction highlights the varying approaches to data querying, emphasizing thoroughness and accepting and accounting for uncertainty in its analysis.

Graph analytics, a technique for analyzing data in a graph format, has attracted substantial attention in contemporary decision-making processes. Analyzing data in this way makes it possible to discover links and relationships that would otherwise escape traditional methods. Graph stores with different models and properties, querying facilities, and analytics libraries with built-in graph analytics algorithms provide tools for exploring graphs. The question is under which conditions the various solutions are better adapted to address different analytics (i.e., intelligent) queries.

4.1 Data Science Pipelines on Graphs

The data science pipeline involves a series of steps to gather raw data from multiple sources, analyse it, and present the results in an understandable format. This process can be applied to constructing, exploring, and analysing graphs. General templates for these tasks can be adapted based on the algorithms and strategies used.

Data Analytics Queries on Graph Stores. like Neo4j, Amazon Neptune, and TigerGraph offer built-in graph operations that implement graph analytics operations such as community detection, centrality (e.g., PageRank and betweenness algorithms), similarity and pathfinding, and search operations. While these systems enable declarative querying of stored graphs, users are responsible for memory management and routing graph components to the execution space when applying graph analysis functions. An example of analyzing a graph built from medical text using PageRank with Neo4j involves constructing a graph representation of the medical text data within the Neo4j database.

```
// Step 1: Create graph nodes and relationships
   from medical text data
LOAD CSV WITH HEADERS FROM 'file:///medical_data.csv' AS row
MERGE (m:MedicalTerm {name: row.term})
MERGE (d:Disease {name: row.disease})
MERGE (m)-[:MENTIONS]->(d);
```

This graph likely includes nodes representing medical terms, diseases, symptoms, treatments, and other relevant entities, with their relationships based on semantic connections extracted from the text. The PageRank algorithm is applied to calculate the importance of each medical term node in the graph based on the relationships between nodes.

```
    // Step 2: Run the PageRank algorithm to
    calculate node importance
CALL algo.pageRank(
  'MATCH (m) RETURN id(m) as id',
  'MATCH (m)-[:MENTIONS]->() RETURN id(m) as source, id() as target',
  {write: true, writeProperty: 'pagerank'}
);
```

The results are retrieved, and the top 10 medical terms with the highest PageRank scores are returned, providing insights into the most significant terms within the medical text data.

```
    // Step 3: Retrieve results and inspect the
    nodes with highest PageRank scores
MATCH (m:MedicalTerm)
RETURN m.name, m.pagerank
ORDER BY m.pagerank DESC
LIMIT 10;
```

The pipeline is thus a sequence of "queries" that operate step by step on the "prepared" graph to produce results. Various factors come into play when considering graph store solutions for applying machine learning algorithms on stored graphs. These factors include performance, scalability, flexibility, built-in algorithms, ease of use, community support, and cost.

Imperative Approaches for Implementing Graph Analytics Pipelines. Data mining and machine learning techniques can be integrated into imperative programs to analyze graphs, where analytics pipelines' control and data flow are implicit. Data analytics environments often utilize libraries with proprietary data structures and built-in functions for graph creation and analysis [31]. Examples include Python Networkx, Spark Graphix, and deep learning models (e.g., graph neural networks) for tasks like node classification, link prediction, and graph clustering, as well as graph processing systems like Pregel and Giraph. A typical pipeline constructs an ad hoc graph representing the data corresponding to data preparation and engineering tasks. Subsequently, the analytics code sequence applies functions with calibrated parameters to analyze the prepared graph. The iterative trial-and-error process for parameter tuning is not explicitly coded but can be automated using other machine-learning techniques.

The pipelines may encompass various tasks to convert graphs into vector representations suitable for training, testing, and evaluating Graph Neural Network (GNN) models. The preprocessing pipeline's complexity can vary depending on the graph type (e.g., heterogeneous or homogeneous, directed or undirected, properties in nodes and/or edges). For example, it may involve extracting nodes and edges separately, along with their properties, and transforming them into vectors using techniques like text2vec. Alternatively, for heterogeneous graphs, one may select graph views where nodes and edges share the same type and compute a global vector to represent the entire graph.

4.2 Related Work in the DOING Community

The study done by the DOING community focuses analytical queries on graphs, seeking to integrate database and machine learning techniques to enable the answering of more sophisticated queries. Imperative approaches rely on libraries and execution environments with no built-in options for managing graph views, resource allocation and graph persistence. In contrast, declarative approaches relying on underlying graph management systems profit from the manager's strategies for managing the graphs on disk and main memory. A collaborative experimental comparison [31] done by the labs LIFO, University of Orléans and LIRIS, CNRS lab in Lyon lead to a tutorial and talk in the contexto of DOING.

We discussed the specification and design of data science pipelines that could process graphs representing textual content. We worked with experts from different disciplines, including databases, artificial intelligence and natural language processing, to address these questions. The work done in Paris at the Leonardo da Vinci School of Engineering (ESILV[10]) by the group of Nicolas Travers [17] and at the LIPADE and LIPN [18,22][11] have developed research with the bottom up approach where graphs are designed from input datasets considering the type of analysis to perform. The work in U. Manchester by Andre Freitas [30][12] relies on graph representations of textual content to perform exploration queries for retrieving factual data and analytics operations for discovering knowledge within the graph. A set of data science pipelines adopting different analytics techniques lead to the construction of knowledge graphs at the LIRIS in Lyon [8]. Experts from the DB and AI communities from the University of Tours in France provided insight into graphs analytics with neural network models [7,19][13].

5 Multidisciplinary Context and Concluding Remarks

This paper summarises insights about the problem of structuring textual content and transforming it into knowledge from a multidisciplinary perspective (within the computer science domain). The discussion conclusions show how the NLP, DB and AI fields reason about textual data. In NLP and DB, the guiding challenge is proposing different forms of abstraction. While the NLP domain focuses on numerous details and variants of unstructured data, the DB domain

---

[10] https://www.esilv.fr.

[11] DOING Webinars: Managing data quality in the age of big data, Salima Benbernou, Université Paris Descartes, LIPADE, France, 5th June 2020; Building Scientific Knowledge Graphs from Scholarly Data, Davide Buscaldi, LIPN, Université Sorbonne Paris Nord, France, 17th May, 2021.

[12] DOING Webinar: From Deep Learning to Deep Semantics, Andre Freitas, University of Manchester, UK, 8th July 2020.

[13] DOING Panel, *Representing content and extracting knowledge from texts: automatic language learning, graph management systems, machine learning for graph analysis and semantic web approaches*, Symposium MADICS, Lyon, 2022.

typically necessitates abstractions that consolidate information into unified concepts or relation types. For both the DB and AI domains, the emphasis lies in reasoning-exploring, correlating facts, and uncovering implicit knowledge -from data extracted from textual content. As AI methods are employed for querying, there is a shift in perspective within the DB domain, moving from certainties to possibilities and from general deduction to a reliance on samples and instantiations to yield results. Throughout the transformation of textual data into knowledge, database management systems play a mediating role. (G)DBMS facilitate the structuring of textual content in NLP practices and applying AI techniques on structured textual content for analysis and knowledge extraction.

The challenge lies in blurring the boundaries among the three disciplines and integrating database concepts into both areas while simultaneously incorporating their respective principles. For example, leveraging database techniques for efficient storage, retrieval, and manipulation of textual data can significantly improve the performance and scalability of NLP and AI systems. Ultimately, this integration of databases, NLP, and AI principles drives innovation and paves the way for more robust and practical solutions in data-driven endeavors.

The study conducted by the DOING coordination action has identified a primary requirement: the development of generic tools to facilitate communication between different system components in collaboration with other scientific domains, particularly in the medical and environmental fields. Such tools involve reasoning on a metamodel, which serves as a framework onto which information can be initially mapped and reconfigured into a more specific model. An example of this principle is the approach proposed in [16], which utilizes a general abstraction to guide textual data into a tree structure corresponding to a formal grammar parse tree, adaptable to different database models. Despite the natural inclination towards graph models as a mode of representation, institutions like BRGM[14] are seeking metamodels to postpone adhering to specific models for modeling content. This trend is evident in other recent works such as [2,21].

The DOING coordination action has enabled a collaborative and multidisciplinary exploration of transforming information into knowledge, making data intelligent. DOING highlights the significance of fostering synergy between diverse scientific disciplines by identifying potential interactions among methods and strategies in NLP, DB, and AI. Such an approach underscores the importance of technological advancements in effectively addressing real-world challenges in medicine, environment, geography and beyond.

Acknowledgements. This work was partially supported by the DOING project, a regional project funded by the council of the Centre Val de Loire Region (APR-IA).

References

1. Amarilli, A., Bourhis, P., Mengel, S., Niewerth, M.: Constant-delay enumeration for nondeterministic document spanners. In: ICDT. LIPIcs, vol. 127, pp. 22:1–22:19. Schloss Dagstuhl - Leibniz-Zentrum für Informatik (2019)

---

[14] French geological survey https://www.brgm.fr/en.

2. Balalau, O., et al.: Statistical claim checking: statcheck in action. In: Hasan, M.A., Xiong, L. (eds.) Proceedings of the 31st ACM International Conference on Information & Knowledge Management, Atlanta, GA, USA, 17–21 October 2022, pp. 4798–4802. ACM (2022)
3. Banko, M., Cafarella, M.J., Soderland, S., Broadhead, M., Etzioni, O.: Open information extraction from the web. In: Proceedings of the 20th International Joint Conference on Artificial Intelligence, pp. 2670–2676, IJCAI 2007. Morgan Kaufmann Publishers Inc., San Francisco (2007)
4. Bonifati, A., Fletcher, G., Voigt, H., Yakovets, N., Jagadish, H.: Querying Graphs, vol. 10. Springer, Cham (2018)
5. Buitelaar, P., Olejnik, D., Sintek, M.: A Protégé plug-in for ontology extraction from text based on linguistic analysis. In: Bussler, C.J., Davies, J., Fensel, D., Studer, R. (eds.) ESWS 2004. LNCS, vol. 3053, pp. 31–44. Springer, Heidelberg (2004). https://doi.org/10.1007/978-3-540-25956-5_3
6. Cimiano, P., Hotho, A., Staab, S.: Learning concept hierarchies from text corpora using formal concept analysis. J. Artif. Intell. Res. 24, 305–339 (2005)
7. Conte, D.: Graphs in pattern recognition: successes, shortcomings, and perspectives. J. Electron. Imaging 32(2), 020701–020701 (2023)
8. Coste, L., Helmers, F., Kheddouci, H., Le Nestour, L., Niazi, M., Vargas-Solar, G.: Strategies for creating knowledge graphs to depict a multi-perspective queer communities representation. In: Workshops of the EDBT/ICDT 2023 Joint Conference, vol. 3379 (2023)
9. Dessí, D., Osborne, F., Reforgiato Recupero, D., Buscaldi, D., Motta, E.: SCICERO: a deep learning and NLP approach for generating scientific knowledge graphs in the computer science domain. Knowl.-Based Syst. 258, 109945 (2022)
10. Drummond, L., Girard, R.: A survey of ontology learning procedures. In: Proceedings of the 3rd Workshop on Ontologies and their Applications (2008)
11. Fagin, R., Kimelfeld, B., Reiss, F., Vansummeren, S.: Document spanners: a formal approach to information extraction. J. ACM 62(2), 12:1–12:51 (2015)
12. Farokhnejad, M., Pranesh, R.R., Vargas-Solar, G., Mehr, D.A.: S_covid: an engine to explore COVID-19 scientific literature. In: Proceedings of the 24th International Conference on Extending Database Technology (EDBT), Nicosia, Cyprus, pp. 23–26 (2021)
13. Faure, D., Nédellec, C.: Knowledge acquisition of predicate argument structures from technical texts using Machine Learning: the system ASIUM. In: Fensel, D., Studer, R. (eds.) EKAW 1999. LNCS (LNAI), vol. 1621, pp. 329–334. Springer, Heidelberg (1999). https://doi.org/10.1007/3-540-48775-1_22
14. Florenzano, F., Riveros, C., Ugarte, M., Vansummeren, S., Vrgoc, D.: Constant delay algorithms for regular document spanners. CoRR abs/1803.05277 (2018)
15. Grabar, N., Claveau, V., Dalloux, C.: CAS: French corpus with clinical cases. In: Proceedings of the Ninth International Workshop on Health Text Mining and Information Analysis, pp. 122–128. Association for Computational Linguistics, Brussels, Belgium, October 2018
16. Hiot, N.: Phd. thesis (in preparation)
17. Lefebvre, P., Moal, S.L., Azough, A., Travers, N.: NeoSGG: a scene graph generation framework for video-surveillance tasks. In: Proceedings 27th International Conference on Extending Database Technology, EDBT 2024, Paestum, Italy, March 25 - March 28, pp. 838–841. OpenProceedings.org (2024)

18. Lovera, F., Cardinale, Y., Buscaldi, D., Charnois, T.: A knowledge graph-based method for the geolocation of tweets. In: Workshop Proceedings of the 19th International Conference on Intelligent Environments (IE2023), pp. 53–62. IOS Press (2023)
19. Maekawa, S., Sasaki, Y., Fletcher, G., Onizuka, M.: Benchmarking GNNs with GenCat workbench. In: Amini, M.R., Canu, S., Fischer, A., Guns, T., Kralj Novak, P., Tsoumakas, G. (eds.) ECML PKDD 2022. LNCS, vol. 13718, pp. 607–611. Springer, Cham (2022). https://doi.org/10.1007/978-3-031-26422-1_40
20. Magnini, B., Altuna, B., Lavelli, A., Speranza, M., Zanoli, R.: The E3C project: collection and annotation of a multilingual corpus of clinical cases. In: Proceedings of the Seventh Italian Conference on Computational Linguistics, CLiC-it 2020, Bologna, Italy, 1–3 March 2021. CEUR Workshop Proceedings, vol. 2769 (2020)
21. Mali, J., Ahvar, S., Atigui, F., Azough, A., Travers, N.: A global model-driven denormalization approach for schema migration. In: Guizzardi, R., Ralyté, J., Franch, X. (eds.) RCIS 2022. LNBIP, vol. 446, pp. 529–545. Springer, Cham (2022). https://doi.org/10.1007/978-3-031-05760-1_31
22. Mammar Kouadri, W., Benbernou, S., Ouziri, M., Ben Amor, I.: WSSA: weakly supervised semantic-based approach for sentiment analysis. In: Proceedings of the 34th International Conference on Scientific and Statistical Database Management, pp. 1–4 (2022)
23. Minard, A.L., Ligozat, A.L., Grau, B.: Apport de la syntaxe pour l'extraction de relations en domaine médical. In: TALN 2011, Montpellier, France, p. 383, June 2011
24. Minard, A., Roques, A., Hiot, N., Halfeld Ferrari, M., Savary, A.: DOING@DEFT: cascade de CRF pour l'annotation d'entités cliniques imbriquées (DOING@DEFT: cascade of CRF for the annotation of nested clinical entities). In: Actes de la 6e conférence conjointe Journées d'Études sur la Parole (JEP, 33e édition), Traitement Automatique des Langues Naturelles (TALN, 27e édition), Rencontre des Étudiants Chercheurs en Informatique pour le Traitement Automatique des Langues (RÉCITAL, 22e édition). Atelier DÉfi Fouille de Textes, Nancy, France, 8–19 June 2020, pp. 66–78. ATALA et AFCP (2020)
25. Peterfreund, L.: Grammars for document spanners. In: ICDT. LIPIcs, vol. 186, pp. 7:1–7:18. Schloss Dagstuhl - Leibniz-Zentrum für Informatik (2021)
26. Peterfreund, L., ten Cate, B., Fagin, R., Kimelfeld, B.: Recursive programs for document spanners. In: ICDT. LIPIcs, vol. 127, pp. 13:1–13:18. Schloss Dagstuhl - Leibniz-Zentrum für Informatik (2019)
27. Prevoteau, H., Djebali, S., Laiping, Z., Travers, N.: Propagation measure on circulation graphs for tourism behavior analysis. In: Proceedings of the 37th ACM/SIGAPP Symposium on Applied Computing, pp. 556–563 (2022)
28. Savary, A., Silvanovich, A., Minard, A., Hiot, N., Halfeld Ferrari, M.: Relation extraction from clinical cases for a knowledge graph. In: Chiusano, S., et al. (eds.) ADBIS 2022. CCIS, vol. 1652, pp. 353–365. Springer, Cham (2022). https://doi.org/10.1007/978-3-031-15743-1_33
29. Toledo-Alvarado, J.I., Guzman-Arenas, A., Luna, G.L.M.: Automatic building of an ontology from a corpus of text documents using data mining tools. J. Appl. Res. Technol. 10, 398–404 (2012)

30. Valentino, M., Ferreira, D., Thayaparan, M., Freitas, A., Ustalov, D.: Textgraphs 2022 shared task on natural language premise selection. In: Proceedings of TextGraphs-16: Graph-Based Methods for Natural Language Processing, pp. 105–113 (2022)
31. Vargas-Solar, G., Marrec, P., Halfeld Ferrari Alves, M.: Comparing graph data science libraries for querying and analysing datasets: towards data science queries on graphs. In: Hacid, H., et al. (eds.) ICSOC 2021. LNCS, vol. 13236, pp. 205–216. Springer, Cham (2021). https://doi.org/10.1007/978-3-031-14135-5_16

Building Model-Driven Knowledge Graphs via Large Language Models

Vaaruni Desai[✉], Yinglan Chi[✉], Jon Stephens[✉],
and Amarnath Gupta[✉] [iD]

University of California San Diego, La Jolla, CA 92093, USA
{v1desai,ychi,jcstephe,a1gupta}@ucsd.edu

Abstract. We consider a special case of knowledge graph construction from text, where the target knowledge graph is structured as specific Directed Acyclic Graph and the input text has the form of a recipe. The intention of this paper is to present a case study that uses a large language model (LLM) for the knowledge extraction process. We formulate knowledge extraction as a model-driven structure recovery process and demonstrate that LLMs can be effectively used in the process. We demonstrate through extensive experiments that using LLMs a zero-shot process produces a wide range of errors. To remedy them, we propose two different *model-driven prompting strategies* by which LLMs can be used to improve the accuracy of knowledge graph construction. We demonstrate that a *state memoization* technique introduces an accuracy-efficiency tradeoff that demands further research.

Keywords: Knowledge Graph Construction · Graph Structure Recovery · Model-driven Construction

1 Introduction

Construction of knowledge graphs from unstructured text is not a new problem (see [15] for a recent survey). The basic strategy of knowledge graph construction is to identify the primary entities that form the nodes, and to extract the relationships between pairs of these entities via text analytics to construct the edges of the knowledge graph. Entity identification is often accelerated in the presence of a domain ontology [5,9]. The relationship extraction process is more nuanced and domain-specific – the set of allowed relationships can be pre-determined at design-time [16], from a distant supervision source like Wikipedia [14], or custom-designed for a specific domain [2]. More complex methods including joint extraction of entities and relationships based on neural models have also been explored in recent literature [3,7].

Supported in part by the NOURISH project grant 2024-68015-41700 from USDA and NSF.

J. Tekli et al. (Eds.): ADBIS 2024, CCIS 2186, pp. 160–172, 2025.
https://doi.org/10.1007/978-3-031-70421-5_14

Beyond entities and relationships, a knowledge graph construction method must also extract properties (attributes) of the extracted entities and relationships. More recently, neural models are being used for automated attribute extraction for entities that have been already recognized [1,11,13].

The motivation for this paper stems from the need to construct knowledge graphs from *process-oriented text* like a laboratory experimental protocol, a "how-to" manual for building an artifact, or a business process. This class of text have the following characteristic features.

1. They start with a set of initial entities that participate in a sequence of processes that represent a temporal progression.
2. There is a terminating process that completes the protocol. Most often, there is a single terminating process.
3. A sequence of steps is often associated with an explicit or implicit state transition process, and the terminating process is associated with an "end state" designating the completion of the process.

We call this process sequence a **recipe** (see Sect. 2 for a formal definition), and develop a method to extract recipe knowledge from unstructured text.

Challenges. Processes are fundamental entities in recipes, but except for very specific domains [10], we cannot assume that publicly available process ontologies exist. Secondly, extraction of process properties (e.g., conditions under which a process will be executed) is an under-researched problem. Third, temporal relationship extraction from process-describing text is an under-researched problem. Finally, we see the development of a fully automated end-to-end system for constructing such graphs as an implementation challenge.

Approach. Our approach comes from 1) the observation that any recipe can be modeled as a directed acyclic graph over processes, and the intuition that 2) the domain knowledge latent in suitable pretrained large-language models can compensate for the lack of domain knowledge needed from recipes. Our working hypothesis is that these two principles are sufficient guidelines to construct recipe knowledge graphs.

Contributions. To this end, the paper makes the following contributions. 1) It demonstrates that the DAG structure of the resultant recipe graph can be used as a *model* that drives the knowledge extraction process via an LLM. 2) It demonstrates the error landscape when an LLM is used directly on text. Further, it shows the impact of two measures of textual complexity on these errors. 3) It presents two new *model-driven prompting strategies*, and measures their relative performances via an extensive set of experiments.

2   A Semantic Model for Recipes

A *recipe* is a multi-stage method that starts with initial ingredients and through the steps (i.e., processes) transforms the ingredients progressively in a prescribed order to ultimately produce an end product. Many documents including food preparation instructions, a manufacturing workflow, a materials synthesis process, and an experimental science procedure, follow this general definition of a recipe. Based on the above description, we can create a formal model of a recipe.

Let I (ingredients), P (processes), T (tools), F (final products), and M (intermediate products) be mutually exclusive sets. We assume we have only a single final product f^r in one recipe r, i.e., $|F^r| = 1$. We make the following assertions.

Assertion 1. *If $P_1^r, P_2^r \ldots P_k^r \in P_r$ are the processes for recipe r, then there exists a partial order $\prec_T$ over P_i^r, where $p_i^r \prec_T p_j^r$ implies p_i^r **occurs earlier than** p_j^r in time.*

Definition 1 (Recipe). *A recipe r is a directed acyclic graph G over nodes $N = \{N_I^r, N_P^r, N_T^r, N_C^r, N_M^r, N_f^r\}$ such that the following conditions hold. In the following a lower case n represents a member of the corresponding node class N.*

1. *Ingredient nodes N_I^r and tool nodes N_T^r used in the recipe are always root nodes*
2. *The end-product node N_f^r is the only leaf node*
3. *$\forall n_P^r[i] \in N_P^r \; \exists \{n_i \in (N_I^r \cup N_M^r)\}$ such that edges labeled $PREREQ\_FOR$ from $\{n_i\}$ are incident on $n_P^r[i]$ and $\exists x \in (N_M^r \cup N_f^r)$ such that a single edge from $n_P^r[i]$ emanates to x.*
4. *The edge $(n_P^r[i], FOLLOWED\_BY, n_P^r[j])$ exists if $p_i^r \prec_T p_j^r$ is true and there exists no $p_y^r \in N_P^r$ such that $p_i^r \prec_T p_y^r \prec_T p_j^r$ holds. We use the term **process path** to refer to any path in the graph induced by the FOLLOWED_BY edges.*
5. *Each process node $n_P^r[i]$ has a collection of necessary tools $T_i = \{t_1, ..., t_{k_i}\} \subset N_T$ satisfying $\forall t \in T_i, \; \exists (t, USED\_IN, n_P^r[i])$ an edge from t to $n_P^r[i]$ indicating the tool t is used in process $n_P^r[i]$.*
6. *The edge $(n_P^r, SATISFIES, n_C^r)$ represents a 1:M relationship between process and condition nodes. No condition node n_C^r is connected to more than one process node.*
7. *Each ingredient node $n_I^r[i]$ and possibly N_f^r has a path to an ontology*

Condition 3 states that a process node can have multiple incident edges (labeled PREREQ_FOR to identify a prerequisite) from some combination of incident nodes and intermediate nodes but a single outgoing edge to a node which can be either an end-product node or an intermediate node. The recipe graph has an unusual edge from *a set of tools* (e.g., a knife and a fork) to a process. The condition node represents a "when", "while" or "until" condition - for example, "*while the water is boiling*, pour the masala mix into the water" represents the condition under which the "pour" process will be executed.

Implicit Constraints. We note that the above definition implies some implicit structural constraints on the recipe DAG. For instance, there can never be an edge between two tool nodes, a tool node and an ingredient node, between two ingredient nodes; two process nodes can only be connected via a FOLLOWED_BY edge. There are also a set of implicit semantic constraints. For example, a process (or action) node will be atomic - thus "wash and peel the potatoes" will result in two atomic process nodes for "wash" and "peel" with the edge FOLLOWED_BY(wash, peel). Similarly, if multiple ingredients are connected to a single process node via a PREREQ_FOR edge, then the process needs every one of these ingredients. If an LLM-generated graph fails to satisfy all of these constraints, then the generation procedure has low quality.

Conceptually, a **Recipe Knowledge Graph** is a collection of recipe DAGs connected via one or more ontologies. Hence, the recipe graphs for potato salad and Spanish omelette will be connected because they both use potatoes and eggs, which are concepts in Food Ontology. However, pragmatically, this step requires an *entity extraction* step such that an ingredient node n with n.name = '2 medium potatoes, scrubbed' will map to the basic entity potato. In our model, we mandate that each ingredient node n has a *non-null* uni-valued attribute n.entity whose value will be matched with the ontology. In case n.entity does not match any concept in the ontology, we assign it a distinguished symbol ϕ. Process nodes are similarly endowed with nullable process-related attributes such as duration. We observe that this means the Recipe Knowledge Graph does not have explicit edges from all potato recipes to an ontology node called potato (FOODON:03315354) – instead, the ontology reference is encoded in the node property called ontology-ref associated with every qualifying node. Though ontological annotation is part of our system, in this paper, we focus on the instruction part of recipes.

Our goal is to utilize the model of the recipe as a means to extract individual recipe graphs from recipe text and combine them strategically to generate the knowledge graph.

3   The Generation Task

A textual recipe r_i is a 3-tuple recipe(name, ingredients[], directions[]). The ingredient list is an unordered list (i.e., a set) of all ingredients, whereas the directions list is ordered, and represents a sequence of preparatory operations as often found in cookbooks. The Knowledge generation process is as follows. Given a recipe r_i, an ontology o, and an LLM, construct a sequence of prompts P_i such that the LLM returns a complex, hierarchical recipe that is structurally correct and semantically faithful to the textual content of r_i. Though not formally stated, a tacit expectation is that even for a complex recipe, the number of prompting steps will be low and manageable.

Role of the LLM. Since the generated recipe is a property graph, we expect the LLM to generate a CYPHER code consisting of primarily update statements (i.e., CREATE, MERGE, and SET). Hence, in addition to maintaining the structural and semantic constraints from Sect. 2, *we expect the LLM to generate the knowledge graph as correct and deployable Cypher code.* Further, since most LLMs are not specifically fine-tuned with ontologies, we decouple the task of ontological annotation of the recipes from the LLMs. For this study, we expect the LLM to generate the recipe graphs without any ontology reference.

3.1   Prompting Strategies

In this section, we discuss three different prompting strategies for directly generating the knowledge graph, structured as a property graph that must be generated as a sequence of Neo4J Cypher code. The overall analysis of the prompts, the errors in LLM responses, together with the impact of the remedial measures are discussed in Sect. 4.

```
A recipe is a directed acyclic graph (DAG) with heterogeneous nodes
and edges. Nodes represent entities such as ingredients, utensils,
or tools, actions (e.g., chop, boil), and possibly conditions
(e.g., for 5 minutes, until golden brown), and final product. Edges
represent relationships or dependencies between these entities.
Every ingredient and process must be atomic. The graph must represent
intermediate states (like the output of mixing spices), optional
steps, and recipe variants. The graph should model XORs (e.g.,
either use dry chilies or green chilies, either bake at 400 degrees
for 10 minutes or broil on high for 20 minutes). It should capture
''measures'' e.g., one cup of milk, temperature at 400 deg. F, bake
for 20 minutes etc., and ''conditions'' like "fry until light brown"
or "check for uniform consistency". The graph needs to represent
sub-recipes (like preparing a gravy) as a hierarchical subgraph.
Create a set of FOLLOWED_BY edges between successive process
nodes. For every process, generate a set of incoming PREREQ (for
prerequisite) edges to connect ingredients to processes. Create USES
edges between process nodes and tool nodes. For each node/edge,
add an attribute called "referenceText" where you identify the
phrase from the original text that relates to the node/edge. Produce
the result using Cypher CREATE, MERGE, and SET statements. Do not
invent any ingredients, tools, or utensil nodes. Ensure the graph is
connected. Ensure each sub-recipe is a small recipe graph whose end
node is the name of the sub-recipe. The sub-recipe end nodes must be
connected to the main recipe.
```

Fig. 1. The baseline prompt

Baseline Prompt (BP). Our baseline prompt has two parts - a complex instruction followed by the recipe text. The recipe text is a single string generated by concatenating the title, the ingredients, and the directions in sequence. The instruction part of the baseline prompt, shown in Fig. 1, is definitional, agnostic of the nature and complexity of the recipe, and is used for all recipes. Notice that the prompt is not zero shot, because it contains several examples of node type instantiation, edge specification, and attribute generation. Furthermore, the prompt specifies hard constraints, and contains "non-hallucination" instructions. We call the baseline prompting strategy *stateless* [8] because the knowledge graph manager that is responsible for invoking the LLM, sends requests and receives responses *without using any context (e.g., conversation history) or other intermediate states to be used for reprompting* [12].

Model-Driven Prompting + Conversation History (MPCH). In this form of prompting, the prompting agent does not give complete instruction in one shot. Instead, it provides the recipe and builds the graph through a set of instructions., executed one at a time. A possible MPCH protocol for recipe DAGs is shown in Fig. 2.

```
1. Use Cypher syntax with CREATE, MERGE and SET. Use MATCH when
   updating a node.
2. Create nodes for ingredients and tools. For each ingredient, add
   attributes for measures (e.g., meat, 2 lbs).
3. Create process nodes. For each process, add an attribute called
   referenceText that specifies the part of the recipe used to
   create the node.
4. Update the process nodes with attributes like duration where
   applicable
5. Create condition nodes and connect them to their corresponding
   process. A condition node will capture information like ''fry
   until light brown''. The edge label from the action node to the
   condition node should be SATISFIES.
6. Create edges to connect processes to their prerequisite
   ingredients or tools. The process-to-tool edges should have the
   label USES and the ingredient-to-process nodes should have the
   label PREREQ_FOR.
7. Create FOLLOWED_BY edges between successive process from the
   recipe
8. Create the "FinalProduct" node and connect it to the last process
   via an edge with the label CREATES.
9. Show the complete graph that you produced from the above steps.
```

Fig. 2. Prompt example for MPCH and MPSM. The difference lies in the way the algorithm interacts with the LLM

As each instruction is executed the prompting agent updates the conversation context and LLM produces an intermediate response. The final answer is collected by the prompting agent as the complete response. Since the target graph is constructed incrementally and fed back to the LLM via the conversation history, which is part of the total prompt, this method is limited by the token restrictions imposed by the LLM. In order to avoid "token limit exceeded" errors, optionally, a mid-prompt requesting a complete script that was produced by the previous steps, is added. Once the mid-flight script is collected, prompting agent will continue with the remaining prompts. The MPCH prompting strategy is captured by Algorithm 1.

Model-Driven Prompting + State Management (MPSM). The MPSM strategy can be viewed as an optimization over the MPCH strategy, where the

---

Algorithm 1: Recipe KG Construction with MPCH

Data: D, textual description of a culinary recipe.
Prompt: P_i, instruction to LLM for extracting N_i (nodes) and E_i (edges), starting with P_0 from Fig 5.
Response: R_i, LLM's response to P_i. viz., partial Cypher code.
Result: S_{cypher}, Neo4j cypher script for D as $G = (N, E)$.

1 $f_{\text{LLM}}(D, P_0)$: Init LLM conversation for D, using P_0 from Fig 5.;
2 **for** P_i *where i in [0, 1, .., 8]* **do**
3     $R_i \leftarrow f_{\text{LLM}}(P_i)$.;
4   Identify missing N_i, E_i.;
5     $P_{i+1} \leftarrow$ Next prompt from Fig 5.;
6   **if** P_{mid} *needed* **then**
7       $P_{mid} \leftarrow$ 'Request a complete script.';
8       $f_{\text{LLM}}(P_{mid})$;
9   **end**
10    $f_{\text{LLM}}(P_{i+1}, R_i)$.;
11 **end**
12 **return** S_{cypher}

---

goal of the optimization is to stay within the token limit of the LLM for long recipes. To accomplish this, the prompting agent uses ***memoization*** to store the intermediate results every time the LLM produces the i-th stage of the graph construction process, and then feeds only the part of the conversation context that is relevant for the next instruction. This obviates the need to execute the last step (Step 9 in Fig. 2) because the memoization data structure already has all the graph components produced so far and can reconstruct the target graph from its parts. Additionally, the prompting agent can check for ***integrity constraint violations*** in the graph and request remedial actions from the LLM. The MPSM strategy is presented in Algorithm 2.

4 Experimental Results

Dataset. All experiments were conducted on the FoodKG dataset [4] which has around 2.2 million uniquely named recipes, each containing an ingredient list and a sequence of cooking instructions. The length of the cooking instructions varied from under 10 sentences to over 140 sentences. We treat the length of a recipe as a measure of its complexity. To systematically address the variance in recipe complexity, we categorized the recipes into 15 groups based on the number of sentences to evaluate the quality of the generated knowledge graph under our three prompting strategies.

Computation. We employed the GPT-3.5 model as our large language model accessed through the OpenAI API, setting the 'temperature' and 'top p' both to 0.5 to optimize output balance. We used Neo4J as the destination of our knowledge graphs.

Algorithm 2: Recipe KG Construction with MPSM

Data: D, textual description of a culinary recipe.
Prompt: P_i, instruction to LLM for extracting N_i (nodes) and E_i (edges),
        starting with P_1 from Fig 5, ending with P_8 from Fig 5.
Response: R_i, LLM's response to P_i viz. partial Cypher code.

1   $f_{\text{LLM}}(D)$: Init LLM conversation for D, using P_0 from Fig. 2.;
2   **for** P_i *where* i *in* $[0, 1, .., 7]$ **do**
3   **if** P_i *creates new nodes* **then**
4         $S_i \leftarrow split(D, k)$;
5   **for** *each* s *in* S_i **do**
6             $R_i \leftarrow f_{\text{LLM}}(P_i)$;
7             $G.execute(R_i)$;
8   **end**
9   **end**
10   **else if** P_i *updates existing nodes of type A* **then**
11         $N_i \leftarrow G.execute\_and\_fetch(A)$;
12         $R_i \leftarrow f_{\text{LLM}}(P_i, N_i)$;
13         $G.execute(R_i)$;
14   **end**
15   **else if** P_i *creates new edges between nodes of type A and type B* **then**
16         $N_i, M_i \leftarrow G.execute\_and\_fetch(A, B)$;
17         $R_i \leftarrow f_{\text{LLM}}(P_i, N_i, M_i)$;
18         $G.execute(R_i)$;
19   **end**
20 **end**
21 **return**;

Evaluation Strategy. Recall that output for our process is a set of Cypher statements generated from the LLM. We evaluate the prompting strategies based on the following error categories in these statements.

1. **Missing Content:** Key details (e.g., a process node) are omitted in the LLM output
2. **Wrong Ranking:** LLM fails to prioritize the most relevant result initially, requiring additional prompts.
3. **Ignored Instruction:** LLM does not follow explicit directions provided in the prompt, e.g., creating Cypher-only output.
4. **Wrong Format:** The response format does not match the specified guidelines. Typically, this results in malformed Cypher statements.
5. **Incorrect Specificity:** LLM provided responses that were more generic than requested.
6. **Incomplete:** LLM was unable to offer a complete response, potentially due to token limitations, especially for long recipes.
7. **Wrong Document Context:** LLM incorrectly specified or failed to indicate the document source for the retrieved information. For example, it misses the name of the recipe in the knowledge graph.

These errors categories are measured against two different metrics of recipe complexity. The first, *recipe length*, measured by the number of instruction sentences, distinguishes between simple recipes like a fruit salad to complex recipes with many steps like Chicken Biryani. The second, the Flesch Reading Ease (FRES) score [6] measures how easy something (in our case a sentence) is to read. For recipe length, we divide our corpus into 15 groups in increments of 10 sentences (i.e., 1–10, 11–20 etc.) till the longest recipes with 150 sentences.

4.1   Results from the Baseline Prompt Strategy

We first illustrate an example error for a simple recipe of a fruit salad, where the ingredients consist of a collection, and the instructions simply state "Wash and cut fruit into chunks. Assemble in bowl. Add lemon juice and fruit concentrate. Mix. Chill salad". Despite the simplicity of the "Fruit Salad" recipe, the result misses the fact that a "mixing bowl" is a tool node, and the graph does not include a "Final Product" node as instructed. The bottom figure shows the proportion of errors with increasing recipe length. The dominant errors are "ignored instruction" and "missing information", followed by the "wrong format" error that represents malformed Cypher statements (Fig. 3).

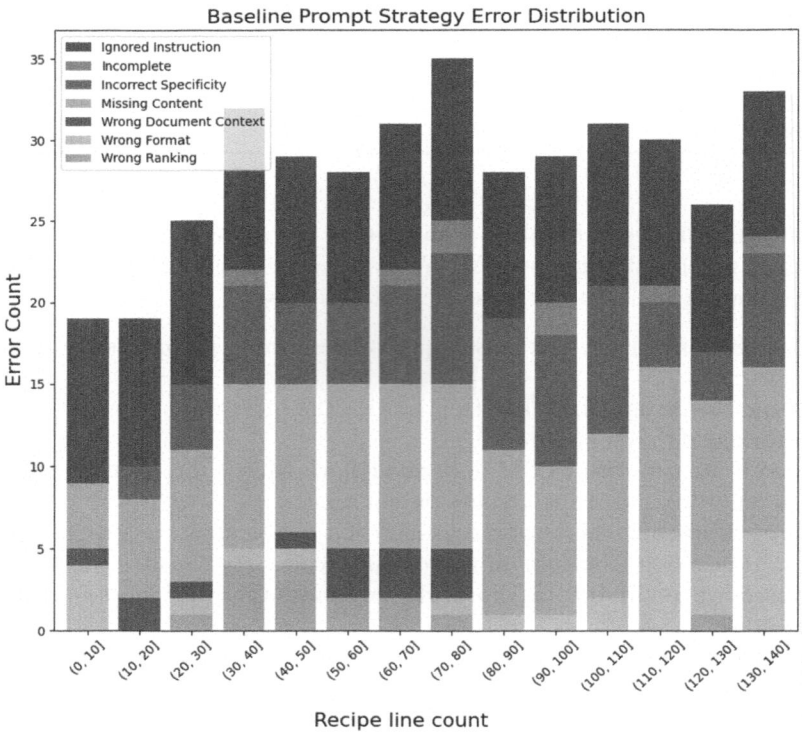

Fig. 3. The error distribution for the baseline prompt with increasing recipe length.

4.2   Impact of Prompting Strategies

The MPCH strategy reduces errors compared to the baseline (Fig. 4). Missing Content and Wrong Ranking errors decreases drastically. The dominant errors now are: Ignored Instruction, Wrong Format, Incorrect Specificity errors and Incomplete Responses. In contrast, the MPSM shows a clear reduction in errors for Missing Information, Wrong Ranking, Ignored Instructions, Incorrect Specificity, and Incomplete Data. Notably, errors from ignored instructions and incorrect specificity categories have decreased significantly, indicating that the MPSM strategy effectively generates more comprehensive knowledge graphs with increased text coverage. This improvement is particularly evident for recipes that are longer than 80 statements. The most dominant error for the MPSM strategy is still the malformed Cypher (Wrong Format) that can be further classified into three categories: (i) the output Cypher code contains typographical errors or misplaced characters; (ii) the output Cypher code has re-declared variable names; (iii) the LLM does not obey the expected structure of the Cypher code delineated in the prompt. Further, the rise in the error may be attributed to the higher volume of queries produced in MPSM compared to MPCH.

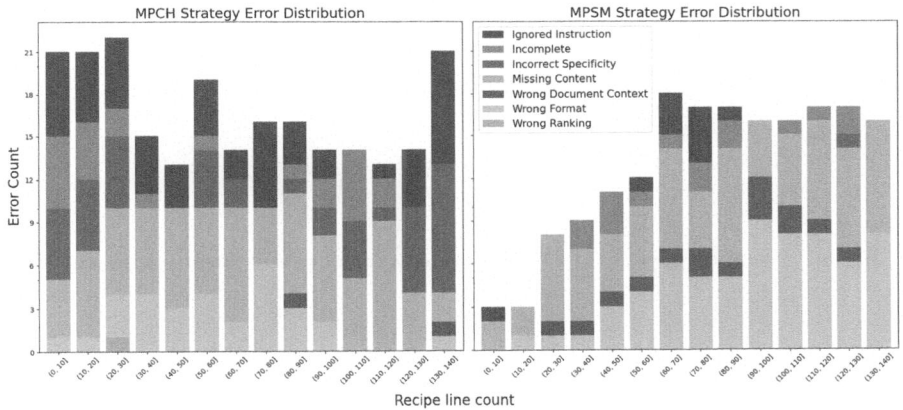

Fig. 4. MPCH and MPSM error distributions show significant improvement over the baseline. MPSM shows better performance compared to MPCH. The majority of errors shift to incorrect Cypher syntax problems.

A different metric to evaluate the error rate is the impact of sentence readability on the error rate. Figure 5 demonstrates that more readable text significantly reduces the error.

Execution Cost. The improvement introduced by the prompting strategies are associated with additional execution cost, primarily because they are multi-stage prompting strategies where the result of one step are fed back to the prompt of the next step. We only present the result of the MPSM prompt due to lack of space, and show that the execution cost per recipe record rises to about 400 s

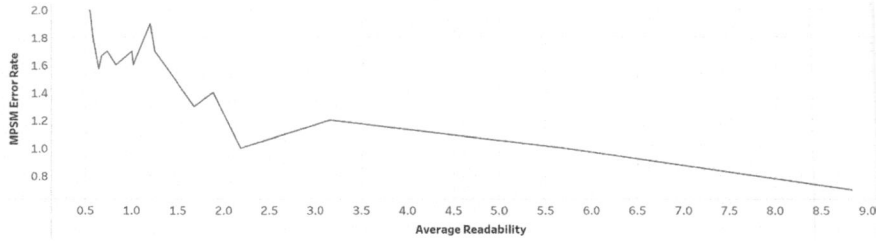

Fig. 5. Lower readability produces more errors. We show the MPSM case for lack of space.

for very long recipes. Not shown here is the observation that over 40% of this cost is spent on generating temporal edges - an observation that needs further investigation (Fig. 6).

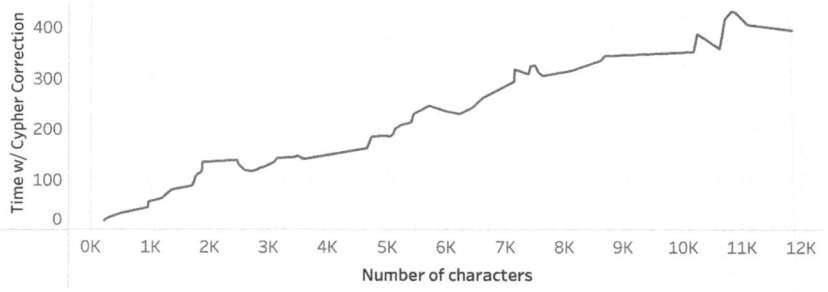

Fig. 6. Execution cost of the MPSM strategy varies nearly linearly with input.

5   Conclusion

This paper investigates how effectively knowledge graphs can be constructed when the topological structure of the graph is well-characterized. It further explores how effective LLMs are in generating the knowledge graphs via prompting strategies that are aware of this topological structure. Our primary finding is that fully automated construction of knowledge graphs for recipe text is possible only when the text is smaller and less complex. However, our experiments revealed the spectrum of errors and demonstrated that our prompting strategies shift the error from bad or incomplete content to correct content delivered via improper form. To our knowledge this is first paper to explore the error landscape for model-driven LLM-prompting toward knowledge graph construction.

As our experiments show, the interaction between the prompting strategy and the LLM behavior needs further exploration. The performance of our knowledge graph construction could be further optimized by having customized query generator. Additionally, the cost-accuracy tradeoff for multi-stage LLM-based knowledge graph construction must be investigated more thoroughly, leading to, what we believe is a new research frontier in automated knowledge graph construction.

References

1. Cheng, Y., Liu, Q., Jiang, C.: Research on equipment entity recognition and attribute extraction for knowledge graph construction. In: 4th International Conference on Signal Processing and Computer Science (SPCS 2023), vol. 12970, pp. 352–358. SPIE (2023)
2. Guan, K., Du, L., Yang, X.: Relationship extraction and processing for knowledge graph of welding manufacturing. IEEE Access **10**, 103089–103098 (2022)
3. Han, J., Jia, K.: Entity relation joint extraction method for manufacturing industry knowledge data based on improved BERT algorithm. Cluster Comput., 1–14 (2024). https://doi.org/10.1007/s10586-024-04386-7
4. Haussmann, S., et al.: FoodKG: a semantics-driven knowledge graph for food recommendation. In: Ghidini, C., Gandon, F. (eds.) ISWC 2019. LNCS, vol. 11779, pp. 146–162. Springer, Cham (2019). https://doi.org/10.1007/978-3-030-30796-7_10
5. Kejriwal, M.: Domain-Specific Knowledge Graph Construction. Springer, Cham (2019). https://doi.org/10.1007/978-3-030-12375-8
6. Kincaid, J., Fishburne, R., Rogers, R., Chissom, B.: Derivation of new readability formulas (automated readability index, fog count, and flesch reading ease formula) for navy enlisted personnel. Research Branch Report 8-75, University of Central Florida, Chief of Naval Technical Training: Naval Air Station Memphis (1975)
7. Li, Q., Yao, N., Zhou, N., Zhao, J., Zhang, Y.: A joint entity and relation extraction model based on efficient sampling and explicit interaction. ACM Trans. Intell. Syst. Technol. **14**(5), 1–18 (2023)
8. McCusker, J.: LOKE: linked open knowledge extraction for automated knowledge graph construction. arXiv preprint arXiv:2311.09366 (2023)
9. Tang, X., et al.: Construction and application of an ontology-based domain-specific knowledge graph for petroleum exploration and development. Geosci. Front. **14**(5), 101426 (2023)
10. Vrijens, G.: Knowledge graph construction to facilitate chemical compound hazard assessment in the toxin project. Ph.D. thesis, Master's thesis School of Engineering and Computer Science-University of Liège (2023)
11. Wang, Y., Ye, F., Li, B., Jin, G., Xu, D., Li, F.: UrbanFloodKG: an urban flood knowledge graph system for risk assessment. In: Proceedings of the 32nd ACM International Conference on Information and Knowledge Management (CIKM), pp. 2574–2584 (2023)
12. Xu, W., Banburski-Fahey, A., Jojic, N.: Reprompting: automated chain-of-thought prompt inference through Gibbs sampling. arXiv preprint arXiv:2305.09993 (2023)
13. Yang, M., Chen, K., Sun, S., Han, Z., Kong, L., Meng, Q.: A pattern driven graph ranking approach to attribute extraction for knowledge graph. IEEE Trans. Industr. Inf. **18**(2), 1250–1259 (2021)

14. Yu, H., Li, H., Mao, D., Cai, Q.: A relationship extraction method for domain knowledge graph construction. World Wide Web **23**(2), 735–753 (2020)
15. Zhao, Z., Luo, X., Chen, M., Ma, L.: A survey of knowledge graph construction using machine learning. CMES-Comput. Model. Eng. Sci. **139**(1) (2024)
16. Zhou, J., Chen, X., Zhang, H., Li, Z.: Automatic knowledge graph construction for judicial cases. arXiv preprint arXiv:2404.09416 (2024)

MADEISD 2024: 6th Workshop on Modern Approaches in Data Engineering and Information System Design

Process Mining in Croatia's Judicial Auctions

Kornelije Rabuzin[1]([✉]) [iD], Maja Cerjan[1] [iD], and Mislav Jakšić[2] [iD]

[1] Faculty of Organization and Informatics, University of Zagreb, Pavlinska 2, Varaždin, Croatia
{krabuzin,macerjan}@foi.unizg.hr
[2] Valicon d.o.o., Draškovićeva Ulica 54, Zagreb, Croatia
mislav.jaksic@valicon.net

Abstract. This work aims to provide a comparative overview of judicial auctions in Croatia and other European countries, focusing on the application of process mining to analyze the process. The study employs process mining to analyze the work of court auctions. It includes a comparative analysis of process models derived from legal requirements and actual process models. Differences exist between the legally prescribed process model issued and the actual process model. The practical part of the study offers insights into various aspects of the sales process, facilitating a better understanding of these differences and enabling improvements in future processes. This pioneering work has practical implications that can significantly contribute to enhancing judicial auction procedures.

Keywords: process mining · judicial auctions · judicial system · Disco · Minit · Croatian judicial auctions · private auctions · Croatian Financial Agency · FINA

1 Introduction

In Process mining, a technique that uses event log analysis to find, examine, and enhance business processes, has gained significant attention in recent years. Its capacity to derive insightful conclusions from extensive volumes of data generated during process execution has been a key factor in its popularity.

According to Van der Aalst et al. (2011), process mining entails the "finding, analysis, and improvement of processes by extracting knowledge from event logs, using software that includes statistical and machine learning approaches."

While large businesses recognize the value of Big Data, they often neglect to establish connections between event data and process models. Process mining plays a crucial role in bridging the gap between Big Data analysis and business process management. By connecting event data and process models, organizations can assure compliance and evaluate performance effectively. Additionally, linking events to process models injects dynamism into traditionally static diagrams. As event data continues to grow rapidly and the demand for smarter and more efficient processes increases, the use of process mining becomes even more enticing.

Croatian court auctions are integral to the country's legal system, facilitating dispute settlement and the execution of judgments (Orlić, 2016). These auctions, overseen by

J. Tekli et al. (Eds.): ADBIS 2024, CCIS 2186, pp. 175–194, 2025.
https://doi.org/10.1007/978-3-031-70421-5_15

court-appointed auctioneers in public settings, enable interested parties to place bids on the properties being sold. The auction process in Croatia is regulated by the Civil Procedure Act of 2008 and the Law on Enforcement Proceedings of 2012. These statutes establish guidelines for conducting court auctions, including rules for bidding, auction fees, and the distribution of the sale proceeds. Croatian court auctions cover a range of property types, such as real estate, automobiles, and personal belongings like furniture and electronics (Orlić, 2016). In certain cases, the court may even order the sale of intangible assets like stocks or intellectual property.

The procedures for Croatian judicial auctions are meticulous and comprehensive. When a debtor fails to repay their debt, the creditor can request the court to initiate an enforcement action. If the debtor lacks sufficient funds, the court can enforce the sale of an asset. Since debtors and creditors have conflicting interests in asset valuation, a court conducts an auction to determine the market price and find a buyer. The court appoints an auctioneer responsible for conducting the sale and ensuring the accurate recording of bids (Marušić, 2018). The auctioneer sets a minimum bid price and notifies interested parties about the auction. During the auction, the property is open for bidding, and the highest bidder becomes the successful buyer. Judicial auctions must be conducted fairly and promptly as they involve the compulsory transfer of assets between citizens. The law provides a framework for conducting judicial auctions. However, the practical implementation of judicial auctions may not be immediately evident based on the broad auction frameworks. On the other hand, process mining offers an analytical method that bridges the gap between business process management and process execution. By comparing the discovered process model with the one prescribed by the law, differences between the intended process and its implementation can be identified. Further analysis can shed light on potential improvements to the judicial auction process, and this is covered within the paper.

This paper focuses on process mining in Croatian court auctions, aiming to enhance the preparation of subsequent auctions by providing a thorough understanding of each step in the sales process based on collected auction data. The rest of the papers is organized as follows: in section two related papers are analyzed, and in section three judicial auctions across Europe are explored. In section four Croatia's Financial Agency (FINA) and court auctions are analyzed, and in section five the dataset which was extracted, stored and prepared for the analysis in given. Section six presents the basics of process mining, and section seven presents how process mining has been utilized on the Croatian judicial auction dataset. After that, the discussion section elaborates some interesting findings and in the end the conclusion is given.

2 Related Papers

As already described, court auctions in Croatia can be conducted transparently and fairly by adhering to the regulations outlined in the Civil Procedure Act and the Law on Enforcement Proceedings, allowing all parties the opportunity to participate in property sales. There are only a few papers which investigate the topic, and their contribution is analyzed in the section below. A recent analysis by Houlberg and Schmitz (2021) investigated the factors influencing real estate values in judicial auctions held in Denmark,

Germany, and the Netherlands. The authors found that the opening offer, number of bidders, and auction duration significantly impacted the final sale price of properties. They also observed variations in auction practices across the three nations, such as mandatory deposits in Germany and the role of auctioneers in the Netherlands.

2.1 Judicial vs. Private Auctions: Better Without Protection

Courts often appoint less effective auctioneers, resulting in net prices received by creditors and debtors in judicial auctions being about 18% to 33% less than those obtained in private auctions. The appointment of auctioneers in judicial auctions aims to limit abuses and promote transparent processes, given the heightened incentives when auctions are mandatory. However, the effects of lifting formal restrictions on entry are unclear when competition is hindered by other regulatory frameworks. When a debtor obtains a loan, they commit to repayment by pledging collateral. If the collateral is insufficient to cover the debt, a court may compel the debtor to pay with their assets. The challenge lies in determining the price of the asset, as debtors and creditors have conflicting incentives. Judicial auctions aim to provide an objective valuation of the asset by obtaining its market value, thereby avoiding subjective pricing when a party is forced to sell. However, mandatory asset sales create additional agency problems. If selling all the debtor's assets still does not fully repay the debt, obtaining a higher price only benefits the creditor. Consequently, the debtor lacks the incentive to monitor auctioneers and may engage in collusive behavior, leading to lower prices and asset retention (Paredes et al., 2014). The misalignment of incentives in judicial auctions is supported by anecdotal evidence, suggesting collusion between auctioneers and bribers, resulting in low asset prices that harm both creditors and debtors. In Honduras, the World Bank (2003) recommended discontinuing the practice of judges having discretion in selecting auctioneers to reduce corruption. Similarly, in Peru, the same bank advised establishing mechanisms for private resolutions to avoid judicial auctions.

2.2 Effective Processes and Enforcement Agents

The paper examines the procedural and practical options available to courts and associated agencies for identifying assets to satisfy a judgment debt. It explores court-supervised asset seizure and sale through public auctions, considering the efficiency of these processes. Factors that can lead to sub-optimal returns on sold assets and higher incentives for corrupt practices are analyzed. The paper also discusses the growing use of private agents to reduce cost enforcement delays. Alternative approaches to public auctions are presented, offering improved prospects for ensuring full payment of a judgment debt (Gramckow, 2012).

2.3   Empirical Analysis: Revisions of Judicial Real System

This paper investigates the impact of system revisions on judicial real estate auctions in the 2000s. It focuses on three key factors: bid acceptance ratio, number of bidders, and highest bid. Using multiyear auction data from various district courts, the study estimates the improvements resulting from the revisions. Furthermore, it compares the auction data with that of voluntary sales and general real estate trading. The analysis reveals that, despite the system revisions, the selling price in auctions remains significantly lower than in voluntary sales. Based on these empirical findings, the paper recommends the necessity of further system revisions (Asami et al., 2021).

2.4   Embedding Process Mining into Financial Statement Audits

This study explores how financial statement audits are changing in the age of digitization and greater transaction processing automation. Also, authors investigate the possibility of process mining as a novel data analysis approach to improve the audit process in light of the challenges faced by auditors in analyzing sizable amounts of mechanically or semi-automatically processed financial transactions. Process mining, which is defined by its automated examination of business processes, is examined with regard to how it may be included into modern audits, in line with accepted audit standards and procedures. In conclusion, the authors illustrate the viability of implementing process mining, which not only improves audit reliability but also fortifies audit evidence by substituting manual methods, through a detailed assessment of audit standards and data from a field study (Werner et al., 2021).

2.5   Process Mining in Auditing: Limits, Future Challenges

This paper explores how process mining and the field of auditing overlap. It acknowledges the wide range of applications that this approach may provide and improves on Wil van de Aalst's original description of process mining as a way to extract process-related insights from event data. The study highlights the critical role of auditors in fostering trust and assurance regarding the veracity of financial accounts in an era of rapidly rising digital data and automated procedures. Because it is a specialist subject, auditing depends on auditors acting as impartial examiners who inspire stakeholders' trust.

The article emphasizes the need of this trust while also exploring how process mining may improve auditing procedures, adding to the continuing conversation about the changing nature of financial supervision and technology-driven improvements in auditing methodology (Jans, 2012).

3   Judical Auctions in Europe

The availability of standardized data sources, like the European Judicial Network (EJN) database, further enhances the research potential of studying judicial auction data in Europe.

European judicial auction data is a valuable resource for scholars interested in understanding asset market values. This database provides details on auctions held in EU member states, serving researchers and professionals alike. These auctions, conducted in person or online, facilitate the sale of diverse assets such as real estate, cars, and artwork. Examining judicial auction data from Europe offers the advantage of exploring the variety of legal systems and auction procedures present in the region. This allows researchers to analyze how different countries handle asset sales and identify common patterns or distinctive traits. However, challenges exist in analyzing European judicial auction data due to inconsistencies in data gathering and auction procedures across countries and regions. Comparing and consolidating data from various sources while ensuring accuracy and reliability can be demanding. Among Spain, France, Sweden, Poland, and Croatia, only Croatia and Spain conduct electronic real estate judicial auctions. Notably, Croatia allows unrestricted access to judicial auction data through the Register of Real Estate and Movable Property Sold in Enforcement Proceedings. Despite these challenges, studying judicial auction data in Europe has the potential to provide crucial insights into asset market dynamics and the broader economic environment.

We have also checked auction websites for several countries, including France, Spain, Poland and Sweden, but due to space limitations we omit the findings.

4   Croatian Financial Agency - FINA

FINA facilitates the sale of real estate and movable property in enforcement and bankruptcy proceedings through electronic public auctions. These auctions aim to improve access to information, enhance transparency and increase the efficiency of enforcement proceedings. The public announcement calls, notices, requests, and related documents about enforcement and bankruptcy proceedings conducted by FINA are published in the application. Publication in the application is considered to be delivered as delivery to all relevant parties eight days after the publication date. This ensures that anyone can monitor the progress of these proceedings. The Register of real estate and movable property sold in Enforcement proceedings serves as a repository for data on properties sold in Enforcement or bankruptcy proceedings. The Register includes information on all real estate sold in these proceedings and movable property with an estimated value exceeding HRK 50,000 (~6636 EUR). Interested individuals can access and review these records within the Register.

To participate in auctions, users must have their access request approved by FINA. The e-Dražba application enables users to access, participate in, and submit bids for auctions, as well as electronically sign relevant documents. Additionally, users can act on behalf of another person through the e-Dražba service based on a power of attorney. The procedure begins after the necessary documentation for sale or registration in the Register is delivered by a court, public notary, tax authority, or another relevant authority. Upon receiving a request for sale or registration, FINA publishes a reimbursement request for procedural costs . If the costs are not reimbursed, FINA will not proceed with the sale or registration and will notify the court of the non-payment. The court will then determine the next steps in the proceedings. Once the procedural costs are reimbursed, FINA publishes an invitation to participate in the auction.

The auction invitation provides details on the sale method and conditions, the start and end date and time of the auction, property viewing times for interested buyers, and other relevant auction information. There must be a minimum of 60 days between the publication of the invitation and the start of the bidding process. Users can place bids in an auction after paying a deposit and submitting a request to participate through the e-Dražba application. Users can only see auctions for which they paid a deposit and submit a participation request through the e-Dražba application. Users can only view and participate in auctions for which they have paid a deposit and submitted a request. The deposit is held in a dedicated account until an order is received from a competent authority. Once a participation request is completed, the auction system assigns a unique identifier to the bidder for each auction. Bids are collected electronically during the tender period, which lasts for 10 working days. After the auction concludes, FINA provides a report to the court regarding the conducted auction. If no valid bids were received during the first auction, FINA publishes a second call for participation on the first working day after the first auction's end to initiate a second auction. If the item being sold is successfully auctioned, FINA publishes the decision of the competent authority overseeing the proceedings and carries out the corresponding court order, settlement order, deposit recovery order, suspension of proceedings decision, or other relevant court orders.

The e-Dražba application allows users to electronically sign participation requests for electronic public auctions and submit bids online during the tender period. The Public announcement application allows users to view calls, notices, requests, and other relevant documents related to the sale of real estate and movable property in enforcement and bankruptcy proceedings. The Register application allows users to access records of real estate and movable property with a value greater than HRK 50,000 (~6636 EUR) sold in enforcement or bankruptcy proceedings. Enforcement against real estate involves the following steps: recording enforcement in the land register, determining the value of the property, conducting the sale, and using the proceeds to settle the claims of enforcement creditors. The value of real estate is established through a discretionary court conclusion based on the findings and opinions of authorized expert witnesses or appraisers. This valuation considers any existing rights and encumbrances that may diminish the property's value even after the sale. Once the property valuation procedure is completed, the court issues a conclusion specifying the property's value and the manner and conditions of the sale (Fig. 1).

Parties involved can agree on the property value either through judicial or extra-judicial means, which forms the basis for establishing a lien or other relevant property rights for securing the settlement claim. Real estate is sold through online judicial auctions conducted by FINA upon the request of a competent authority. The request for sale and other necessary documents are submitted to the regional FINA centres with territorial jurisdiction corresponding to the location and jurisdiction of the enforcement court. An online judicial auction commences with a call for participation. During the first online judicial auction, the real estate cannot be sold below four-fifths of its established value. For the second online judicial auction, the minimum sale price is set at three-fifths of the established value. Upon the successful completion of the real estate sale through an online judicial auction, the court, upon receiving notification from FINA, issues a decision adjudicating the real estate. This decision confirms that, once it becomes final and the buyer submits the purchase price, the right of ownership to the adjudicated property will be entered in favour of the buyer in the land register. Additionally, any rights and encumbrances on the property that ceases upon its sale will be deleted. After the decision adjudicating the real estate to the buyer becomes final, the court will schedule a hearing to divide the purchase price.

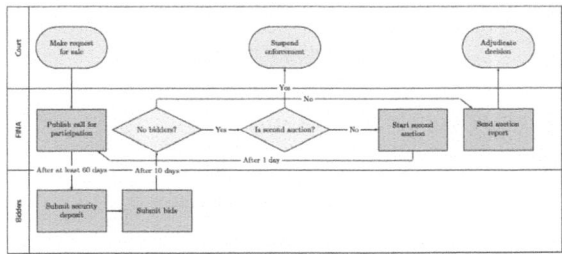

Fig. 1. Croatia's Judicial Auction Process

5   Data Collection and Engineering

Table 1 provides an example of the data available in a Croatian judicial auction dataset. A remote server downloads the judicial auction data from FINA's Register of real estate and movable property sold in enforcement proceedings each day at 18:00. The data is then bundled, compressed, and uploaded to a workstation for process mining every month.

```
0 16 * * * Wget --Output-Document = /root/ocevidnik-$(Date
+"\%Y-\%m-\%d").Csv https://ponip.fina
```

Each file contains approximately 10,000 rows and includes all electronic public auctions, encompassing movable and immovable property, as well as rights. Data recording the entire auction process from start to finish was collected over four months in 2022. Since the official currency in the Republic of Croatia was the kuna, and it changed to the euro in 2023, prices are presented in both currencies in the table. The database containing all the data has been implemented in MS Access database management system (Table 2).

Table 1. Sample data regarding auction process

Column Name	English Column Name	Data Type
Nadležno tijelo	Competent authority	TEXT
Poslovni broj spisa	Business file number	TEXT
Opis	Description	TEXT
Vrsta predmeta prodaje	Type of sale item	TEXT
Složenost PP/opseg imovine	Complexity of PP/scope of assets	TEXT
Utvrđena vrijednost	Determined value	REAL
Napomena uz detalje predmeta prodaje	Note with the details of the item of sale	TEXT
Način prodaje	Method of sale	TEXT
ID nadmetanja	Contest ID	INTEGER
Oznaka EJD	Mark EJD	TEXT
Datum odluke o prodaji	Date of sale decision	TEXT
Datum i vrijeme početka	Start date and time	TEXT
Datum i vrijeme početka nadmetanja	Date and time of the start of the bidding	TEXT
Datum i vrijeme završetka nadmetanja	Date and time of the end of the bidding	TEXT
Mogućnost produljenja nadmetanja (za dodatnih 10 minuta)	The possibility of extending the competition (for an additional 10 min)	TEXT
Ostali uvjeti prodaje	Other conditions of sale	TEXT
Minimalna zakonska cijena ispod koje se predmet prodaje ne može prodati	The minimum legal price below which the object of sale cannot be sold	REAL
Početna cijena za nadmetanje	Starting price for bidding	REAL
Iznos dražbenog koraka (iznos u kunama)	Amount of the auction step (amount in kuna)	REAL
Rok u kojem je kupac dužan položiti kupovinu	The term in which the buyer is obliged to deposit the purchase	TEXT
Iznos jamčevine	Bail amount	REAL
Ostali uvjeti za jamčevinu	Other bail conditions	TEXT
Razgledavanje	Sightseeing	TEXT
Napomena uz uvjete prodaje	Note with terms of sale	TEXT
Stanje na dan	Balance on day	TEXT

Table 2. Sample data regarding auction process (sensitive data are hidden)

Description	Value - determined	Other conditions of sale	Minimum legal price below which the object of sale cannot be sold	Starting price for bidding	Term in which the buyer is obliged to deposit the purchase amount	Deposit amount	Viewing	Note
Property owned by the debtor registered in zk.ul.br. 155, cadastre municipality Vodinci	236.000,00 kn 31322.58 EUR (2024.)	N/A	188.800,00 kn 25058.07 EUR (2024.)	188.800,00 kn	30 days after the completion of the electronic public auction	23.600,00 kn 3132.26 EUR (2024.)	Every Thursday from 11:00 to 13:00	There is an easement of servitude and passage on the property
Properties registered in cadastre municipality Požega, zk.ul.br. ...	523.000,00 kn 69414.03 EUR (2024.)	N/A	261.500,00 kn 34707.01 EUR (2024.)	261.500,00 kn	30 days from the finality of the award decision	20.000,00 kn 2654.46 EUR (2024.)	Viewing by prior arrangement	The property is not free from occupants and belongings

6   Process Mining

The concept of process mining was introduced by Van der Aalst (2012) as an innovative technology enabling process analysis based on evidence. The author presented three fundamental categories of process mining: discovery, conformance, and enhancement, using simple and extensive examples to demonstrate their practical applications (Fig. 2.). However, there are still unresolved scientific challenges, and many organizations have yet to fully recognize the benefits of process mining. According to the author, the primary and most significant category is discovery, where a model is built solely based on event records. It is remarkable to businesses that modern approaches can identify real processes by analyzing patterns captured in event logs. The second category is conformance, which involves comparing an event log from the actual process with an existing process model. The third category is an enhancement, which aims to use data from the recorded event logs to expand or improve existing process models.

Process mining enables the discovery, analysis, and improvement of business processes based on event data (Van der Aalst, 2012). Rather than relying on subjective opinions or outdated experience, businesses should manage, support, and enhance their processes based on objective event data. However, traditional data mining techniques such as classification, clustering, regression, association rule learning, and sequence/episode mining, are not tailored specifically to process analysis. Process mining research aims to bridge the gap between data mining and business process management (BPM), which often relies on manually crafted models. It connects data mining and business process modeling by starting with an event log that captures activities and phases of a process for specific cases (Van der Aalst, 2012). In the context of judicial auctions, the process

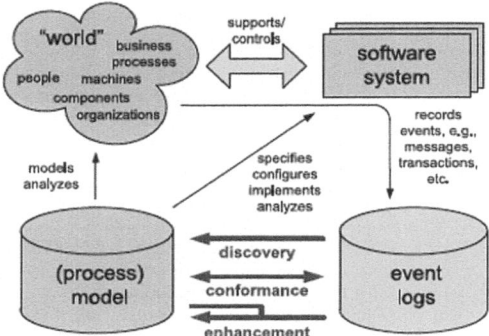

Fig. 2. Basic types of process mining (source: van der Aalst, 2012)

mining approach involves analyzing an event log specific to the auction process. Each event in the log corresponds to an activity or step in the process and is associated with a particular case. The events within a case follow a chronological sequence, representing a single run of the process. However, acquiring a judicial auction event log can be challenging. To conduct online judicial auctions, a country must have both the necessary legislation and infrastructure in place. In the absence of either, auctions are typically announced physically on a bulletin board outside the courthouse. Even if the data access barriers are overcome, open data laws are required to enable citizens to access the judicial auction data. However, even if citizens can access the data, it does not guarantee that the data will be available in the required event log format required for process mining.

7   Analyzing Croatian Judicial Auction

For the practical aspect of this work, an approach was used that allows users to gain insight into the sales execution process and access detailed statistics. Due to the complexity and extensive range of data display options, only the most important components will be presented and discussed below. For process analysis, two popular tools were selected: Minit Process Mining and Disco. Although both are commercial in nature, access is also possible through an academic license. The following is a comparison of process analysis using both tools.

7.1   Minit Process Mining

Minit Process Mining software is a great option since it can import data, show how the process is carried out, and assist in identifying issues and the underlying causes of inefficiency. Additionally, it allows AI-powered root cause analysis, and its dashboards are user-friendly with a variety of sophisticated options, including the creation of BPMN models based on data. The technology is used by many of the top 500 hundred companies, and Microsoft recently bought it.

The following procedure illustrates the steps after loading the data within the process mining tool. The process starts with the choice to sell, as depicted in Fig. 3. Once the decision to sell is made, the start date of the sale is established. Next, the date for the auction to begin is determined. Subsequently, the final date is set, concluding the actual sales procedure.

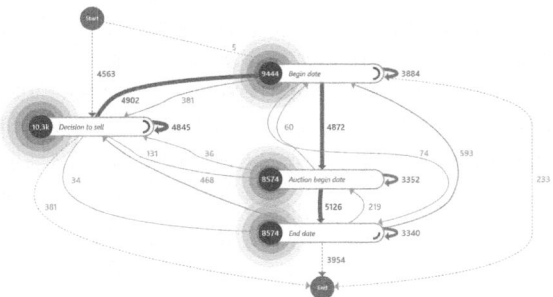

Fig. 3. Process execution steps

Figure 3 provides a comprehensive overview of all possible scenarios and events, encompassing 100% of the gathered data. It presents every conceivable situation, allowing viewers to observe the frequency of cases at each stage of the actual process. For example, when considering the "Choice to sell," which comprises 10.3k instances, the majority proceeds to the next stage, which is determining the start date (9444 instances). However, there are alternative routes, including a direct path to the end date for 34 instances and a route leading to the very end for 381 instances. These alternative paths must be taken into account.

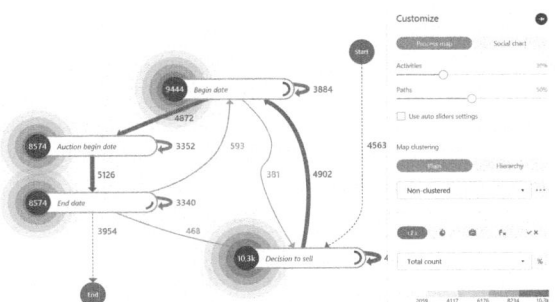

Fig. 4. Graph customization

The graph in Fig. 4 is customizable, allowing users to apply various filters to obtain the desired data. It can be presented as either a process map or a social graph. Users can select activities and pathways using sliders, ranging from 0% to 100%, with preset options available. The second component of the filter enables map clustering, offering plan, and hierarchy options. By default, non-clustered is selected, but users can choose clustering from the drop-down selection. The third section of the filter offers a range of display choices, such as focusing on the largest actions, the duration of specific procedures, or using a unique formula. To aid user understanding, a legend is provided at the bottom of the filter.

Figure 5 showcases the window for presenting the sales process animation. It displays the execution path and key components of the sales process on the process map. Users can view the planned procedure and the actual process through two sequences. The lower portion of the display features a timeline that allows for a simultaneous and accurate comparison within a specific period with the top part representing the planned activities and the lower part representing the accomplished ones. This display is crucial as it enables users to identify any errors in sales planning in advance and make necessary adjustments.

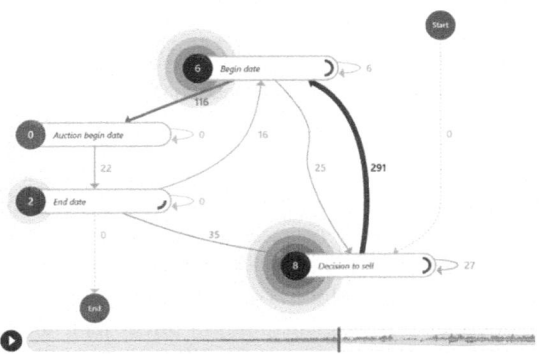

Fig. 5. Process animation view

An alternative view is the data view, comprising three primary components. The first section includes buttons for searching and filtering data, enabling users to easily personalize the presentations according to their needs. Next, a table displays the variations, showing the variant name, number of cases, occurrence quantity, and the proportion of events represented by visuals. The third section visually represents the occurrences within each variation. Cubes are used to depict the occurrences, with the initials of the corresponding procedures inscribed on them. Each procedure is assigned a unique color (Fig. 6).

The first variant (Fig. 6) covers over 50% of all situations and includes four events. The corresponding statistics can be obtained by selecting the first variant (Fig. 7).

When a specific variant is chosen, a comprehensive overview with multiple sections is displayed. In the first section, users can select the data presentation format, including variations display, table cases, and Gantt cases. On the left side of the screen, a timeline of steps is available, showing the duration of each step based on the applied. For instance, the image displays a 5.35-month time interval between steps 1 and 2. The right side of the display consists of three sections. The upper section includes a graph and KPIs that provide specific statistics. Users can access filtered and absolute coverage of cases and occurrences. The second section displays timestamps for crucial time-related data related to the selected variation, such as start and finish timestamps, beginning and ending

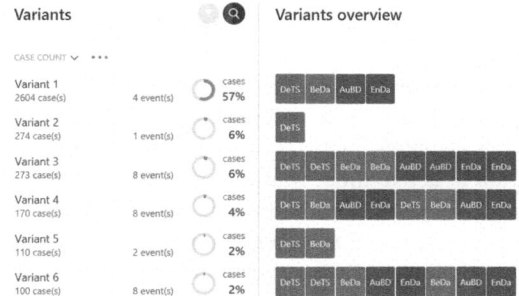

Fig. 6. Variants overview

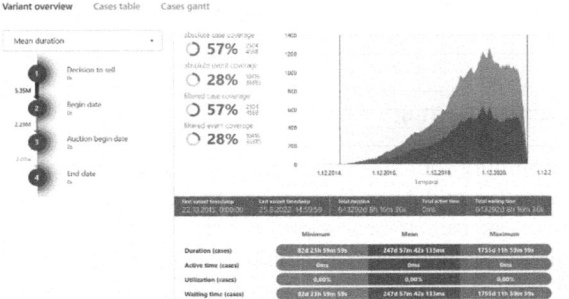

Fig. 7. Detailed overview of a selected variant

dates, and total waiting time. In the third section, various components are graphically represented, including waiting time, duration, etc. Each component is categorized into a minimum, mean, and maximum values, each represented by a distinct color. One can examine different cases individually (Fig. 8):

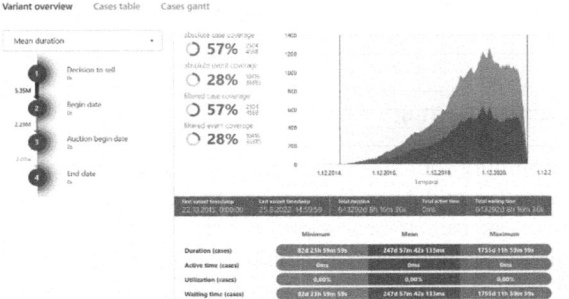

Fig. 8. Case overview

The user can examine each example independently. The display in Fig. 8 is divided into several sections. The first section presents a graphical representation of instances over time. Using a slider, the user can select a specific period. Filters are available on the right side of this section. The user can choose the instances to view from the drop-down menu (active cases are shown by default). The appearance of the x-axis on the graph can be customized as linear or logarithmic. The entire display can be exported in the chosen format or as a PNG image.

In the second section, the most important parts of the selected case are highlighted, along with timestamps and other crucial components. The third section provides a table of the selected instance's elements, including case ID, number of cases, start timestamp, finish timestamp, duration, active time, waiting time, utilization, self-loop percentage, loop, and rework. Additionally, the tool can generate a Business Process Modeling Nation (BPMN) model. A model is a visual representation that shows the planned phases of a business process. It visually illustrates the sequential business operations and information flows required to complete a process, making it a crucial component in BPM. Its objective is to simulate strategies for improving productivity, considering changing conditions, or gaining a competitive edge. A detailed view of the data organized by the individual case is available to the user. While extensive filtering will not be explored for this study, all data can be filtered.

7.2 Disco Process Mining Tool

The Disco process mining tool is another great tool that deserves to be highlighted. Drs. Christian W. Günther and Anna Rozinat created the tool in 2009. The tool's objective is to enable business users to access process mining. Disc's framework, which has applications in many different industries including user journey analysis, audits, process improvement, and optimization, is founded on validated scientific research. With its display and filtering features, the tool makes process discovery easy and versatile (Viner, 2021).

The steps after inserting the data into the process mining tool are shown in the following procedure, which is similar to the Minit process mining tool example shown before.

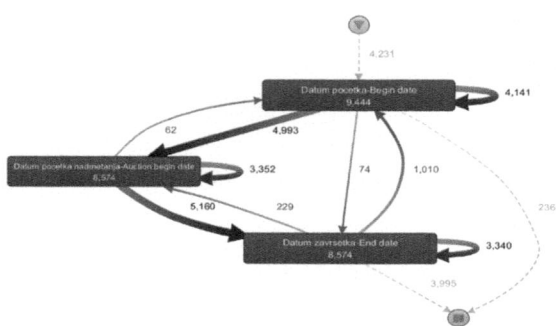

Fig. 9. Disco Process execution steps

A thorough view of all the scenarios and events that are added to the imported data is provided in Fig. 9. The process stages, the number of occurrences in each phase, and each scenario are visible to the user directly. In addition, one can view other options and the likelihood of success. There is a distinct interaction for each piece in Fig. 9.

As a result, when the user clicks on an element, an information window displaying the frequency and performance statistics opens. To make the chosen piece stand out, it can also be put directly to the filter (Fig. 10).

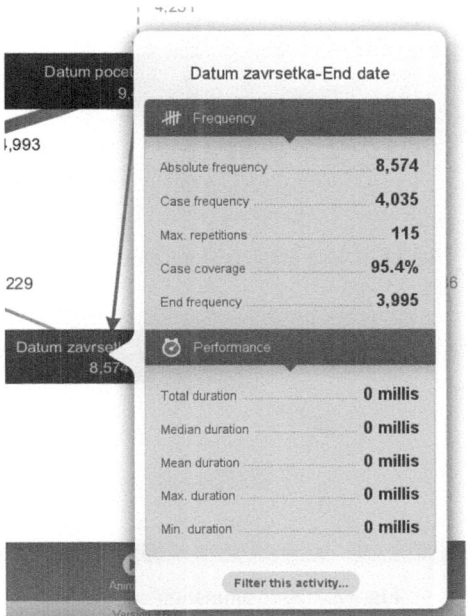

Fig. 10. Disco Info view about selected graph activity

The application itself has various filters that can be applied to alter the visual display seen in Fig. 9. The user can adjust a number of parameters that impact how data is displayed on the screen under the "Detail" section located on the right side of the screen. Two sliders that control the percentage of shown activities and paths make up the first section of the options. The frequency settings are shown in the second section of the settings. The display of absolute frequency, case frequency, maximum repetitions, and case coverage are all customizable by the user. Performance choices are located in the third part of the settings. The following choices are available to the user: mean duration, max. Duration, min. Duration, total duration, and media duration (see Fig. 11).

Additionally, the user has the option to monitor the status of the sales process itself (Fig. 12). This comprehensive animated view shows you how each step of the sales process unfolds in detail. It also highlights specific trends and potential mistakes that could have a big influence on decisions made in the future.

The dashboard, or data display section, of the Disco application is extremely comprehensive. The viewer can obtain a global summary of particular statistical data in this area. The activity overview, status overview, and main statistics window are the three key components that make up the dashboard section. A number of options are available for showing data, as shown in Fig. 13: events over time, active cases over time, case

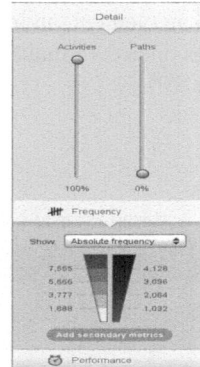

Fig. 11. Disco graph customization options

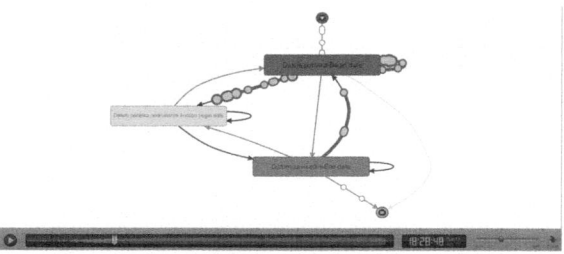

Fig. 12. Disco animation options

variants, events per case, and case length. On the right side of the screen, each option provides both an educational description and a comprehensive graphic overview.

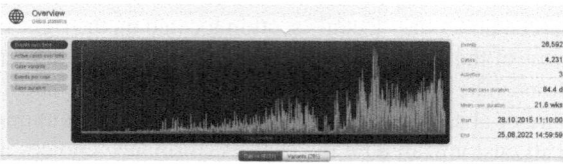

Fig. 13. Disco global statistics overview

Figures 14, 15 and 16 depict the appearance of the module. A tabular presentation of variants and cases with the number of events, unpredictability, and success % is located on the left side. Only the selected cases' results are displayed immediately to the user upon clicking on the preferred variant and case. A graphic overview of the chosen cases and variants can be seen on the right side of the screen. Selected parameters are displayed in the red portion of the graph for instant visibility. There are two methods to see the data in the bottom portion of the screen.

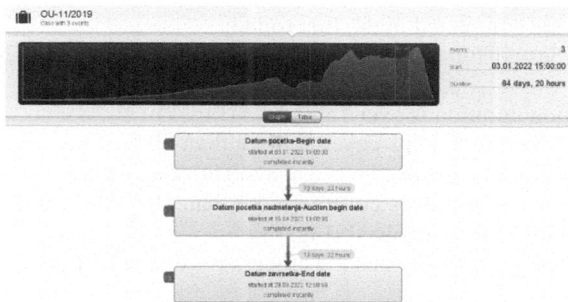

Fig. 14. Disco graph variant overview

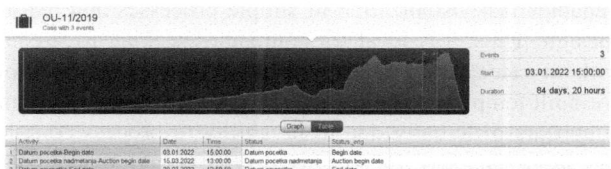

Fig. 15. Disco table variant overview

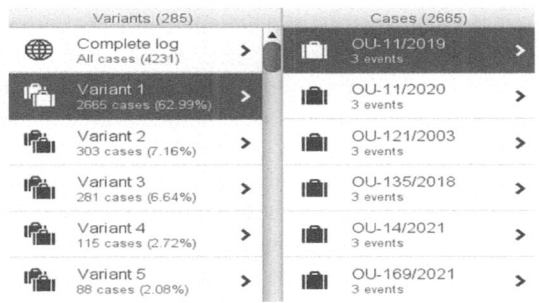

Fig. 16. Disco variants and cases selection view

8 Discusion

The Croatian Financial Agency (FINA) facilitates electronic judicial auctions for the sale of real estate and movable property in enforcement and bankruptcy proceedings. To support the proceedings, FINA undertakes the following actions:

- All calls, notices, requests, and relevant documents about judicial auctions are published in the Public Announcement.
- The status of all electronic judicial auctions is provided through the Register of Real Estate and Movable Property Sold in Enforcement Proceedings.
- Bids for judicial auction are collected through the e-Dražba (e-Auction) application.

Based on the analysis above we can conclude that:

- Some auctions are completed in the first round;
- Some auctions lasted for a very long period of time, more than 200 or 300 days (Fig. 7);
- Maximum auction duration is over 1500 days (Fig. 7);
- While Variant 1 with four steps covered more than 50% of the cases, there are also some variants with only one or two steps involved (Fig. 6);
- While Variant 1 with four steps covered more than 50% of the cases, there are also some variants with 8 steps involved (Fig. 6);
- By looking at the Fig. 9, we see that the process is executed differently than described above, etc.

Regarding the tools, a comparative study of Minit and Disco has revealed that both tools possess valuable features for the mining process but differ in other aspects. Minit is popular as it enables fast visualization of simple processes and real-time data with a user-friendly design. Thus, it is ideal for companies seeking to streamline and gain quick insights without a steep learning curve. Moreover, Minit's library of best practice standards and pre-built templates can help fresh businesses on process mining greatly. However, Disco appears differently due to its ability to process complex procedures as well as manage large data sets. Its advanced filtering options are the reason why it is the preferred option for deep research, especially in industries where processes are very intricate or require thorough examination of the variances of the process. Disco's scripting and customization tools may be beneficial to experienced process mining professionals in their more detailed analysis and complete process optimizing. It also suits organizations that handle diverse data systems, given that it interconnects well with most data analytical applications and allows multiple forms of data integration. Finally, the choice between Minit and Disco is determined by the specific attributes of the company, the intricacy of data that is processed, and the level of expertise in working with a user base. Minit may be more suitable for smaller businesses starting out with process mining. However, larger organizations with complex processes and knowledgeable analysts may prefer the advanced features and customizable nature of Disco. To derive maximum benefits from process mining, it is important for organizations to understand the minute differences between BPM and workflow. Additionally, this includes the cost, scalability and support of process mining projects, so companies can choose between them depending on how these factors will work.

However, it is important to acknowledge the limitations of the mining procedure in court auctions. Access to the auction data may not always be available, and the data from other sources may lack accuracy. External factors that influence the auction process and outcomes, such as market demands and the state of the economy, must also be considered. Additionally, the mining process prioritizes the auction process itself, potentially placing other activities on hold. Further research could explore these aspects to gain a deeper insight into Croatia's judicial auctions.

9 Conclusions

Process mining involves analyzing event logs from judicial auctions to understand the implementation process. However, obtaining such logs can be challenging due to the broad framework for judicial auctions. By comparing the discovered process model with the legally prescribed process, areas for improvement can be identified. In Croatia, the EU Judicial Auctions Website facilitates online judicial auctions for real estate. The process begins with a call for participation and the court determines the property value. The property cannot be sold below four-fifths of its established value at the first auction or three-fifths at the second auction. If no valid bids are received, the court suspends enforcement. Movable property can be sold through oral judicial auctions, direct dealing, or online judicial auctions, with a minimum selling price set at half of the estimated value. The sale of movable property may also be done through an online judicial auction. In France, creditors set starting prices for seized assets, and immovable property auctions are not conducted online. Spain conducts judicial auctions through the BOE Auction Portal, and the Asset Recovery and Management Office is responsible for selling assets from criminal activities through electronic auctions. In Poland, judicial auctions are not conducted online, and only auction notices are published on the National Bailiffs' Council website. Sweden has proposed legislative amendments to allow electronic auctions for real estate.

In the paper practices from different countries have been analyzed, and the procedure in Croatia has been scrutinized and well described. Dataset containing judicial auction data has been created based on the data published online. Relevant attributes have been extracted and prepared to be used within the process mining tool. Process mining has been used to identify discrepancies between actual processes and legal regulations, and to help to improve and advance court auction procedures in the future. To the best of author's knowledge, this is the first attempt to apply process mining in judicial court auction process.

References

Asami, Y., Higano, Y., Fukui, H.: Frontiers of Real Estate Science in Japan. Springer (2021). https://doi.org/10.1007/978-981-15-8848-8

Civil Procedure Act. Official Gazette of the Republic of Croatia, No. 53/91, 91/92, 112/99, 129/00, 88/01, 117/03, 88/05, 02/07, 84/08, 123/08, 57/11, 25/13, 89/14, 36/15, 123/17, 98/19 (1991)

European Judicial Network (EJN) Database. (n.d.). Retrieved from https://e-justice.europa.eu/content_judicial_auctions-391-en.do

Gramckow, H.: Court Auctions Effective Processes and Enforcement Agents. The Justice and Development Working Paper Series, Washington DC (2012)

Houlberg, K., Schmitz, M.: Determinants of real estate prices in judicial auctions: evidence from Denmark, Germany, and the Netherlands. J. Real Estate Financ. Econ. **62**(2), 227–253 (2021)

Jans, M.J.: Process mining in auditing: From current limitations to future challenges. Business Process Management Workshops, pp. 394–397 (2012). Retrieved from https://doi.org/10.1007/978-3-642-28115-0_37

Law on Enforcement Proceedings. Official Gazette of the Republic of Croatia, No. 112/12, 25/13, 93/14, 55/16, 73/17, 34/19, 115/20 (2012)

Marušić, V.: Croatian enforcement law reform: One step forward, two steps back? Eur. J. Law Reform **20**(3), 303–317 (2018)

Orlić, M.: Enforcement proceedings in Croatia. Int. J. Soc. Sci. Humanit. Res. **4**(3), 326–332 (2016)

Paredes, R.D., Crisosto K,A., Martí C.P.: Judicial versus private auctions: Better without protection?. Estudios De Economía **41**(2), 171–186 (2014)

Van der Aalst, W.: Process mining. ACM Trans. Manag. Inf. Syst. **3**(2), 1–17 (2012)

Viner, D.: Disco - process mining software comparison. Process Mining Software Comparison (2021). Retrieved from https://www.processmining-software.com/tools/disco

Werner, M., Wiese, M., Maas, A.: Embedding process mining into financial statement audits. Int. J. Acc. Inf. Syst. **41**, 100514 (2021). Retrieved from https://doi.org/10.1016/j.accinf.2021. 100514

Towards the Utilization of AI-Powered Assistance for Systematic Literature Review

Marija Đukić$^{(\boxtimes)}$ ⓘ, Milica Škembarević ⓘ, Olga Jejić ⓘ, and Ivan Luković ⓘ

Faculty of Organizational Sciences, University of Belgrade, Belgrade, Serbia
{marija.djukic,milica.skembarevic,olga.jejic,
ivan.lukovic}@fon.bg.ac.rs

Abstract. Systematic literature review (SLR) serves as a cornerstone of evidence-based research, yet its execution often proves time-consuming and labor-intensive. The advent of ChatGPT has sparked interest in harnessing artificial intelligence to streamline various tasks. This paper aims to evaluate the efficacy of two GPT-based tools, Elicit and SciSpace, in automating the SLR process. We assess their usability and reliability in two SLR steps: literature search and citation screening. Through testing via conducted SLR, we examine the benefits and limitations of these tools. Our findings suggest that while Elicit and SciSpace offer significant assistance in literature retrieval and study selection, their integration with human expertise is essential to ensure the thoroughness and accuracy of the review process.

Keywords: systematic literature review · automatization · artificial intelligence · generative AI · literature search

1 Introduction

Scientific papers are based on the discussion, confirmation, and extension of previous research. By reviewing the existing literature, researchers gain insights into current scientific knowledge and identify research gaps to develop new research ideas. As stated by [11], a systematic literature review (SLR) is a "means of identifying, evaluating, and interpreting all available research relevant to a particular research question or topic area". The goal of SLR is to provide trustworthy and unbiased insights into research topics. SLRs have become an important research method to summarize existing evidence. As a result, the number of SLRs has considerably increased over the last few years [23]. Although SLRs offer great value, they require a lot of resources. The collection, extraction, and synthesizing of the required data are known to be highly manual, error-prone, and labor-intensive tasks. A SLR was found to take 1.72 years for a single reviewer and would cost \$141,194.80 [16].

Fortunately, various aspects of the SLR process are amenable to automation [22]. Different studies have focused on automating specific steps of the SLR process, such as study selection [20], data extraction [10], query writing [13], as well as research synthesis [15]. Automation is a method that can operationalize the manual tasks of the SLR by using specific tools and computer systems. The integration of automation into

J. Tekli et al. (Eds.): ADBIS 2024, CCIS 2186, pp. 195–205, 2025.
https://doi.org/10.1007/978-3-031-70421-5_16

SLR promises to enhance the review process by saving time [6] and increasing accuracy [19]. Automation tools are being developed to aid in completing the essential stages of SLRs, employing techniques such as machine learning [10, 15], visual text mining [20], and specialized software packages [7].

The advent of ChatGPT has marked a new era of artificial intelligence (AI) and the growth in the number of GPT-based tools dedicated to processing text-based data. A recent study [18] outlines the increasing importance of AI in research, with researchers recognizing its value and efficacy in data analysis and literature review. Moreover, numerous GPT-based tools have been developed to aid in automating the SLR process [21]. However, not much research is currently present on the application of these tools for SLR automation. Researchers often encounter barriers to the adoption of automation tools due to a limited understanding of available options [19]. In this paper, we aim to expand the current understanding of utilizing GPT-based tools to automate the SLR process and introduce two tools that may be relatively unfamiliar to researchers. By comparing Elicit and SciSpace, the objective of the paper is to evaluate the usability and reliability of such tools in automating specific steps within SLR processes. We intend to spark a discussion on whether GPT-based tools should be integrated into the SLR process or if traditional methods should be maintained. Should they prove valuable, an important question for researchers is how such tools can be best leveraged in different stages of the SLR process and how they can complement human researchers. We anticipate offering insights grounded in the practical implementations of Elicit and SciSpace at different stages of the SLR process. Furthermore, we strive to highlight the potential opportunities and risks associated with their adoption, particularly for researchers who may not have extensive experience in conducting SLRs.

The paper is structured as follows: Sect. 2 presents a discussion of related works. To ensure comprehension, Sect. 3 describes the SLR process. An overview of the compared tools is presented in Sect. 4, while the methodology and research questions are outlined in Sect. 5. Section 6 describes the conducted SLR, along with results and analysis. Section 7 discusses the primary findings. Finally, Sect. 8 concludes the paper and provides concrete answers to each research question.

2   Related Work

Deploying automation into SLR promises to enhance the review process. Results from [14] highlighted the predominance of visualization and text mining techniques in supporting various SLR stages. [4] presented a system for the semi-automation of the SLR process based on machine learning. Additionally, [2] identified several tools used to support the SLR process, noting that StArt, SLuRP, and SLR-Tool have demonstrated usefulness. A new approach proposed in [1] leverages generative AI to extract keywords from a provided abstract, query a search engine, and generate a related work section based on the retrieved results.

ChatGPT has led to the advent of GPT-based tools tailored for processing text-based data. Researchers increasingly recognize the value of AI for scientific research, with a growing trend toward utilizing AI to assist in automating the SLR process.

In [24], the authors outlined how AI can expedite the SLR process and how AI-based tools can help further automate individual steps of literature reviews. Authors of [21] investigated the efficacy of several AI-based tools developed using ChatGPT technology to automate a specific step within a defined context. ChatGPT's performance in completing several SLR tasks was evaluated in [17], focusing on tasks relevant to language interpretation. The effective utilization of ChatGPT as a research tool in SLR was demonstrated in [3], while [9] offered a practical guide on conducting literature searches using SciSpace. In [5], authors incorporated SciSpace as one of the databases utilized for literature search in conducting an SLR.

As evidenced in this section, researchers have already experimented with various AI-based tools to expedite different stages of SLR, including ChatGPT as a precursor to other presently available tools. Some researchers have even integrated these tools into the SLRs they have conducted, as demonstrated by [7]. Building upon the findings of previous studies, we aim to investigate the potential of GPT-based tools as it can be argued that AI has the potential to significantly streamline various aspects of SLR. In contrast to prior studies, our research centers on the comparison of Elicit and SciSpace, aiming to assess their effectiveness in automating specific steps within the SLR process. In the example of these two tools, we aim to provide insights into whether GPT-based tools are suitable for SLRs and how such tools can be leveraged across various stages of the process. Besides being less discussed in the existing literature and their increasing prominence, Elicit, and SciSpace are selected due to their documented suitability for specific SLR steps, as outlined in their respective documentation.

3 Systematic Literature Review

Authors of [11] introduced an SLR method tailored to the software engineering domain. Researchers rely on this structured framework to systematically identify relevant literature with minimal bias and high rigor. The SLR methodology described in [11] is selected for this study due to its specific adaptation to the software engineering domain. This methodology is particularly well-suited for our research needs, having been widely utilized and validated in numerous prior studies within the field of software engineering. Its extensive application in previous research underscores its reliability, making it the preferred choice for conducting systematic reviews in this area.

To familiarize readers with the concepts and facilitate comprehension of the subsequent sections of the paper, we outline the proposed method, which has garnered widespread acceptance in the field (Table 1). It is important to introduce the necessary terminology as these terms will be utilized throughout the remainder of the paper.

Specifying the research questions (SLR2) drives the systematic review methodology. A review protocol (SLR3) outlines the methods used to conduct an SLR. A predefined protocol is essential to minimize the possibility of researcher bias and prevent analysis driven by researcher expectations. The literature search (SLR5) step aims to identify studies that address the research questions. The goal is to find as many relevant primary studies as possible using an unbiased search strategy. Study selection criteria (SLR6) is designed to identify primary studies that provide direct evidence related to the research questions. This step is typically divided into the first screening based on titles and abstracts and a more restrictive second screening based on full texts. The data extraction (SLR8) step involves extracting the data items needed to answer the research questions, while the data synthesis (SLR9) step analyzes the data in a way that allows the research questions to be answered.

Table 1. Steps in the SLR process as proposed by [11]

ID	Category	SLR Step
SLR1	Planning the review	Commissioning the review
SLR2		Specifying the research question(s)
SLR3		Developing a review protocol
SLR4		Evaluating the review protocol
SLR5	Conducting the review	Identification of research (Literature search)
SLR6		Selection of primary studies (Citation screening)
SLR7		Study quality assessment
SLR8		Data extraction and monitoring
SLR9		Data synthesis
SLR10	Reporting the review	Specifying dissemination mechanism
SLR11		Formatting the main report
SLR12		Evaluating report

The SLR process encompasses both creative and manual tasks, offering opportunities for AI to reduce the effort for time-consuming and repetitive tasks, thereby enabling authors to dedicate more time to creative ones that demand human interpretation, and expertise.

4   AI-Powered Tools

ChatGPT has paved the way for numerous tools designed to assist with various tasks, and some of these tools can be leveraged to support certain stages of the SLR process. In this section, we present an overview of two GPT-based tools tailored for the research discovery process: Elicit and SciSpace. The selection is based on tools suitability for the literature search (SLR5) and citation screening (SLR6) steps of the SLR process. The selected tools provide a wider range of research information in comparison to other presently available tools. Moreover, their respective documentation highlights their convenience for application in the aforementioned steps of the SLR process.

Both tools leverage GPT-3 language models to identify relevant papers based on the user's research question, condense key insights from these papers, and organize essential information into a research matrix. According to available information, SciSpace has not yet adopted GPT-4, whereas a new version of Elicit utilizing GPT-4 was released in 2023 but is still in private beta. Semantic search functionality enables the discovery of publications addressing the research question, even without precise keyword matches [9]. Users have the flexibility to select specific results from a paper to extract only essential insights. Moreover, users can employ keyword-based filtering for further refinement, but neither of the tools is intended for keyword-based searches utilizing search syntax such as logical operators. Equipped with a copilot feature, SciSpace allows users to engage with research papers via a chatbot interface, export the workspace in multiple formats, and share it with teams or research groups. Elicit possesses its repository, while SciSpace retrieves publications available in Semantic Scholar. Semantic Scholar collaborates with academic publishers to create a reliable source of research works, aiming to enhance the discoverability of research content. With over 50 direct partnerships with publishers and data providers, including ACM Digital Library, IEEE, Springer, Microsoft Academic, and Wiley, it maintains a robust database. Elicit imports papers from Semantic Scholar into its database, focusing exclusively on academic papers and excluding other research documents like books. The Elicit database is updated weekly. Elicit explores the citation graph within the selected papers, in search of further relevant results and uses a process-based machine learning system that provides "better differential capabilities" [12]. This feature enables users to provide feedback to Elicit, illustrating how AI research assistants improve with human input.

5 Research Method

A SLR is a resource-intensive process, with the selection of primary studies, data extraction, and literature search being the most time-consuming tasks [19]. A challenge in developing an efficient search strategy is the use of different search models across various digital libraries. To address this challenge, search strings must be customized for each library requiring adjusting search syntax and keywords according to the library's requirements to ensure result accuracy. One of the high-priority requirements, yet to be fulfilled by existing automation tools, is the capability for integrated search across different databases [2]. GPT-based tools leverage AI technologies, particularly machine learning, and natural language processing, to facilitate the process of integrated search across multiple digital libraries.

In the paper, we compare the potential of the selected tools in the literature search (SLR5) step. The SLR5 step is chosen because, in addition to being one of the most time-consuming and resource-intensive steps of the SLR process [19], it significantly influences the SLR6 step, which is the most demanding stage. The search query in SLR5 directly impacts the number of results retrieved and subsequently screened, affecting the overall efficiency and effectiveness of the review process. The need for AI support becomes evident considering the rapid growth of research output and the inefficiency of allocating time for academic experts to complete repetitive tasks. To evaluate the tools in the SLR5 step, we consider criteria such as the number of retrieved papers, the number of papers relevant to the research question, and the number of missed papers based on reference screening. We anticipate that the tools will demonstrate their utility in constructing the literature sample, but only as complementary aids in this process.

A search system must enable high recall and precision. Recall is the percentage of relevant article records retrieved from the total number of relevant records available, while precision is the percentage of retrieved records that are relevant [8]. High recall indicates that the search captures many relevant items, whereas high precision means the search retrieves a few irrelevant records.

Furthermore, we evaluate the reliability of the selected tools in the citation screening (SLR6) step. The SLR6 step is selected because it is the most time-consuming and resource-intensive part of the SLR process [19]. Consequently, automation and AI support in this step would be highly beneficial. During this step, authors work with the search results to distinguish relevant papers from those that should be excluded from the review. The criteria for comparing the tools in the SLR6 step include the provided key insights and section summaries, reliability, and clarity of this information, which determines the need to read the abstract and subsequently the full text of the paper. Typically, this step is divided into the first screening based on titles and abstracts, and the second screening, which is more restrictive and based on full texts. We anticipate that the potential for AI support will be high for the first screening, where irrelevant papers can be excluded, and moderate for the second screening stage, where more exclusion criteria are applied.

5.1 Research Questions

The objective of this research is to analyze and assess two GPT-based tools, Elicit and SciSpace, that can be utilized to assist in conducting an SLR. This paper aims to give answers to the following research questions:

RQ1: How reliable are GPT-based tools when used for SLR?

RQ2: What are the benefits and limitations of using GPT-based tools in SLR?

6   Case Study

To demonstrate the practical application of the selected tools, we delve into the domain of "systematic literature review automation using artificial intelligence". It's important to clarify that while this paper outlines the methodology for conducting an SLR, it does not delve into the specific findings of the SLR process. The main objective is to evaluate the reliability and effectiveness of two selected tools in the SLR5 and SLR6 steps of the SLR process. The performed SLR process aims to identify papers related to the use of artificial intelligence in automating SLRs. Given that the chosen tools utilize a question-based approach for retrieving relevant papers, both tools are presented with the same question to enable a direct comparison of the results obtained. The question posed for the literature search is: "How does the utilization of artificial intelligence automate the process of systematic literature review?".

Elicit retrieved a total of 60 papers spanning from 2013 to 2023, offering a concise summary in response to the posed question based on insights from the four most relevant papers. However, some papers were duplicated, as one version was a preprint while the other was the published version, resulting in a total of 56 unique papers. Papers were then analyzed based on the abstract summary, and three irrelevant papers were excluded. On the other hand, SciSpace retrieved 52 papers from the years 2018 to 2023, generating an answer to the research question based on insights derived from the top five relevant papers. After excluding preprints and retaining only the published versions, 47 papers remained for analysis. Further exclusion based on whether the paper answers the research question and abstract summary resulted in the removal of six more papers. However, an

Table 2. Results of the SLR5 and SLR6 steps performed in each tool

Tool	Papers retrieved	Duplicates removed	After first screening	After reference search	**Total number**
Elicit	60	56	53	53	**53**
SciSpace	52	47	41	43	**43**

additional two papers were discovered while screening the references, making a total of 43 papers (Table 2).

Elicit retrieved more papers compared to SciSpace, which primarily found more recent references. Elicit achieved a precision of 88.3%, compared to SciSpace's 78.8%. Additionally, SciSpace omitted two papers that were later discovered during the reference search of the retrieved papers. Both tools offer similar filtering options based on year and full-text PDF availability, with a distinction regarding keywords. Elicit filters based on keyword appearance in abstracts only, while SciSpace scans the full paper for keywords and allows for the inclusion or exclusion of papers containing specified keywords. SciSpace offers a broader range of export options compared to Elicit and provides the functionality of saving search processes for future use. Elicit lacks this functionality but allows the creation of a search through predefined steps. Another advantage of SciSpace is its integrated chatbot, SciSpace Copilot, which enables users to interact with each paper through subsequent questions. On the other hand, Elicit offers a significantly wider spectrum of insights through additional columns such as research gap, methodology, and paper section summaries. Elicit performs better in extracting key insights from papers, offering enhanced reliability, and even providing a warning in the form of an exclamation mark to signify low confidence in the provided insight.

7 Discussion

GPT-based tools mark a substantial advancement in AI research, offering potential applications across various domains, including SLRs. However, it's crucial to recognize that as generative models, these tools cannot ensure the absolute accuracy of their outputs. Hence, this section will delve into the potentials and limitations of GPT-based tools in the realm of conducting SLRs.

Database search: Elicit and SciSpace both provide integrated search functionality, removing the necessity of searching across different databases. Elicit retrieves research papers from its database, whereas SciSpace accesses papers from the integrated Semantic Scholar database. However, these databases contain fewer papers compared to well-known databases like Google Scholar and may potentially overlook some relevant references. Elicit's database contains approximately 125 million papers [12], whereas the Semantic Scholar database holds around 218 million papers [9]. In contrast, Google Scholar offers a significantly larger collection with about 389 million papers. While a larger number of papers does not necessarily guarantee a higher number of relevant references, broader coverage is generally considered advantageous. Searching larger databases, all other factors being equal generally results in higher search recall [8].

Search strategy: Elicit and SciSpace streamline the literature search step through question-based searches, eliminating the need for query formulation and query refinement as required in traditional digital library searches. However, this approach may pose limitations as researchers must formulate the right questions to ensure relevant papers are returned. Although the tools can retrieve papers without exact keyword matching, there's a possibility of missing relevant references because of discrepancies in the variations of the research question's keywords. In contrast, the search query typically encompasses all relevant synonyms.

Time-saving: Elicit and SciSpace demonstrate significant potential in saving time during both SLR5 and SLR6 steps, which are known to be time-consuming and resource-intensive. Researchers can leverage their capability to quickly retrieve relevant studies as the tools incorporate both forward and backward citation searches. The SLR6 step presents a great opportunity for utilizing GPT-based tools, allowing researchers to quickly analyze and summarize extensive literature volumes. By extracting key information, researchers can identify relevant studies more efficiently leading to significant time and effort savings. Some of the information that can be extracted include main findings, paper section summaries, methodology, research contributions, and limitations. Initially, the main findings and abstract summary columns are displayed, but users can add more details by selecting additional predefined columns provided by the tool. Of the two tools, Elicit offers a wider range of options for extracting paper information and a possibility of creating custom columns. Particularly noteworthy is the support provided for the first screening based on titles and abstracts. However, it is essential for human experts to meticulously review the tool-generated summaries.

Accuracy and efficiency: The tools show promising potential to improve the accuracy and efficiency of SLRs. Their proficiency in natural language processing enables precise content analysis, reducing the likelihood of errors and omissions in research interpretation. However, despite the ability to filter and extract paper insights, they may encounter limitations in extracting all relevant information, particularly if the information is presented in non-textual formats. Although the documentation claims that the tools generate summaries based on paper text, they still operate using large language models. A significant limitation of these models is their potential to fabricate facts, generating statements that appear legitimate but are false. In our case study, we did not encounter non-existent papers, which is expected since the databases source papers from academic publishers, university presses, and scholarly societies. Elicit customizes large language models to ensure that extracted information is based on the paper's text. However, since these models are not explicitly trained to be entirely faithful to the text, they might generate non-factual information and overlook the latest research not included in their training data. Elicit claims that around 90% of the provided information is accurate.

Reproducibility: The search results in both Elicit and SciSpace are influenced by user prompts, establishing a basis for reproducibility. Consistent outcomes can be obtained by following the same guidelines. SciSpace provides the option to save the results of the search, including all retrieved papers for a given research question and the extracted information. The "search workspace" can also be shared with other researchers. On the other hand, Elicit allows users to create a "notebook" and save individual retrieved papers within it.

7.1 Threats to Validity

Despite our thorough examination of the application of GPT-based tools for SLR automation, there are several potential threats to the validity of this research that must be acknowledged. One concern is the possibility of algorithm biases that may arise from factors such as the training data used or the tool design choices, which could impact the results the tools generate. Although we found information on the sources of papers, we did not find detailed information on the tool's functioning or the ability to check for algorithmic bias. Consequently, the "black box" review of the tool represents a potential threat to the validity of the research. Furthermore, technological advancements or changes in tool functionality could invalidate previous findings or conclusions. Findings derived from research on specific contexts may not necessarily generalize to different settings, particularly since the documentation of the tools indicates their best performance with empirical research questions.

8 Conclusion

AI is anticipated to become increasingly integrated into the research workflow across various scientific disciplines, offering opportunities to enhance various research tasks. In this study, we explored how AI can support the SLR process and sketched potential directions for future research.

Through a performed SLR, we evaluated the practical application of two selected tools, Elicit and SciSpace, in automating the SLR5 and SLR6 steps. In response to the first research question concerning reliability, the tools can help automate the SLR process, but these contributions require human validation and interpretation. Therefore, they should be regarded as complementary to human expertise. By combining the capabilities of these tools with human insight, researchers can enhance the productivity and overall quality of their research and scientific writing endeavors.

In response to the second research question, the tools undeniably offer valuable assistance, but they also have notable limitations. These tools promise to improve search efficiency by retrieving papers without exact keyword matches, relying on the research question rather than search query. However, they may overlook relevant studies by omitting keyword synonyms. Both tools show significant potential for time savings as they efficiently analyze and summarize large volumes of data. Of particular significance is the potential for automating the first screening phase within the SLR6 step. The tools offer substantial assistance by extracting key insights from papers, thereby minimizing the risk of researcher omissions and errors. However, a significant limitation lies in the possibility of generating false statements.

The integration of AI necessitates careful attention to ethical considerations related to data privacy, transparency, and the responsible use of AI technologies. By attending to these ethical concerns, researchers contribute to an environment that prioritizes responsible and ethical utilization of AI.

Our findings demonstrate that AI has the potential to revolutionize traditional methods and can be utilized to automate labor-intensive and repetitive tasks in the SLR process. As a part of the future work, the research can be expanded to include more test cases to analyze these two tools or to evaluate this test case using alternative methods, such as a question-answering approach. Additionally, the research can be broadened to include other available GPT-based tools. Another direction for future research and further automation of the SLR process involves evaluating the usability of these tools for other stages of the SLR, such as study quality assessment, data extraction, or research synthesis. While Elicit and SciSpace may not be best suited for these stages, some other GPT-based tools may be employed to automate these aspects of the SLR. In doing so, these tools could enable human workers to concentrate on higher-level tasks that necessitate human skills, such as critical thinking, empathy, and creativity.

Acknowledgments. This paper is funded by the Faculty of Organizational Sciences, University of Belgrade.

Disclosure of Interests. The authors have no competing interests to declare that are relevant to the content of this article.

References

1. Agarwal, S., Laradji, I.H., Charlin, L., Pal, C.: LitLLM: A Toolkit for Scientific Literature Review. arXiv e-prints, arXiv-2402 (2024)
2. Al-Zubidy, A., Carver, J.C., Hale, D.P., Hassler, E.E.: Vision for SLR tooling infrastructure: prioritizing value-added requirements. Inf. Softw. Technol. **91**, 72–81 (2017)
3. Alshami, A., Elsayed, M., Ali, E., Eltoukhy, A.E., Zayed, T.: Harnessing the power of Chat-GPT for automating systematic review process: methodology, case study, limitations, and future directions. Systems **11**(7), 351 (2023)
4. Bacinger, F., Boticki, I., Mlinaric, D.: System for semi-automated literature review based on machine learning. Electronics **11**(24), 4124 (2022)
5. Barros Adamantino De Oliveira, B., Gerken Brasil, B.: Influence of the overweight and obesity on Systemic lupus erythematosus: A review. In incbac.org. INCBAC Institute (2023). https://www.incbac.org/wp-content/uploads/2024/04/UNIGOU-Training-Public ation-2023-Bruno-Barros-Adamantino-de-Oliveira.pdf
6. Clark, J., McFarlane, C., Cleo, G., Ramos, C.I., Marshall, S.: The impact of systematic review automation tools on methodological quality and time taken to complete systematic review tasks: case study. JMIR Med. Educ. **7**(2), e24418 (2021)
7. Cleo, G., Scott, A.M., Islam, F., Julien, B., Beller, E.: Usability and acceptability of four systematic review automation software packages: a mixed method design. Syst. Rev. **8**, 1–5 (2019)
8. Gusenbauer, M., Haddaway, N.R.: Which academic search systems are suitable for systematic reviews or meta-analyses? Evaluating retrieval qualities of Google Scholar, PubMed, and 26 other resources. Res. Synth. Methods **11**(2), 181–217 (2020)
9. Jain, S., Kumar, A., Roy, T., Shinde, K., Vignesh, G., Tondulkar, R.: SciSpace literature review: harnessing AI for effortless scientific discovery. In: Goharian, N., et al. Advances in Information Retrieval. ECIR 2024. LNCS, vol. 14612. Springer, Cham (2024). https://doi.org/10.1007/978-3-031-56069-9_28
10. Jonnalagadda, S.R., Goyal, P., Huffman, M.D.: Automating data extraction in systematic reviews: a systematic review. Syst. Rev. **4**, 1–16 (2015)
11. Kitchenham, B., Charters, S.: Guidelines For Performing Systematic Literature Reviews in Software Engineering. Keele University (2007)

12. Kung, J.Y.: Elicit. J. Can. Health Libr. Assoc. J. **44**(1), 15–18 (2023). https://doi.org/10.29173/jchla29657

13. Li, H., Scells, H., Zuccon, G.: Systematic review automation tools for end-to-end query formulation. In: Proceedings of the 43rd International ACM SIGIR Conference on Research and Development in Information Retrieval, pp. 2141–2144 (2020)

14. Marshall, C., Brereton, P.: Tools to support systematic literature reviews in software engineering: a mapping study. In: 2013 ACM/IEEE international symposium on empirical software engineering and measurement, pp. 296–299. IEEE (2013)

15. Marshall, I.J., Wallace, B.C.: Toward systematic review automation: a practical guide to using machine learning tools in research synthesis. Syst. Rev. **8**, 1–10 (2019)

16. Michelson, M., Reuter, K.: The significant cost of systematic reviews and meta-analyses: a call for greater involvement of machine learning to assess the promise of clinical trials. Contemp. Clinical Trials Commun. **16**, 100443 (2019)

17. Qureshi, R., Shaughnessy, D., Gill, K.A., Robinson, K.A., Li, T., Agai, E.: Are ChatGPT and large language models "the answer" to bringing us closer to systematic review automation? Syst. Rev. **12**(1), 72 (2023)

18. Ray, P.P.: ChatGPT: A comprehensive review on background, applications, key challenges, bias, ethics, limitations and future scope. Int. Things Cyber-Phys. Syst. **3**, 121–154 (2023). https://doi.org/10.1016/j.iotcps.2023.04.003

19. Scott, A.M., Forbes, C., Clark, J., Carter, M., Glasziou, P., Munn, Z.: Systematic review automation tools improve efficiency but lack of knowledge impedes their adoption: a survey. J. Clin. Epidemiol. **138**, 80–94 (2021)

20. Shakeel, Y., Krüger, J., Nostitz-Wallwitz, I.V., Saake, G., Leich, T.: Automated selection and quality assessment of primary studies: a systematic literature review. J. Data Inf. Q. (JDIQ) **12**(1), 1–26 (2019)

21. Souifi, L., Khabou, N., Rodriguez, I.B., Kacem, A.H.: Towards the use of AI-based tools for systematic literature review. ICAART **2**, 595–603 (2024)

22. Tsafnat, G., Glasziou, P., Choong, M.K., Dunn, A., Galgani, F., Coiera, E.: Systematic review automation technologies. Syst. Rev. **3**, 1–15 (2014)

23. Van Dinter, R., Tekinerdogan, B., Catal, C.: Automation of systematic literature reviews: a systematic literature review. Inf. Softw. Technol. **136**, 106589 (2021)

24. Wagner, G., Lukyanenko, R., Paré, G.: Artificial intelligence and the conduct of literature reviews. J. Inf. Technol. **37**(2), 209–226 (2022)

Estimating Information Efficiency of Bitcoin Inscriptions

Niki Hrovatin[1,2]([✉]) [iD] and Aleksandar Tošić[1,2] [iD]

[1] Faculty of Mathematics, Natural Sciences and Information Technologies,
University of Primorska, Glagoljaška 8, 6000 Koper, Slovenia
`niki.hrovatin@famnit.upr.si`
[2] InnoRenew CoE, Livade 6a, 6310 Izola/Isola, Slovenia
`niki.hrovatin@innorenew.eu`

Abstract. This paper investigates the information efficiency of text based Bitcoin inscriptions. We estimate the entropy of text-based inscriptions using the ZIP compression algorithm, which is a common technique in information theory. Our analysis reveals that despite the economic costs associated with inscribing data on the Bitcoin blockchain, inscriptions exhibit a surprisingly high compression ratio compared to common text-based data. Moreover, our analysis of a 10-day moving average yields similar compression ratios even in presence of huge spikes in raw data inscribed. Finally, we conclude that the results are rather unexpected considering inscriptions are not free, as the limited block size must is contested by other participants including transactions.

Keywords: Bitcoin · Inscriptions · Compression · Entropy

1 Introduction

The practice of inscribing data into the Bitcoin blockchain dates back to the genesis block, which contains a headline inscribed by Satoshi Nakamoto. However, the number of data inscriptions was low due to the high transaction fees associated with these operations [13].

In the recent years, the Bitcoin protocol [9] has undergone several modifications, referred to as Bitcoin Improvement Proposal (BIP)[1], mostly aimed at overcoming certain limitations related to scalability, transaction throughput, privacy and security of smart contracts, as well as extending the supported complexity upon their execution. Although a detailed overview of the many changes Bitcoin went through is out of the scope of this paper, we refer the readers to existing work (e.g. [6]) for a detailed description.

Notably, in 2017, the controversial BIP141 introduces the Segregated Witness structure (SegWit)[2] that designs a new structure called witness that is

---

[1] https://github.com/bitcoin/bips/tree/master.
[2] https://github.com/bitcoin/bips/blob/master/bip-0141.mediawiki.

A. Tošić: The authors would like to thank Michael Mrissa for his support in the preparation of this paper.

© The Author(s), under exclusive license to Springer Nature Switzerland AG 2025
J. Tekli et al. (Eds.): ADBIS 2024, CCIS 2186, pp. 206–212, 2025.
https://doi.org/10.1007/978-3-031-70421-5_17

committed to blocks separately from the transaction merkle tree. SegWit is understood as a "soft-fork", meaning that it is a backward-compatible evolution of the original protocol. Combined in 2021 with the TapRoot[3] and TapScript[4] extensions, they allow multiple conditions encoded in a Merkle tree to enable different alternative ways to validate a spending.

In a similar way to non-fungible tokens (NFTs) proposed for the Ethereum protocol, the extension, primarily meant for smart contract and to alleviate the blockchain from the taxable part of the transaction, led to the development of the Ordinal protocol [1]. This allows data to be permanently stored in the blockchain for social, artistic, and other purposes unrelated to transaction validation, and at a low transaction fee.

This situation raises a number of problems, the most prominent arguably could be the place taken in the mempool[5] with this data, which recently reaches new highs and also stimulates the research community to address the problem [11]. In this paper, we follow this line of work and explore a burning question: how meaningful is the data encoded in inscriptions, in particular text inscriptions? We cover existing work in Sect. 2, which demonstrates the relevance of our approach for evaluating the meaningfulness of inscriptions. In order to guarantee the repeatability of the obtained research results, we describe in Sect. 3 our original methodology that takes advantage of the properties of lossless compression to evaluate data entropy and draw conclusions about the nature of the data stored in inscriptions, with a specific focus on text inscriptions. Section 4 discusses our interpretation of the coding entropy measurement using moving average over a large period of collected data. Section 5 summarizes the results obtained and discusses the many opportunities that this explorative work opens for future research in this domain.

2   Related Work

A large amount of research has been realized around blockchain since its creation, related to different topics such as peer-to-peer systems, network usage, consensus protocols, etc. For more details, we refer the reader to the extensive bibliographic analysis of research trends related to blockchain between 2014 and 2020, published in 2021 [7]. This study shows the growth of the research activity in this domain and the connections with other fields such as cloud or fog computing, security, or privacy.

More specifically related to this paper's focus, in [3], the author studies the characteristics of ordinals and how they affect transaction fees. Ordinals is a theory that makes every satoshi[6] uniquely identifiable, if we accept to comply to a small number of constraints[7]. Ordinals connect to inscription in the sense that it provides a unique identification system for the later.

---

[3] BIP341: https://github.com/bitcoin/bips/blob/master/bip-0341.mediawiki.

[4] BIP342: https://github.com/bitcoin/bips/blob/master/bip-0342.mediawiki.

[5] The term mempool refers to the in-memory space that is used during the elaboration of a block, before consensus has been reached.

[6] A satoshi is the basic unit behind bitcoin where 1 BTC = 1 million satoshis.

[7] https://github.com/casey/ord/blob/master/bip.mediawiki.

In [6], the authors give an overview of the underlying technologies described in the introduction of this paper, compares the drawbacks and advantages with classical Ethereum NFTs, and discuss challenges and opportunities for the research and industry communities. The first challenge identified in the paper is blockchain bloat, due to the increase in size of the blocks and the impact on network usage and computational requirements.

Galaxy Research and Mining's latest report highlights the explosive growth of the Ordinals ecosystem on Bitcoin, noting over 33 million inscriptions and significant market activity [10]. The report also delves into the impact of Ordinals on Bitcoin transaction fees and the emergence of new inscription techniques like recursion and reinscription.

In [14], the authors study the inscription based BRC-20 token standard, analyzing its mechanisms and market impact, comparing it to Ethereum's ERC-20 standard and noting its role in introducing non-fungibility to Bitcoin. Their findings suggest that while BRC-20 expands Bitcoin's functionality, it still lacks the extensive decentralized application ecosystem that Ethereum offers.

As explained in [12], the amount of non-transactional data stored since the apparition of the OP_RETURN operator in 2014 has also been widely studied, quantified and evaluated. As shown in this review of the related work, although non-transactional data on the blockchain has been studied, to this day, none of the existing work has focused on the meaning that one could evaluate from such data. This is the gap we are addressing in this paper.

3   Methodology

Data Acquisition: Inscriptions are arbitrary data cryptographically coupled to a satoshi, the smallest subdivision of a bitcoin. The content of inscriptions is stored entirely on-chain, within Taproot spend scripts. To collect this data, we first synchronized a Bitcoin Core full node[8] with the Bitcoin mainnet. The full node synchronization with transaction indexing took 3 days. Following this, we set up the open source ORD server[9], connecting it to our Bitcoin full node. ORD is an index, block explorer, and command-line wallet written in rust specifically developed for the tracking of sats and managing bitcoin inscriptions. For the purpose of this study, the ORD server provides an API for accessing inscription data. The ORD instance required building an index of all existing sats; syncing with the Bitcoin full node took 24 h. With both the Bitcoin full node and the ORD server fully synchronized, we created a python script to fetch all inscriptions corresponding to the MIME type *text/plain* using the ORD server's API. This script retrieved inscriptions from block #771573 on 12-01-2024–when the first text inscription was made[10]–up to block #836403 on 26-03-2024. The script ran for 12 days and produced 13GB of output.

---

[8] Bitcoin-Core full node: https://bitcoin.org/en/full-node.

[9] ORD: https://github.com/ordinals/ord.

[10] Inscription Id: f58ad8178e7fe78624bcd814cf4b655dab8a6d5f293d4a395a8f24c49aaba78ai0.

Data Preprocessing and Cleansing: We began by closely examining the data and noticed that most of the inscriptions are JSON objects of the format: *{"p":"brc-20","op":"transfer","tick":"ordi","amt":"12"}*. These represent BRC-20 token transfers and minting activities, inscribed with the MIME type *text/plain*. Since our interest is solely in plain text inscriptions, we decided to remove all inscriptions containing JSON content, resulting in an output of 1 GB. After further inspection, we noticed that there are many inscriptions containing single emojis, artistic artifacts consisting of unusual symbols, and a lot of text written in languages using non ascii characters. We decided to remove all this instances using a filter that removes all inscriptions where non ascii characters are the majority. The filtering produced an output of 817 MB, which after metadata removal resulted in 158 MB of plain text inscription content.

Data Analysis: Estimating entropy using compression algorithms like ZIP is a common technique in information theory. The basic idea is that if the data has low entropy (i.e., it is predictable or has a lot of redundancy), it will compress well, resulting in a smaller compressed file size [2]. Conversely, if the data has high entropy (i.e., it is unpredictable or has little redundancy), it will not compress well, and the compressed file size will be closer to the original size [4].

Based on this assumption, we designed two experiments. In the first experiment, we compressed all inscription content into one zip file, adding one inscription at a time, and at each step, we documented the compression ratio as given in the following equation:

$$compression\_ratio = 100 * (1 - \frac{compressed\_size}{uncompressed\_size}) \quad (1)$$

To form the baseline for comparing with the inscriptions compression rates we sourced plain text data from the following datasets: Harry Potter 1 - Sorcerer's Stone[11], News Category Dataset [8], and Amazon Product Reviews [5]. We obtained compression ratios by compressing the whole Harry Potter book, all short descriptions of the News Category Dataset, and 2 GB of 5-core Amazon reviews from the clothing, shoes and jewelry category. The second experiment was designed to assess how the compression ratio evolves over time. To accomplish this, we defined a moving window function that progresses through the inscription data in increments of 1440 Bitcoin blocks, equivalent to a time span of 10 days. At each step, we compressed all inscriptions within the window and documented the compression ratio. Both experiments were conducted using Python scripts relying on the *zipfile*[12] library for compression using the DEFLATE compression algorithm.

4 Results and Discussion

Before interpreting the results it is important to note that inscriptions on the Bitcoin network are constrained by the block space, which is contested by other

---

[11] Harry Potter: https://github.com/amephraim/nlp/tree/master/texts.

[12] zipfile: https://docs.python.org/3/library/zipfile.html.

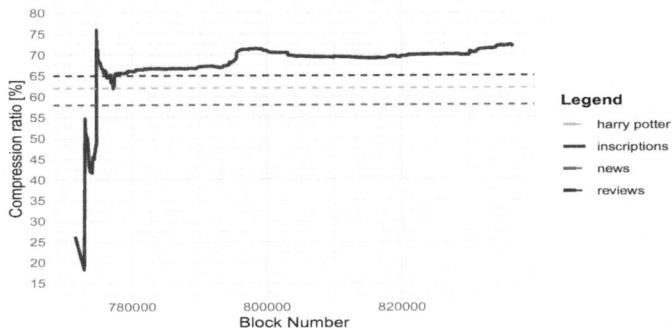

Fig. 1. A comparison of compression ratios between inscriptions and common texts that serve as a baseline. Bitcoin inscriptions generally have a significantly higher compression ratio then common texts.

market participants and hence can be very costly. We therefore make a fairly intuitive assumption that what users choose to inscribe (and consequently pay for) is not meaningless as it would be colloquially refereed to as throwing money away. Consequently, one would expect the compression level to be lower as conciseness is desired, and repetition is avoided. However, Fig. 1 contradicts this expectation. We provide common, non-random text as comparison that can serve as the baseline of what typical text based data compression factors might look like. We observe that regardless of economical costs of inscription, the compression of data is higher then the average. Naturally, news are expected to have a lower compression rate given that what is new, should generally not be repetitive (should carry new information), while reviews are observed to have a higher compression factor as users are reviewing a small subset of purchased items which increases the repetitiveness of certain words. We conclude the Bitcoin blockchain now stores a considerable amount of textual data, which is repetitive. A possibly undesired effect given transactions are contesting the same limited block space potentially hindering the primary goal of the network, a peer to peer electronic cash system.

Taking a closer look at Bitcoin's inscriptions we observe the initially, the compression ratios were fairly low as observed in Fig. 2. As more data was inscribed, the compression factor significantly increases hinting at the repetitiveness or lack of new information being added. Moreover, with time, the on-chain size of these inscriptions follows a linear growth while the compression ratio remains fairly stable. Additionally, the compressed size hints at the inefficiency of block space currently used by inscriptions. Theoretically, at least, the storage requirement on nodes is growing disproportional to the actual information being added. This is not surprising, as more data is added, the compression ratio will stabilize. Possibly a more informative approach is offered in Fig. 3, where instead of looking at the entirety of inscriptions, we compute the ratio of a moving average. A sufficiently long time-frame was chosen to avoid a noisy signal and Fig. 3 shows

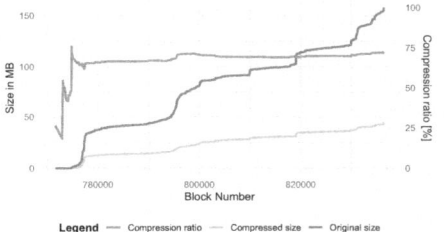 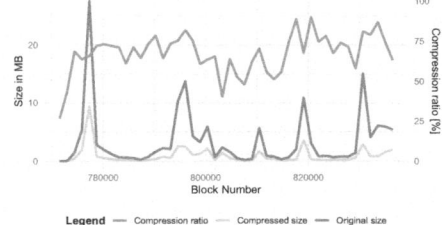

Fig. 2. A time-series comparison between actual size, compressed size, and compression ratio of inscriptions on the Bitcoin network.

Fig. 3. Compression ratio, original size, and compressed size observed in a 1440-block (cca 10 days) moving average.

the moving averages of compression, compressed, and original sizes of inscriptions. We observe that while the moving average is quite small, the compression ratio remains quite high. Moreover, there are interesting spikes in the original size (note that moving averages smoothen such spikes to an extent) while the compression ratio did follow suit. Moreover a couple of such spikes in inscribed data are observed, which could only mean the inscribed text was very repetitive or had a very low coding entropy.

5   Conclusion

In conclusion, our experiments show that inscriptions represent a significant problem contesting the limited block size of the Bitcoin protocol. Inscriptions can be valuable as they can serve as a fully transparent and immutable log of time-stamped information. However, a distinction must be made between data, and information. The compression ratios, commonly analog to entropy and used as a measure of information show that Bitcoin's text-based inscriptions have a below average/expected entropy, and hence a higher compression factor. Moreover, the 10-day moving averages were used to reduce the signal to noise ratio and the results conform with the previously observed. Finally, growing compression factors may indicate the repetitiveness of inscriptions, which is unexpected given the costs associated. This is evident in comparing compression factors with other known literature as well as a moving average both of which confirm that inscriptions are growing in similarity. Future research should focus on topic detection and similarity measures to establish weather this is just correlation, or causation. This study is limited to text based inscriptions, where coders used in popular loss-less compression algorithms are known to be close to the theoretical lower bound. Future research should focus on analyzing other data types by potentially using binary coders or computing the actual entropy.

References

1. Ordinal theory. https://docs.ordinals.com/. Accessed 03 July 2024
2. Avinery, R., Kornreich, M., Beck, R.: Universal and accessible entropy estimation using a compression algorithm. Phys. Rev. Lett. **123**(17), 178102 (2019)
3. Bertucci, L.: Bitcoin ordinals: determinants and impact on total transaction fees. Res. Int. Bus. Financ. **70**, 102338 (2024). https://doi.org/10.1016/j.ribaf.2024. 102338. https://www.sciencedirect.com/science/article/pii/S0275531924001314
4. Henriques, T., Gonçalves, H., Antunes, L., Matias, M., Bernardes, J., Costa-Santos, C.: Entropy and compression: two measures of complexity. J. Eval. Clin. Pract. **19**(6), 1101–1106 (2013)
5. Hou, Y., Li, J., He, Z., Yan, A., Chen, X., McAuley, J.: Bridging language and items for retrieval and recommendation (2024)
6. Li, N., Qi, M., Wang, Q., Chen, S.: Bitcoin inscriptions: foundations and beyond (2024)
7. Luo, J., Hu, Y., Bai, Y.: Bibliometric analysis of the blockchain scientific evolution: 2014–2020. IEEE Access **9**, 120227–120246 (2021). https://doi.org/10.1109/ ACCESS.2021.3092192
8. Misra, R.: News category dataset (2022)
9. Nakamoto, S.: Bitcoin: A peer-to-peer electronic cash system (2008). https:// bitcoin.org/bitcoin.pdf. Accessed 22 Apr 2024
10. Galaxy Research Mining: Bitcoin inscriptions & ordinals: A maturing ecosystem (2023). https://www.galaxy.com/insights/research/bitcoin-inscriptions-and-ordinals/
11. Sohan, M.S.H., Mahmud, M., Sikder, M.A.B., Hossain, F.S., Hasan, M.R.: Increasing throughput and reducing storage bloating problem using IPFS and dual-blockchain method. In: 2021 2nd International Conference on Robotics, Electrical and Signal Processing Techniques (ICREST), pp. 732–736 (2021). https://doi.org/ 10.1109/ICREST51555.2021.9331254
12. Strehle, E., Steinmetz, F.: Dominating OP returns: the impact of omni and veriblock on bitcoin. J. Grid Comput. **18**(4), 575–592 (2020)
13. Sward, A., Vecna, I., Stonedahl, F.: Data insertion in bitcoin's blockchain. Ledger **3** (2018)
14. Wang, Q., Yu, G.: BRC-20: Hope or hype (2023). https://arxiv.org/abs/2310.10652

Deep Learning-Based Cellular Nuclei Segmentation Using Transformer Model

Mateusz Erezman$^{(\boxtimes)}$ and Tomasz Dziubich$^{(\boxtimes)}$ ⓘ

Computer Vision and Artificial Intelligence Laboratory, Department of Computer
Architecture, Faculty of Electronics, Telecommunications and Informatics,
Gdańsk University of Technology, Gdańsk, Poland
erezman.mateusz@gmail.com, tomasz.dziubich@eti.pg.edu.pl

Abstract. Accurate segmentation of cellular nuclei is imperative for
various biological and medical applications, such as cancer diagnosis and
drug discovery. Histopathology, a discipline employing microscopic examination of bodily tissues, serves as a cornerstone for cancer diagnosis.
Nonetheless, the conventional histopathological diagnosis process is frequently marred by time constraints and potential inaccuracies. Consequently, there arises a pressing need for automated image analysis tools
to augment medical practitioners' efforts. In this paper, we present a
novel approach utilising Transformer model, originally designed for natural language processing tasks, for automated cellular nuclei segmentation
in whole-slide microscopic images. Specifically targeting cell nuclei, this
methodology holds significance as the initial phase in diagnosing various
illnesses, streamlining the analysis and quantification process.

The study introduces a novel model that combines a U-Net architecture with a Transformer-based network functioning as a parallel encoder.
This model was compared against three other popular architectures in
the literature: U-Net, ResU-Net, and LinkNet-34. The impact of augmentation and colour normalisation techniques was investigated. The
average Dice similarity coefficient for the considered images was found
to be 0.8041. The obtained results seem to be clinically relevant.

Keywords: Cell nuclei segmentation · transformer · convolutional
network · whole-slide image

1 Introduction

Tumours in various forms pose a serious health threat, characterised by high
mortality rates. Research conducted by [16,19] indicates that tumours are the
second leading cause of death. In women, breast cancer and lung cancer are
the most deadly, while in men, prostate cancer and lung cancer are the most
dangerous. In recent years, thanks to scientific advancements in the field of
diagnosis, the mortality rate associated with breast cancer has declined. However,

---

https://cvlab.eti.pg.gda.pl/.

J. Tekli et al. (Eds.): ADBIS 2024, CCIS 2186, pp. 213–224, 2025.
https://doi.org/10.1007/978-3-031-70421-5_18

the diagnostic process carried out by medical personnel is difficult and time-consuming. Additionally, the accuracy of diagnosis significantly depends on the experience and skills of the examiner. Analysis of images, obtained through tissue biopsy in form of Whole Slide Images (WSI), becomes even more complicated due to their large data volume, ranging from several to several dozen gigabytes [14]. Thus, computer systems supporting work become a key area of research [4].

Detecting cell nuclei is a crucial task in image segmentation because observing them is used as the first step in analysing most tissues [6, 14]. This paper focuses on the segmentation of cell nuclei. Over the past decade, there has been significant development in cell nuclei segmentation methods for whole-slide images. These methods generally fall into two categories: based on computer vision methods and deep learning-based approaches. Computer vision techniques can be categorised into two main groups: gray level-based methods and structure-based methods. These encompass various machine learning techniques, including contour-based segmentation, fuzzy c-mean clustering (FCM), the greedy snakes algorithm, adaptive thresholding, watershed transform applied to gradient magnitude, confidence-connected region-growing algorithm, Gaussian mixture models, the fast marching method, random forest, graph cuts, and constrained regions of shape. However, most of these methods are semi-automatic and have been evaluated on small datasets, limiting their solution generality and requiring long computational time. On the other hand, deep convolutional neural networks have shown promising results in medical image segmentation, leveraging the significant advancements in natural image processing achieved by deep learning techniques. The survey of methods can be found in [1, 7, 10].

Yildiz et al. [25] focus on the segmentation of cell nuclei, aiming to classify them into six different types. They utilise UNet-based architecture trained on the Lizard dataset, achieving IoU results at 48.57%, accuracy at 69.61%, precision at 94.23%, and recall at 94.48%. In [24], Wazir et al. propose a coder-encoder network architecture with an attention module, using a multi-loss function. They utilise the MoNuSeg and GlaS datasets, achieving F1-score, IoU, and Dice scores at 75.08%, 71.06%, 96.20% for MoNuSeg, and 98.07%, 76.73%, 99.09% for GlaS, respectively. Belharbi et al. [2] propose image segmentation and classification, utilising a pre-trained ResNet-18 network. They concentrate on regularising the Kullback-Leibler divergence loss. They propose two methods: Explicit Entropy Minimisation (EEM) and its surrogate, Surrogate for Explicit Entropy Maximisation (SEM). The datasets used in the study are GlaS and Camelyon16. The aim of the study [22] is to create a network sensitive to tumours and detect tumour regions. Sun et al. proposed the Cancer Sensitive Cascaded Networks (CSC-Net), which incorporates two UNet networks. They process large-scale images by dividing them into smaller images with dimensions of 512×512px as input to one network and 1024×1024px as input to the other. The outcome of their work expands upon [12]. They once again integrate two models: the Shallow UNet (SU-net) and the Deep UNet (DU-net). Inputs are separately fed into the two networks, and the resulting outputs are merged. Large-scale images are subdivided into images with dimensions of 512×512px, 1024×1024px, and

2048×2048px. The Dice coefficient serves as the metric for testing, yielding a performance level of 79.68%. Kasturi et al. presented a comparison and analysis of methods utilising machine learning techniques for cell nucleus segmentation in medical images, particularly focusing on various architectures, with emphasis on UNet architectures [8]. The study employed the MoNuSeg dataset, with images standardised to a size of 512×512px and augmented by altering their orientation. Results indicated that a pretrained DenseNet-121 achieved the highest Jaccard index of 66.9%, F1 score of 80%, and accuracy of 93.9%. Additionally, a modified UNet exhibited the highest sensitivity at 82.1%, while a pretrained ResNet-50 demonstrated the highest precision at 83.3%.

The research presented in [21] demonstrates that transformer-based networks employing attention mechanisms have garnered significant traction in the realm of text processing, supplanting the conventional use of recurrent neural networks (RNNs) within this domain. Additionally, in recent years, these networks have been adapted into vision transformers for the processing of image data. To the author's understanding, as per the conducted review, this methodology has not yet been applied within the discussed problem domain. Consequently, the proposed novel methodology will be centred around network leveraging transformers. Research outlined in [27] illustrates that transformer-based networks are well-suited for segmentation tasks and are implemented across various modalities. Additionally, architectures composed entirely of attention layers and transformers have demonstrated notable performance in segmentation tasks [20]. The contributions of this study are as follows: (1) introducing an innovative utilisation of the Transformer model for cell nuclei segmentation on whole-slide images; and (2) comparison study between automatic methods for cell nuclei segmentation: LinkNet, Res-UNet and Image Transformer based segmenter, with the reference point being the UNet model, which is commonly used as a baseline.

The structure of this paper is as follows: Sect. 2 provides a review of methods utilising convolutional networks for cell nuclei segmentation. Section 3 describes the proposed segmentation method, while Sect. 4 focuses on detailing the experiment comparing the performance of selected models on the available dataset and the results of the conducted experiments are presented and discussed. The paper concludes with a summary and outlines for future work.

2   Related Work

In this section, we focus on several relevant studies that utilise image processing and machine learning techniques to segment histopatological whole-slide images. During the systematic literature review, the IEEE Xplore scientific database was utilised (April–June 2023). The keywords and phrase used for article search were: *"Segmentation" AND "Image" AND "Histopathology"*. The phrase returned 344 different articles. After applying exclusion criteria (all papers with a publication year older than 2018), 183 articles remained.

The significant majority of methods for medical image segmentation utilise machine learning methods and deep learning techniques. Most of these methods employ convolutional neural networks (CNN), to create more advanced neural network architectures such as LinkNET-34, Fast-SCNN, or UNet. The UNet [17] architecture significantly dominates the literature, appearing in almost every publication and serving as the foundation for creating other architectures used in the field. Various variations of the UNet architecture emerge, such as combining the UNet model with residual networks.

Benmabrouk et al. [3] focus on semantic segmentation of breast cancer on histopathological images. Authors use the U-Net architecture and employ metrics including accuracy, precision, recall, F1-score, and IoU to evaluate the solution. The BreaKHis dataset is utilised; however, it lacks segmentation masks, hence automatic data annotation and mask generation are crucial aspects of the article. Authors employ colour detection and colour separation algorithms for this purpose. Nevertheless, the reliability of such annotated data remains uncertain. Ultimately, utilising UNet, authors achieved accuracy of 0.92, precision of 0.94, recall of 0.89, F1-score of 0.91, and IoU of 0.95 for the created dataset.

Natarajan et al. [14] introduced a deep neural network model named LinkNET-34 for cell nuclei segmentation. This architecture is of the encoder-decoder type. The architecture resembles the UNet model; however, it is not symmetrical. Successive levels of the model consist of a varying number of layers. The encoder and decoder parts differ from each other, and connections between them are made by summing tensors instead of concatenation. The Data Science Bowl 2018 dataset was used for training and evaluation. Authors compared the results obtained for both UNet and LinkNet architectures. It was found that the LinkNet model outperformed, achieving an accuracy of 0.972, DSC (Dice similarity coefficient) of 0.89, and IoU of 0.898.

In [11], Li et al. focus on the detection and segmentation of breast cancer areas in histopathological images. The article also describes the workflow with large-scale images. The images are divided into 256×256px and 512×512px resolutions, utilising the Camelyon16 dataset. To achieve faster and more accurate processing, the authors employ a model that combines classification and segmentation. Using the ResNet101 and MobileNetV2 networks, divided image parts are examined for the presence of tumours, and if a tumor is detected segmentation is performed using the UNet architecture. For the MobileNetV2 network an accuracy is at the level of 0.972 and a sensitivity - 0.956, while the ResNet101 network attained 0.983 and 0.968, respectively. The Dice metric for the UNet network is 0.814 for images divided into 512×512px and 0.846 for 256×256px.

Samudrala et al. [18] aimed to perform colorectal tumour region segmentation. They utilised the GlaS dataset to train a model with a hybrid encoder-decoder architecture. The DlinkNet network was employed as the encoder for feature extraction, utilising the ResNet-50 network as its core. For the decoder, the U-Net model was utilised. Additionally, the authors proposed an Attention Gate Module (AGM) to facilitate attention focusing between network layers. The accuracy achieved by the proposed model is 0.992, with a Dice coefficient of 0.92, IoU of 0.839, and a sensitivity of 0.917.

Rashid et al. proposed a new method, MD-Unet (Multiscale Dilated UNet), for detecting and segmenting cell nuclei in multiple organs [15]. In the proposed network, the skip connection blocks, which connect the corresponding encoder layer to the decoder layer, were expanded. Additionally, new blocks were added. Large-scale images are divided into 256×256px patches, which are passed as input to the network. For validation on MoNuSeg dataset, the results obtained by the authors are: 0.921 accuracy, 0.801 F1-score, 0.80 Dice coefficient, and 0.077 loss function value.

Zhan et al. [26] aimed to detect cell nuclei in histological images using two connected architectures: UNet and a residual UNet (ResUNet). Furthermore, the authors investigated the influence of colour normalisation on the segmentation results of stained medical images. They utilised three normalisation methods: Reinhard, Macenko, and Vahadane. The evaluation metrics used on the training set MoNuSeg included IoU, Dice coefficient, accuracy, and F1-score. In experiments with both deep learning models, it was found that the Macenko method yielded the best results in most metrics. The standard U-Net model outperformed Res-UNet.

3   Proposed Method

According to the conducted review, it appears that only a minority of studies make use of the attention module. One of the emerging architectures in this regard is the Transformer model [23] and hybrid architectures which linking convolutional networks with transformers [27]. In contrast to [28], which implemented the Transformer model with CNN as parallel path in the classification task, our proposal involves extending the architecture of Unet to incorporate the TransUnet model [5], where a transformer-based network is added as the final stage of the UNet encoder. In the current approach to the UNet model, an additional parallel encoder has been incorporated, which functions as a transformer-based network. The architecture is illustrated in Fig. 1.

The input image is passed to both the transformer and the standard encoder of the UNet model. The encoder utilising the transformer first divides the input image into sub-images, which are then transformed into embeddings, and their positions are encoded in the subsequent stage.

The core of the encoder consists of 12 transformer layers, which comprise a self-attention layer and an MLP layer composed of multiple fully connected layers. Additionally, normalisation operations are applied before the MLP layer and the attention layer. Unlike batch normalisation, which normalises over batches of data, layer normalisation focuses on normalisation across channels.

The output from the last transformer block is reshaped to match the shape for integration with the UNet network. The output from the encoder is then concatenated with the lowest layer of the UNet architecture. Subsequent steps involve standard operations typically used in the decoder of UNet architectures.

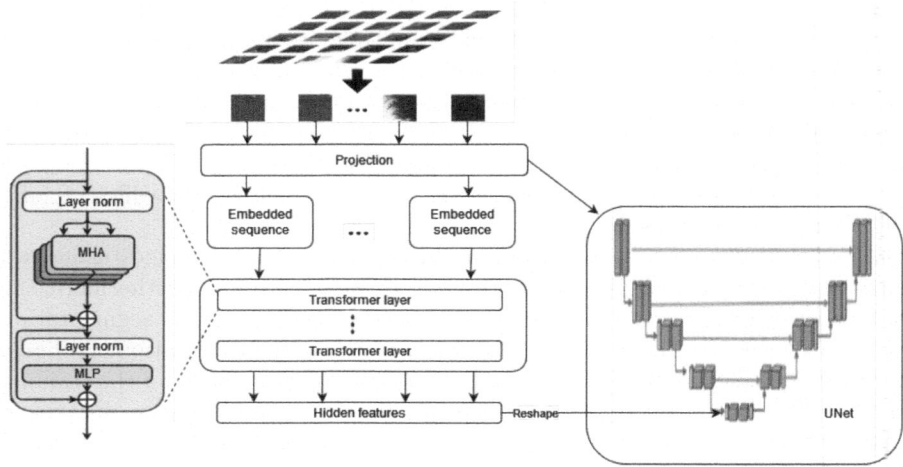

Fig. 1. Architecture of proposed model

4   Experiments and Results

4.1   Selected Architectures

To assess the effectiveness of the proposed method, a series of experiments were conducted, in which our architecture was compared with three architectures presented in Sect. 2, namely: UNet, ResUNet, and LinkNet-34. We conducted training and testing on a MoNuSeg dataset described further below. The impact of preprocessing methods - augmentation and colour normalisation - on the segmentation accuracy was examined. The aim of colour normalisation is to improve the segmentation results performed by the neural network model. Due to variations in colour among microscopic images resulting from various techniques, image acquisition methods, or tissue sampling locations, models may encounter difficulties in learning effectively. To address this issue, colour normalisation of dataset images may prove beneficial. In studies regarding cell nucleus segmentation [26], the authors investigated the impact of several colour normalisation methods on the segmentation outcomes of microscopic images. The research demonstrated that colour normalisation positively affects segmentation results, with the best outcomes achieved using the Macenko method [13]. Hence, in the current study, colour normalisation will also be employed, with the Macenko method selected based on the results reported by the authors [26] These methods were investigated in the context of different loss functions - Binary Cross Entropy (BCE) and Multiloss [24], which were employed for network training. In summary, each of the four architectures was trained in four various variants:

1. BCE - BCE loss function without augmentation or normalisation,
2. Multi - multi-loss function without augmentation or normalisation,

3. Multi_A - multi-loss function and augmentation without normalisation,
4. Mulit_AN - multi-loss function, augmentation, and normalisation.

To evaluate the segmentation results obtained by the trained deep model, three evaluation metrics were employed, namely accuracy (Acc), Dice similarity coefficient (DSC), and Intersection over Union (IoU). For increased experimental credibility, the training and testing procedures were carried out three times, and the final outcome was averaged.

In the context of the described task of cell nucleus segmentation, datasets most closely related to the subject include the MoNuSeg, GlaS, Lizard, Camelyon16, CDC-LungHP, BraKHis, PAIP 2019, and Data Science Bowl 2018 datasets. A literature review conducted revealed that the MoNuSeg dataset appears most frequently in papers. Consequently, this dataset was utilized as the basis for further investigations. MoNuSeg [9] is a collection of histopathological images obtained by sampling tissues from several patients with tumours originating from different organs. It consists of sub-images extracted from whole-slide images. Sample images were acquired using a 40-fold magnification microscope and staining. Cell nuclei were meticulously delineated and annotated as labels and masks on images depicting the positions of cell nuclei. The entirety is publicly available. Training data containing 37 images, each of size 1000×1000px, containing 21623 nuclear boundaries and test data - 14 images with additional 7223 masks. For the purposes of our tests, the original images were divided into sub-images of size 256 × 256 pixels. Following this division, the training set consists of 592 images, while the test set comprises 224 images of the same size.

The models were implemented using the Keras library. A platform with a GPU on an NVIDIA GeForce RTX 2060 card and 16 GB of RAM was used for the training. Training of each network was carried out using the Adam optimiser and batch size = 8. Training lasted for 200 epochs.

4.2   Results and Discussion

The Table 1 contains the results on the test set of all trained models. The learning curves of all models exhibit significant similarity, oscillating around comparable values, with noticeable disparities between training and validation set results in each training iteration. This phenomenon could be attributed to the limited availability of data required for model learning. However, even after employing augmentation, this difference did not significantly improve, suggesting the complexity of the problem and its solutions.

The results of all architectures on the test set closely resemble each other. The ResUNet model achieved the weakest results, while the remaining architectures attained nearly identical outcomes. The introduction of multi-loss functions (Multi) did not substantially improve the results. However, dataset augmentation and normalisation positively impacted the outcomes. Depending on the model, improvements of approximately 2–5% were observed in IoU and DSC. Dataset normalisation and augmentation had the most significant effect on the LinkNet models and the proposed transformer-based method, indicating their pronounced

Table 1. Summary of results obtained. The best results for a given metric are in bold.

Model	Acc	IoU	DSC
Unet - BCE	0.9134	0.7749	0.7656
Unet - Multi	0.9149	0.7788	0.7778
Unet - Multi_A	0.9163	0.7848	0.7794
Unet - Multi_AN	**0.9290**	0.8043	0.7966
Res-Unet - BCE	0.9099	0.7639	0.7714
Res-Unet - Multi	0.9088	0.7625	0.7725
Res-Unet - Multi_A	0.9125	0.7713	0.7806
Res-Unet - Multi_AN	0.9171	0.7790	0.7853
LinkNet - BCE	0.9054	0.7593	0.7708
LinkNet - Multi	0.9035	0.7568	0.7705
LinkNet - Multi_A	0.9168	0.7852	0.7942
LinkNet - Multi_AN	0.9261	0.8046	0.8035
Our model - BCE	0.9064	0.7562	0.7509
Our model - Multi	0.9142	0.7771	0.7655
Our model - Multi_A	0.9123	0.7824	0.7762
Our model - Multi_AN	0.9285	**0.8075**	**0.8041**

influence on large-scale models. Nonetheless, these models ultimately achieved results very close to the UNet model. Therefore, if the selection criterion for the model is its size, it is advisable to opt for the basic UNet, which is considerably smaller in architecture compared to others while achieving very comparable results. In the absence of memory constraints, utilising a transformer-based architecture is suggested.

The results obtained for the baseline UNet model are consistent with those reported in the literature. Some studies, such as [3], demonstrate better outcomes (IoU= 0.95); however, they utilise different training and testing datasets. For the MoNuSeg dataset, the results are comparable to those reported in the literature, for instance in [15], where DSC is reported as 0.80. This is likely due to the better diversity and size of the MoNuSeg dataset.

Figure 2 presents an sample image segmentation result for each architecture in the Multi_AN variant. Marked in white and black are pixels correctly classified as cell nucleus and background (TP and TN). Red colour marked are pixels incorrectly classified as background (FN). Grey colour indicates pixels misclassified as cell nucleus (FP). It can be observed that all architectures performed very similarly, albeit making errors in slightly different cells.

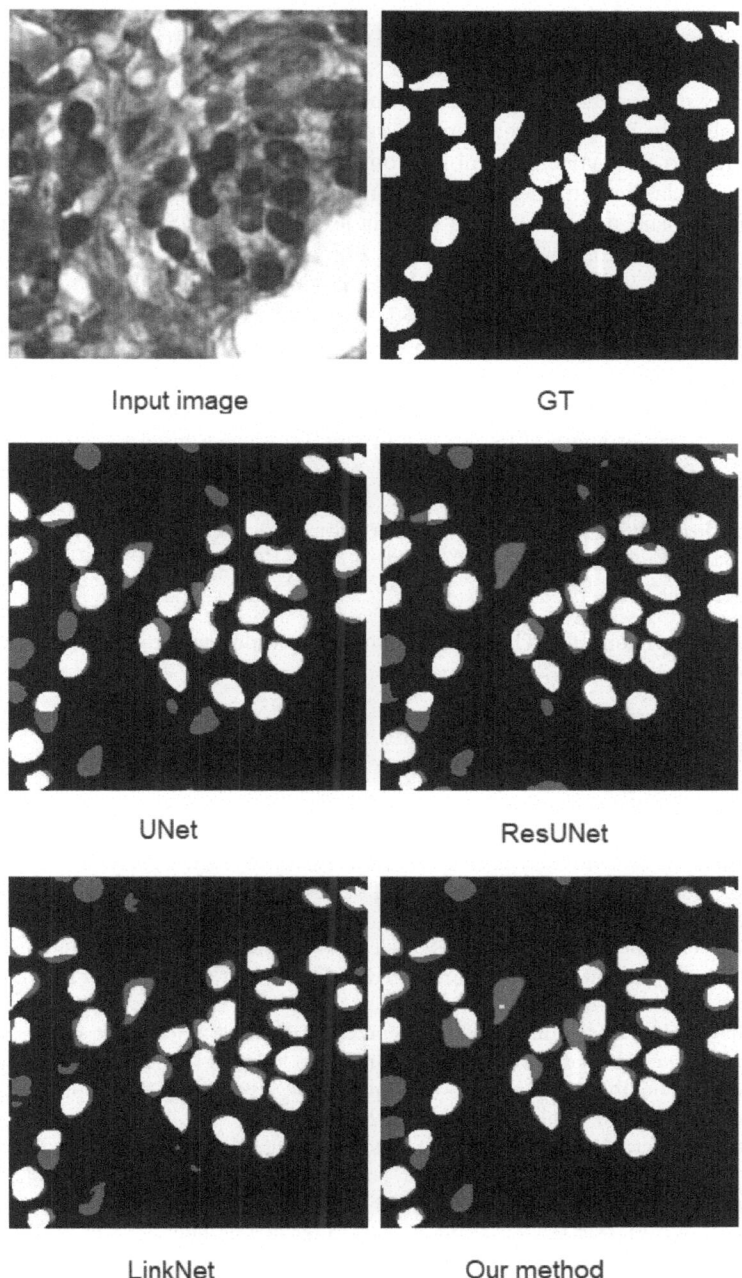

Input image

GT

UNet

ResUNet

LinkNet

Our method

Fig. 2. Segmentation results of the sample image for each tested model.

5   Summary

In this paper, the segmenter based on Transformer architecture is adopted and applied to cellular nuclei segmentation. After implementing the four automatic methods, experiments conducted on the MoNuSeg dataset showed that this model is slightly better at segmenting cell nuclei in WSI images. The improvement is due to the parallel combination of two encoders: the descending path of the Unet classical model and the TransUnet-based module. The proposed transformer-based model received the highest IoU and DSC, but these are very close, to the Unet and Linknet models. The weakest score was achieved by the ResUNet model. In the future, we will enhance the fusion effectiveness of CNN and Transformer with more robust models and strategies to achieve greater accuracy in this medical task. Expansion of the model to include segmentation of a greater number of classes is also planned. After evaluating the capabilities of our segmenter in three-dimensional whole-slide image analysis, its implementation is intended as part of the STOS project in the Tri-City Academic Supercomputer Centre (Poland).

Acknowledgments. The research was partially funded by the *Cloud Artificial Intelligence Service Engineering (CAISE)* project No. KPOD.05.10-IW.10-0005/24, KPO, European Programme IPCEI-CIS.

Disclosure of Interests. The authors have no competing interests to declare that are relevant to the content of this article.

References

1. Amitha, H., Selvamani, I.: A survey on automatic breast cancer grading of histopathological images. In: 2018 International Conference on Control, Power, Communication and Computing Technologies (ICCPCCT), pp. 185–189 (2018). https://doi.org/10.1109/ICCPCCT.2018.8574291
2. Belharbi, S., Rony, J., Dolz, J., Ayed, I.B., Mccaffrey, L., Granger, E.: Deep interpretable classification and weakly-supervised segmentation of histology images via max-min uncertainty. IEEE Trans. Med. Imaging **41**(3), 702–714 (2022). https://doi.org/10.1109/TMI.2021.3123461
3. Benmabrouk, Y., Gasmi, M., Bendjenna, H., Nadjah, A.: Semantic segmentation of breast cancer histopathology images using deep learning. In: 2022 4th International Conference on Pattern Analysis and Intelligent Systems (PAIS), pp. 1–7 (2022). https://doi.org/10.1109/PAIS56586.2022.9946874
4. Cellina, M., et al.: Artificial intelligence in lung cancer screening: the future is now. Cancers **15**(17) (2023). https://doi.org/10.3390/cancers15174344. https://www.mdpi.com/2072-6694/15/17/4344
5. Chen, J., et al.: TransUNet: transformers make strong encoders for medical image segmentation (2021). https://arxiv.org/abs/2102.04306
6. Cuadros Linares, O., Aurea Soriano-Vargas, A., Faiçal, B.S., Hamann, B., Fabro, A.T., Traina, A.J.: Efficient segmentation of cell nuclei in histopathological images. In: 2020 IEEE 33rd International Symposium on Computer-Based Medical Systems (CBMS), pp. 47–52 (2020). https://doi.org/10.1109/CBMS49503.2020.00017

7. Hussain, H., Hujran, O., Nitha, K.P.: Survey on mitosis detection for aggressive breast cancer from histological images. In: 2019 5th International Conference on Information Management (ICIM), pp. 232–236 (2019). https://doi.org/10.1109/INFOMAN.2019.8714696

8. Kasturi, S., Tran, W.T., Shenfield, A.: Accurate nuclei segmentation in breast cancer tumour biopsies. In: 2022 IEEE Conference on Computational Intelligence in Bioinformatics and Computational Biology (CIBCB), pp. 1–8. IEEE Press (2022). https://doi.org/10.1109/CIBCB55180.2022.9863023

9. Kumar, N., et al.: A multi-organ nucleus segmentation challenge. IEEE Trans. Med. Imaging **39**(5), 1380–1391 (2020). https://doi.org/10.1109/TMI.2019.2947628

10. Laxmisagar, H., Hanumantharaju, M.: A survey on automated detection of breast cancer based histopathology images. In: 2020 2nd International Conference on Innovative Mechanisms for Industry Applications (ICIMIA), pp. 19–24 (2020). https://doi.org/10.1109/ICIMIA48430.2020.9074915

11. Li, C., Lu, X.: Computer-aided detection breast cancer in whole slide image. In: 2021 International Conference on Computer, Control and Robotics (ICCCR), pp. 193–198 (2021). https://doi.org/10.1109/ICCCR49711.2021.9349391

12. Li, Y., Xu, Z., Wang, Y., Zhou, H., Zhang, Q.: SU-Net and DU-Net fusion for tumour segmentation in histopathology images. In: 2020 IEEE 17th International Symposium on Biomedical Imaging (ISBI), pp. 461–465 (2020). https://doi.org/10.1109/ISBI45749.2020.9098678

13. Macenko, M., et l.: A method for normalizing histology slides for quantitative analysis. In: 2009 IEEE International Symposium on Biomedical Imaging: From Nano to Macro, pp. 1107–1110 (2009). https://doi.org/10.1109/ISBI.2009.5193250

14. Natarajan, V.A., Sunil Kumar, M., Patan, R., Kallam, S., Noor Mohamed, M.Y.: Segmentation of nuclei in histopathology images using fully convolutional deep neural architecture. In: 2020 International Conference on Computing and Information Technology (ICCIT-1441), pp. 1–7 (2020). https://doi.org/10.1109/ICCIT-144147971.2020.9213817

15. Rashid, S., Fraz, M., Javed, S.: Multiscale dilated UNet for segmentation of multi-organ nuclei in digital histology images. In: 2020 IEEE 17th International Conference on Smart Communities: Improving Quality of Life Using ICT, IoT and AI (HONET), pp. 68–72 (2020). https://doi.org/10.1109/HONET50430.2020.9322833

16. Robin, M., John, J., Ravikumar, A.: Breast tumor segmentation using U-Net. In: 2021 5th International Conference on Computing Methodologies and Communication (ICCMC), pp. 1164–1167 (2021). https://doi.org/10.1109/ICCMC51019.2021.9418447

17. Ronneberger, O., Fischer, P., Brox, T.: U-Net: convolutional networks for biomedical image segmentation. In: Navab, N., Hornegger, J., Wells, W.M., Frangi, A.F. (eds.) MICCAI 2015. LNCS, vol. 9351, pp. 234–241. Springer, Cham (2015). https://doi.org/10.1007/978-3-319-24574-4_28

18. Samudrala, S., Mohan, C.K.: Semantic segmentation in medical image based on hybrid Dlinknet and UNet. In: 2022 International Conference on Computing, Communication, and Intelligent Systems (ICCCIS), pp. 42–47 (2022). https://doi.org/10.1109/ICCCIS56430.2022.10037693

19. Siegel, R.L., Miller, K.D., Jemal, A.: Cancer statistics, 2019. CA Cancer J. Clin. **69**(1), 7–34 (2019). https://doi.org/10.3322/caac.21551. https://acsjournals.onlinelibrary.wiley.com/doi/abs/10.3322/caac.21551

20. Strudel, R., Garcia, R., Laptev, I., Schmid, C.: Segmenter: transformer for semantic segmentation (2021). https://arxiv.org/abs/2105.05633

21. Subakan, C., Ravanelli, M., Cornell, S., Bronzi, M., Zhong, J.: Attention is all you need in speech separation. In: ICASSP 2021 - 2021 IEEE International Conference on Acoustics, Speech and Signal Processing (ICASSP), pp. 21–25 (2021). https://doi.org/10.1109/ICASSP39728.2021.9413901

22. Sun, S., Yuan, H., Zheng, Y., Zhang, H., Jiang, Z.: Cancer sensitive cascaded networks (CSC-Net) for efficient histopathology whole slide image segmentation. In: 2020 IEEE 17th International Symposium on Biomedical Imaging (ISBI), pp. 476–480 (2020). https://doi.org/10.1109/ISBI45749.2020.9098695

23. Vaswani, A., et al.: Attention is all you need. CoRR **abs/1706.03762** (2017). http://arxiv.org/abs/1706.03762

24. Wazir, S., Fraz, M.M.: HistoSeg: quick attention with multi-loss function for multi-structure segmentation in digital histology images. In: 2022 12th International Conference on Pattern Recognition Systems (ICPRS), pp. 1–7 (2022). https://doi.org/10.1109/ICPRS54038.2022.9854067

25. Yildiz, S., Memiş, A., Varl, S.: Nuclei segmentation in colon histology images by using the deep CNNS: a U-Net based multi-class segmentation analysis. In: 2022 Medical Technologies Congress (TIPTEKNO), pp. 1–4 (2022). https://doi.org/10.1109/TIPTEKNO56568.2022.9960188

26. Yıldırım, Z., Hançer, E., Samet, R., Mali, M.T., Nemati, N.: Effect of color normalization on nuclei segmentation problem in H&E stained histopathology images. In: 2022 30th Signal Processing and Communications Applications Conference (SIU), pp. 1–4 (2022). https://doi.org/10.1109/SIU55565.2022.9864814

27. Zheng, S., et al.: Rethinking semantic segmentation from a sequence-to-sequence perspective with transformers. In: IEEE Conference on Computer Vision and Pattern Recognition, CVPR 2021, virtual, 19–25 June 2021, pp. 6881–6890. Computer Vision Foundation / IEEE (2021). https://doi.org/10.1109/CVPR46437.2021.00681

28. Zou, Y., Chen, S., Sun, Q., Liu, B., Zhang, J.: DCET-Net: dual-stream convolution expanded transformer for breast cancer histopathological image classification. In: 2021 IEEE International Conference on Bioinformatics and Biomedicine (BIBM), pp. 1235–1240 (2021). https://api.semanticscholar.org/CorpusID:245934770

Towards a Model-Driven Approach to Enable Uniform Access to Vector Databases

Elena Akik$^{(\boxtimes)}$ ⓘ, Marko Vješticaⓘ, Vladimir Dimitrieskiⓘ, Slavica Kordićⓘ,
and Sonja Ristićⓘ

Faculty of Technical Sciences, University of Novi Sad, Trg Dositeja Obradovića 6,
21000 Novi Sad, Serbia
{elena,marko.vjestica,dimitrieski,slavica,sdristic}@uns.ac.rs

Abstract. Vector databases have recently gained much attention for managing unstructured data in the era of big data. They efficiently store, retrieve, and analyze data in the vector form, facilitating the processing of vast amounts of data. However, challenges arise due to syntax variations among database vendors, resulting in complexities in querying and migrating between vector databases of different vendors. Additionally, the lack of a standardized language for accessing vector databases necessitates programming skills, as end-users usually access them by using programming languages, hindering user experience. The main goal of our research is to provide end-users with straightforward use of vector databases by proposing a uniform approach across various vector databases. To achieve such a goal, the development of a Model-Driven (MD) solution is proposed, with a novel Domain-Specific Language (DSL) as its core component, aiming to overcome identified challenges. As a domain analysis is a prerequisite for creating a DSL, vector database concepts and terminology are analyzed, and in this paper, the results of such a domain analysis are presented using the Feature-Oriented Domain Analysis (FODA) method, alongside a proposal for the MD solution architecture.

Keywords: Vector Database · Domain-Specific Language · Model-Driven Software Development · Big Data · Machine Learning

1 Introduction

A vector database management system [1] constitutes a specialized subclass of database management systems engineered with a principal focus on the proficient administration of high-dimensional vector data. A vector database is designed to store, retrieve, and analyze data represented in the form of vectors, thus enabling efficient processing and leveraging of large amounts of data [2]. It can be used as a key component in Artificial Intelligence (AI) and Big Data applications, addressing the challenges posed by the intricate nature and vast volume of such data.

Big Data is changing how people live and work nowadays, as data is the key driver of innovation and development in various fields. Based on the research concluded by International Data Corporation (IDC), it is predicted that by 2025, the global datasphere will be about 165 zettabytes of data, whereby 80% of generated data will be unstructured

© The Author(s), under exclusive license to Springer Nature Switzerland AG 2025
J. Tekli et al. (Eds.): ADBIS 2024, CCIS 2186, pp. 225–237, 2025.
https://doi.org/10.1007/978-3-031-70421-5_19

[3]. The challenges of managing vast amounts of data include determining storage solutions and processing methods to understand the data's context and generate knowledge from it [4]. The journey from data to formed decisions is intricate and prolonged, with predictive analytics being crucial in identifying trends and patterns that form consequential decisions. Since humans cannot independently process large amounts of data and generate knowledge based on them in an effective and expeditious manner, Machine Learning (ML) has been used to support decision making for decades [5].

To understand the data context, a predominant methodology involves partitioning data into smaller segments, each meticulously represented numerically. This process culminates in the comprehensive representation of the dataset through a series of numbers, i.e., vectors, wherein akin data shares analogous numerical values in corresponding positions. ML algorithms are pivotal in generating such numerical representations from structured and unstructured data. With data exhibiting diverse types, sizes, and formats, vectors proliferate commensurately with data volume. Consequently, the challenge of effectively storing high-dimensional data emerges, underscoring the preference for vector databases over other NoSQL or traditional relational databases.

As vector databases excel in storing structured and unstructured data in the form of vectors, they enable efficient semantic search, indexing, scalability, and query performance. The utilization of vector embeddings generated by advanced ML algorithms, such as Word2Vec [6] and Bert [7], enriches the functionality of vector databases. These embeddings, encapsulating semantic similarities and intricate relationships, refine data retrieval, similarity search, and clustering operations, thereby elevating the capabilities of vector databases across diverse domains. However, despite the benefits, the complexity of vector database usage poses challenges. One primary challenge lies in mastering the concepts of vector databases because of their complexity and diversity from the end-users' perspective. Furthermore, as there is no standardized language for interacting with vector databases, users mostly access them through programming code, necessitating proficiency in interpreting desired actions within the selected programming language.

When transitioning between vector databases from different vendors, it is necessary for end-users to adapt to the way they access the chosen vector database. Additionally, various terms are used across vector databases to express identical concepts, further complicating the transition process. Acknowledging the need for a system supporting various vector databases, the aim of our research is to enhance communication with vector databases by establishing a novel language for uniform access to various vector databases. The proposed language is planned to be declarative and optimized for interaction with vector databases. This language is anticipated to render the system more accessible to end-users, fostering ease of use and comprehension. In this paper, we propose establishing a system around the language to simplify vector database utilization for users of varying technical expertise, minimize learning curves, and simplify data querying. Ultimately, uniform access to diverse vector databases enables seamless utilization across vendors, eliminating the need for users to adapt to differing concepts and interpretations.

A domain analysis should be conducted before creating a language, and the terminology used should be researched. This paper gives an overview of various concepts that vector databases utilize, summarized through a Feature-Oriented Domain Analysis

(FODA) model. In addition, the proposal for the design and development of the system based on Model-Driven (MD) principles is presented to facilitate uniform access to various vector databases. Further development and implementation will be addressed in subsequent works.

Following Introduction, this paper proceeds as follows. In Sect. 2, the challenges related to accessing data in vector databases are addressed, followed by a brief overview of the MD paradigm. In Sect. 3, vector database concepts are examined, while the proposed MD solution architecture is introduced in Sect. 4. Finally, conclusions are drawn, and future work is outlined in Sect. 5.

2 Background

In this section, the approaches and challenges of accessing data in vector databases are analyzed, primarily in Sect. 2.1. Subsequently, in Sect. 2.2, a brief overview of Model-Driven Software Development (MDSD) and Domain-Specific Languages (DSLs) is provided.

2.1 Accessing Data in Vector Databases

Data-driven ML-based applications pervade nearly all industries nowadays, posing challenges in managing predominantly unstructured data and understanding its context. In addressing these challenges, vector databases emerge as a favored solution due to their ability to store structured and unstructured data in the form of vectors.

Vector databases store data in the form of vectors through an approach formally called vectorization [8]. This process involves representing data context numerically as n-dimensional vectors, where each dimension value signifies a feature or part of a feature. Analogous data exhibits similar numerical values in designated vector positions, thereby simplifying intricate data analysis and knowledge generation processes. This approach hinges upon comparing vector similarity, utilizing vector distance metrics.

Despite the advantages vector databases provide, there are challenges when using them. Syntaxes of languages for accessing vector databases vary depending on a database vendor, resulting in complexities in querying and migrating between vector databases of different vendors. There is no standardized or universally accepted language for uniform access to vector databases of different vendors. Accessing vector databases typically entails programming, forcing users to navigate various technologies and interpret actions within specific frameworks. Most software engineers are familiar with relational databases and their standardized declarative Structured Query Language (SQL) aimed to ease database access. Given that SQL offers a straightforward method for managing relational databases, there is a preference for a declarative language that similarly facilitates a convenient approach to vector databases. Albeit such an approach brings potential hazards due to the existence of concepts used both in relational and vector databases, but with different meanings, as discussed in Sect. 3.2. If an approach to vector database pretends to name concepts as they are in SQL, such a syntax can be difficult to understand, prolonging the mastery of concepts and complicating database management tasks from the end-user point of view. The need to create such a language may be caused by the absence of a language that would allow the vector database to be accessed in a straightforward manner. When creating a new language, the attention needs to be paid to the naming of concepts, so end-users can be focused on the very actions with vector databases, and not on the syntax they use. The syntax of language used to access a database needs to be clear, concise, and easy to learn.

A DSL aimed at specifying tasks over various vector databases uniformly is needed to ensure consistent and uniform access to vector databases. The initial step prior to creating a DSL is to establish unique terminology for the concepts employed.

2.2   A Brief Overview of MDSD and DSLs

Model-Driven Development (MDD) is a paradigm that uses models as the primary artifact of the development process, typically involving the automated or semi-automated generation of implementation artifacts from abstract models [9]. Model-Driven Software Development (MDSD) represents a concretization of the MDD paradigm, in which models are used to lead the process of software system development. Being the core of MDSD, DSLs are used to create various models and transformation rules are applied to transform models into other ones or into executable programing code.

DSLs are designed specifically for a certain domain, context, or company, allowing users to specify a solution with familiar concepts. If the language is aimed at modeling, it may also be referred to as a Domain-Specific Modeling Language (DSML). General-Purpose Languages (GPLs) can be applied to any sector or domain, but users may need more time and expertise to specify a solution in the specific domain [10].

The absence of uniform formal language for working with vector databases, as described in Sect. 2.1, necessitates the development of a novel MD solution to establish a standardized approach for interacting with vector databases. Accordingly, in search of an efficient method to address this need, a new DSL aimed at providing a uniform and simplified way to work with various vector databases was set out to be created.

Before introducing a proposal for a dedicated DSL, it is necessary to thoroughly analyze the utilization of vector databases. This analysis encompasses an examination of the fundamental concepts employed, as well as the approaches for accessing vector databases through supported technologies and programming languages. Consequently, in Sect. 3, a comprehensive overview of the concepts and approaches for accessing vector databases is provided.

3   An Overview of Vector Database Concepts

With the aim of the comprehensive understanding and effective utilization of vector databases, detailed examination of the concepts and approaches for accessing vector databases is presented in this section. A clearer understanding of the underlying logic among end-users is facilitated by the examination, fostering increased familiarity with these systems. In Sect. 3.1, a general overview of the workflow with vector databases is given. Subsequently, in Sect. 3.2 and Sect. 3.3, the concepts used when working with vector databases are explored and analyzed in detail. The exploration is conducted on several vector databases widely used at the time of the research: *Pinecone* [11], *Milvus* [12], *Chroma* [13], *Weaviate* [14] and *Qdrant* [15].

3.1   Vector Database Workflow

In general terms, vector databases serve as repositories for the output of embedding model algorithms, housing vectors, referred to as vector embeddings within the domain of vector databases. They also store associated meta-data, thereby facilitating efficient storage and retrieval processes. The database expedites similarity searches by preserving these embeddings, adeptly aligning user queries with corresponding vector embeddings. Vector databases employ ML algorithms to index vector embeddings, mapping them to specialized data structures optimized for swift similarity or distance searches, notably Approximate Nearest Neighbor (ANN) searches.

Querying vectors entails calculating distances between vectors, with similarity metrics, such as the cosine similarity, often employed to gauge the proximity of vectors. A Vector DataBase Management System (VDBMS) computes distance and conducts similarity calculations between query vectors and those stored in the vector database, subsequently furnishing the most akin vectors or nearest neighbors in accordance with similarity rankings. These computations underpin various ML tasks, encompassing recommendation systems, semantic search, image recognition or Natural Language Processing (NLP) endeavors.

3.2   Vector Database Concepts

In this section, universally applicable definitions of core concepts are aimed to be offered, albeit expressed in varied terminologies. This analysis of terminology enhances comprehension and facilitates comparison of fundamental principles central to the operation of vector databases.

In Table 1 is presented terminology of the concepts of analyzed vector databases. For each concept, an insight is given into the terminology used to express it, depending on the vendor of a vector database in which the term is used. The last column of Table 1 represents the term used to denote a particular concept that could be used when constructing a new DSL. Based on the terminology employed across various vector databases, the terms that most accurately describe the concept and distinguish it from others have been selected for the purposes of this work.

Working with vector databases entails the definition of the database structure, the manipulation of data, and the execution of data queries, including both filtering data and vector search. Data definitions and data structures are described in Sect. 3.2.1, while querying and vector search methods are described in Sect. 3.2.2.

3.2.1   Data Structures

The interaction workflow with a vector database unfolds through a series of steps. At the outset, end-users should undertake preparatory measures before data insertion into the vector database, delineating the organization of data and specifying a model to govern its representation within the database.

In support of formulating this organizational model, the *schema* concept emerges as pivotal, serving to elucidate correlations between data and objects within the database framework. Moreover, the schema assumes multifaceted utility, extending beyond mere collection definition to encompass the delineation of fields within the database structure. It should be mentioned that schema is not used in all vector databases, but here it is singled out as a concept as it can potentially contribute to a better organization of data within the database.

In vector databases, the concept of a *collection* is analogous to the relation, i.e., table, in a Relational DataBase Management System (RDBMS), serving as a repository for organizing and managing entities. When creating a collection in certain vector databases, it is necessary to establish a predefined schema outlining its structure, including the fields within and additional constraints applied at the collection level. Likewise, it is noteworthy that in certain VDBMSs, the term index is used instead of the term collection, potentially posing a challenge for users navigating between relational and vector database environments due to the nuanced distinctions in terminology. Furthermore, within the realm of vector databases, the term index carries a unique connotation, as delineated below.

In vector databases, a *limit* refers to a rule or condition that can be defined at the cluster, collection, partition, or field level. The introduction of this concept exerts a

Table 1. The terminology used for concepts in different vector databases.

| | | VDBMS | | | | Chosen term |
	Pinecone	Milvus	Chroma	Weaviate	Qdrant	
Data structures	-	Schema	-	Schema	-	*Schema*
	Index	Collection	Collection	Collection	Collection	*Collection*
	Meta-data	Limit, Constraint	Meta-data	Meta-data	Meta-data	*Limit*
	Meta-data	Field	Document	Property	Payload	*Field*
	Meta-data	Entity	-	Data object	Point	*Entity*
	Vector	Field	Embedding	Vector	Vector	*Embedding*
	Namespace	Partition	-	-	-	*Partition*
	Pod	Shard	-	Shard	Shard	*Shard*
	Cluster	Cluster	Cluster	Cluster	Cluster	*Cluster*
	Pod-index	Vector index	Index	Vector index	Vector index	*Vector index*
	Pod-index	Scalar index	-	Inverted index	Payload index	*Scalar index*
Vectori-zation	Vector	Embedding model	Embedding model	Vectorizer	Model	*Embedding model*
Querying	Metric	Metric	Meta-data	Metric	Metric	*Metric*
	Query	Vector search	Query	Vector search	Vector search	*Vector search*
	Filtering	Filtering	Query	Filtering	Filtering	*Filtering*

significant influence on both data integrity and query optimization processes across the entire database system.

The *field* concept refers to a specific attribute of an entity in a collection. The field may be either vector or non-vector, with non-vector fields accommodating scalar data types. In contrast, vector fields are represented by a one-dimensional array of numbers, generated by ML algorithm, where context of data is represented through each vector dimension. Schemas can also describe the structure of fields within a collection, including defining properties such as field names, data types, and other field constraints like automatic primary key allocation.

An *entity* is represented by a set of fields within a collection, each corresponding distinct entity within the real world. Each entity is uniquely identified by a primary key.

Embedding, a concept rooted in ML, entails mapping data into a high-dimensional space where data exhibiting similar semantic attributes are positioned in proximity. Embedding is stored as the value of the vector fields within the entity.

In vector databases, *partitioning*, a fundamental process involving the division of a collection to establish partitions within physical storage, is a commonly adopted strategy to optimize data organization and access efficiency. This strategy is facilitated through processes like range partitioning, where vector data is allocated based on key column value ranges, and list partitioning, which assigns vector data based on predefined value lists. Notably, it is to be observed that while partitioning is used to optimize the management of vector fields within databases, its applicability to non-vector fields may vary, with some databases demonstrating the capacity for partitioning solely in relation to vector data. Previously described partitioning strategies lead to efficient data distribution, improved performance, scalability and usability of VDBMS.

Sharding encompasses the horizontal partitioning of data across multiple servers or nodes, with each designated shard containing a distinct subset of the overall dataset. By distributing workload across these shards, sharding facilitates parallel processing and alleviates the burden on individual nodes, thereby augmenting the system's overall efficiency and scalability.

A *cluster* typically refers to a logical grouping of similar or related entities in vector databases. This clustering mechanism organizes data points based on their proximity or similarity in feature space, facilitating efficient data retrieval and analysis. Clusters in vector databases serve various purposes, including enhancing search performance, enabling data compression, and supporting data analytics tasks, such as clustering algorithms and nearest neighbor searches. They play a crucial role in database optimization and improving overall system performance.

Indexing procedures are employed to expedite query processing after the vectorization and storage of entities within a vector database. When describing the collection concept, it was noticed that the term *index* is occasionally used as a synonym. However, it is necessary to emphasize that, within this research, the term index is not used as a synonym for collection. Instead, it has a distinct contextual meaning, conforming to its conventional usage in most vector databases. *Scalar indexes* typically facilitate exact match queries and are utilized for indexing non-vector fields. In contrast, *vector indexes* utilize embeddings to enable approximate match searches, enhancing efficiency in exploring large datasets and indexing vector fields. These indexes, constructed based on a chosen similarity metric and methodology, rely on ANN algorithms for querying. A vector index, a data structure derived from raw data, significantly accelerates the process of vector similarity search by organizing the vectors in an optimized manner.

3.2.2 Vectorization and Querying

At the core of operations within the vector database lies the utilization of *embedding models,* which transform raw data into dense vector representations stored as embedding vectors within the database. Typically implemented through Deep Neural Networks (DNNs), such as those derived from BERT or other Transformer families, these embedding models proficiently capture the semantics of various data types, including text and images, by representing them with numerical vectors.

Distances between vectors within a vector space are computed by utilizing distance metrics customized to suit the specific demands of the problem under consideration. *Metrics,* such as Euclidean, Manhattan, cosine, and Chebyshev measures, find prevalent utilization within vector databases, particularly in tasks like vector similarity search. This search mechanism aids in identifying vectors that closely resemble the target vector, thereby facilitating efficient retrieval of pertinent data. Unlike relational databases that primarily support queries based on exact matches or predefined criteria, vector databases offer advanced functionalities for discerning similar or relevant data by assessing vector distance or similarity.

In addition to the core functionality of *vector search,* which serves as the primary feature of vector databases, the capacity to query non-vector data, akin to relational databases, is also viable, designated as *filtering.* This process typically operates autonomously from vector search. However, certain implementations of vector databases restrict filtering capabilities solely within the vector search framework. An alternate term synonymous with this operation is query. Besides vector search and filtering, a VDBMS offers standard operations to manage data in a vector database. These operations include inserting, updating, and deleting data from a vector database, as well as creating, altering and dropping database structures.

Based on a thorough examination of the concepts prevailing in vector databases, the Feature-Oriented Domain Analysis (FODA) model is formulated and presented in Fig. 1, by using Yet Another Feature Modeling Tool (YAFMT) [16]. The FODA method is used for domain analysis and identifying features and their relationships to support software development [17]. The cardinality-based FODA method [18], which represents an extension of standard FODA and supports feature and group cardinalities, has been used in the presented model of vector database concepts.

3.3 Accessing Vector Databases

As the aim of our research is to create a uniform way of working with vector databases, approaches of accessing vector databases are analyzed, which are mostly based on access through program code. Thus, it is explored which programming languages are supported in vector databases, as well as whether they have their own query language. The exploration showed that the most supported programming languages in the researched vector databases are *Python, Java, Node.js, JavaScript/TypeScript* and *Go*. Furthermore, all analyzed vector databases support access through *REST API*. However, no query language was encountered within the analyzed databases. In the following section, a proposed architecture of an MD solution is presented. It includes both a novel approach and a software system, aiming to enable a uniform way of working with vector databases, regardless of their vendor, and outlines access via DSL as a key component. The given solution relies on a DSL that utilizes concepts outlined within presented FODA model.

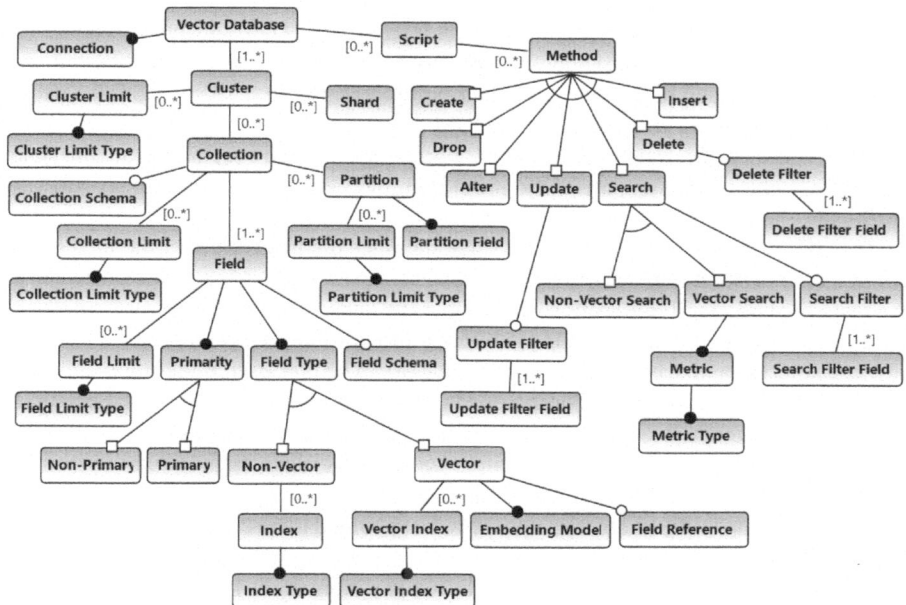

Fig. 1. An FODA model of vector database concepts.

4 A Proposal of an MD Solution for Vector Database Access

In Sect. 4.1, an architecture of a MD solution, which integrates both a novel approach and a software system, is proposed. In the following Sect. 4.2, Design Science Research Methodology (DSRM) [19], on which our research is based, is outlined, with a focus on six basic activities that follow the current and future activities of our research.

4.1 An Architecture of a Solution for Vector Database Access

In this section, an architecture of a solution based on MD principles is proposed to cope with the challenges discussed in this paper, such as: (i) complexity and diversity of vector database concepts; (ii) variety of terminology in different vector databases; (iii) absence of a standardized language for vector databases; (iv) required programming skills for end-users to access vector databases; and (v) difficulty in migrating between different vector databases.

The solution needs to contribute to facilitating a uniform access to vector databases, regardless of the database vendor. Such uniformity is important for end-users across diverse domains, including IT specialists and individuals.

The development of the MD solution is planned in two stages, as presented in Fig. 2. The first development stage serves as a proof-of-concept in which a novel DSL and code generators will be developed and tested by software engineers. They will use the text editor that supports the usage of the DSL, equipped with code validators, to write scripts containing different statements regarding vector databases. These scripts will be sent to code generators that will automatically transform the scripts into executable code for the chosen vector database. That code can be represented in a specific programming language or a language of a particular VDBMS.

However, software developers would not prefer such an approach. Instead, the statements should be written as part of the programming code in the environment software developers choose.

Therefore, the second development stage includes the integration of the novel DSL within a code written in programming language supported in a chosen vector database. The main goal is to facilitate software developers by ensuring that an integrated approach is used, eliminating the need for the environment to be changed when programming the software solution and accessing the vector database. The proposed integration approach is to form a built-in module or library, or to implement integration through a parser. In addition, data migration from the source database of various types to the target vector database is planned, with relevant features selected and data transformation into the target vector database defined by using our DSL.

To develop the DSL, an abstract syntax is first planned to be created by using a meta-meta-model, such as the Ecore language, which is part of Eclipse Modeling Framework (EMF) [20], and a set of rules expressed by Object Constraint Language (OCL) [21]. The DSL aims to incorporate all the specified concepts presented in Fig. 1. Afterward, it is planned to create a declarative textual syntax, which can be done by using Xtext [22]. The main reason for choosing a textual syntax over a graphical one is that the proposed DSL will primarily be used by software engineers familiar with programming languages and textual syntax and thus may better fit their needs. As the final step of the first development stage, code generators are intended to be developed using Xtend [22]. The purpose of code generators is to transform the resulting models into appropriate scripts or program code for the selected vector database. This enables different vector databases to be used easily without requiring end-users to be dependent on the choice of vector database they use. Another advantage is the possibility of testing various vector databases in the same way by setting the same query for different databases to analyze the obtained results relatively quickly. Additionally, as part of the second development stage, integration of the DSL within the Python programming language is intended, considering its frequent usage in ML algorithms and embedding models.

The proposed solution aims to streamline vector database usage for end-users, minimizing the learning curve associated with vector database concepts and facilitating

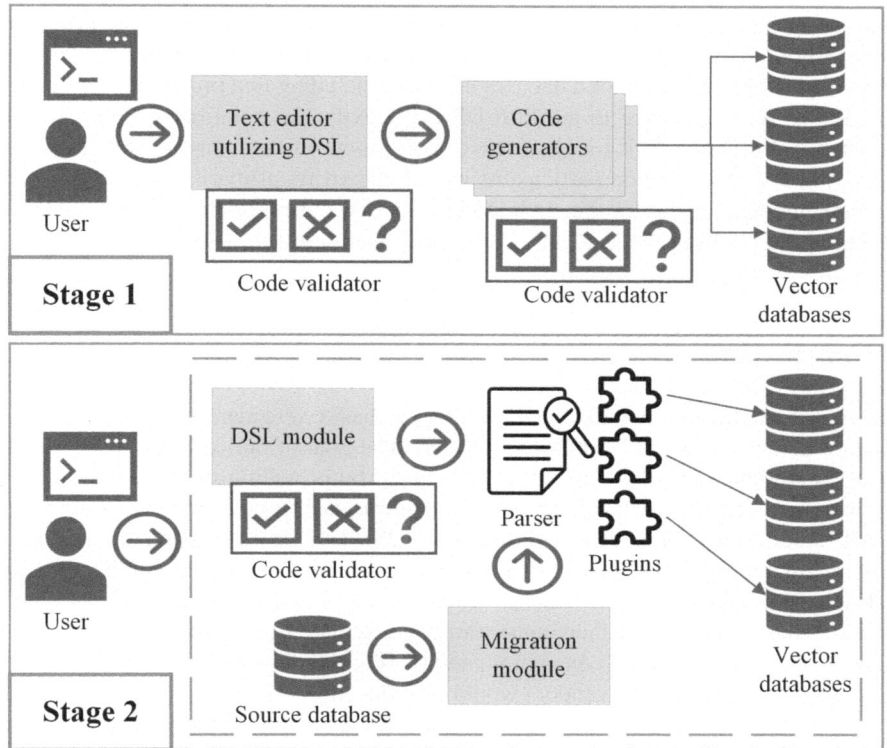

Fig. 2. The architecture of MD solution with development stages.

simpler data querying. Consequently, an approach to accessing diverse vector databases uniformly enables users to seamlessly utilize databases from various vendors.

4.2 Research Methodology

Our research is conducted based on DSRM, formulated on six guiding activities. The first activity pertains to identifying problems and challenges that require resolution. The second activity involves establishing goals that the proposed solution should achieve. The third activity encompasses the design and development of the solution, while the fourth activity pertains to the demonstration of the formulated solution in specific use cases. The fifth activity involves the evaluation of the solution, and the sixth activity concerns presenting the results.

Challenges with the lack of a uniform way of working with vector databases have been identified as part of the first DSRM activity, while objectives that our solution needs to achieve have been defined as part of the second DSRM activity. The third DSRM activity includes the design and development of the solution, where the architecture of the solution prototype has been presented as a part of this paper, while the system implementation will be carried out in the two proposed stages. The development of the first phase is currently underway, and the solution is planned to be presented in the future. As part of the fourth DSRM activity, our solution will be applied with various use cases where vector databases are used, including text processing and image recognition. With reference to the fifth DSRM activity, our solution is planned to be evaluated using Framework for Qualitative Assessment of Domain-specific languages (FQAD) [19] and evaluation participants would be experts primarily oriented towards software engineering. As for the sixth DSRM activity, our research results are planned to be presented both in the academic community and industry, with this paper being a part of this activity.

5   Conclusion

In this paper, challenges in working with vector databases were identified and the domain of vector database core concepts was analyzed. The goal of our research is to provide end-users straightforward use of vector databases, by providing them with a uniform way to access various vector databases. To achieve such a goal, the development of the MD solution was proposed, with the novel DSL as its core component, aiming to overcome identified challenges. In that manner, end-users can have several benefits: (i) work with the selected vector database would be simplified; (ii) time needed to learn vector database concepts and their context would be reduced; (iii) testing different vector databases would be less time-consuming; and (iv) complexity of the migration between vector databases of different vendors would be reduced.

As part of future work, it is planned to extend the analysis of vector database concepts and terminologies, by examining additional vector databases, with the aim of creating a solution that will cover a wide range of vector databases. Furthermore, the proposed solution will be iteratively developed following the two stages of development, applied in various use cases, and evaluated to gain insights into the system's advantages and disadvantages.

Acknowledgments. This research has been supported by the Ministry of Science, Technological Development and Innovation (Contract No. 451–03-65/2024–03/200156) and the Faculty of Technical Sciences, University of Novi Sad through project "Scientific and Artistic Research Work of Researchers in Teaching and Associate Positions at the Faculty of Technical Sciences, University of Novi Sad" (No. 01–3394/1).

References

1. Pan, J.J., Wang, J., Li, G.: Vector database management techniques and systems. In: Companion of the 2024 International Conference on Management of Data, pp. 597–604. ACM, New York, NY, USA (2024). https://doi.org/10.1145/3626246.3654691

2. Taipalus, T.: Vector database management systems: fundamental concepts, use-cases, and current challenges. Cogn. Syst. Res. **85**, 101216 (2024). https://doi.org/10.1016/j.cogsys.2024.101216

3. Altan, Z., ed.: Applications and Approaches to Object-Oriented Software Design: Emerging Research and Opportunities. IGI Global (2020). https://doi.org/10.4018/978-1-7998-2142-7

4. Naimi, A.I., Westreich, D.J.: Big data: a revolution that will transform how we live, work, and think. Am. J. Epidemiol. **179**, 1143–1144 (2014). https://doi.org/10.1093/aje/kwu085

5. L'Heureux, A., Grolinger, K., Elyamany, H.F., Capretz, M.A.M.: Machine learning with big data: challenges and approaches. IEEE Access **5**, 7776–7797 (2017). https://doi.org/10.1109/ACCESS.2017.2696365

6. Church, K.W.: Word2Vec. Nat. Lang. Eng. **23**, 155–162 (2017). https://doi.org/10.1017/S1351324916000334

7. Devlin, J., Chang, M.-W., Lee, K., Toutanova, K.: BERT: pre-training of deep bidirectional transformers for language understanding. In: Proceedings of the 2019 Conference of the North American Chapter of the Association for Computational Linguistics: Human Language Technologies, pp. 4171–4186. Association for Computational Linguistics (2019). https://doi.org/10.18653/v1/N19-1423

8. Wang, J., et al.: Milvus: a purpose-built vector data management System. In: Proceedings of the 2021 International Conference on Management of Data, pp. 2614–2627. ACM, Virtual Event China (2021). https://doi.org/10.1145/3448016.3457550

9. Brambilla, M., Cabot, J., Wimmer, M.: Model-Driven Software Engineering in Practice, 2nd edn. Morgan & Claypool Publishers, San Rafael, CA, USA (2017)

10. Mernik, M., Heering, J., Sloane, A.M.: When and how to develop domain-specific languages. ACM Comput. Surv. **37**, 316–344 (2005). https://doi.org/10.1145/1118890.1118892

11. Pinecone. https://www.pinecone.io/. Accessed 20 Nov 2023

12. Milvus. https://milvus.io/. Accessed 13 Jun 2024

13. Chroma. https://www.trychroma.com. Accessed 13 Jun 2024

14. Weaviate. https://weaviate.io/. Accessed 13 Jun 2024

15. Qdrant. https://qdrant.tech/. Accessed 13 Jun 2024

16. Bossert, A., Pikl, J.: Yet Another Feature Modeling Tool (YAFMT). https://github.com/anb0s/YAFMT. Accessed 08 May 2024

17. Kang, K.C., Cohen, S.G., Hess, W.E., Peterson, A.S.: Feature-Oriented Domain Analysis (FODA) Feasibility Study. Carnegie Mellon University, Software Engineering Institute (1990)

18. Czarnecki, K., Helsen, S., Eisenecker, U.: Formalizing cardinality-based feature models and their specialization. Softw. Process Improv. Pract. **10**, 7–29 (2005). https://doi.org/10.1002/spip.213

19. Kahraman, G., Bilgen, S.: A framework for qualitative assessment of domain-specific languages. Softw. Syst. Model. **14**, 1505–1526 (2015). https://doi.org/10.1007/s10270-013-0387-8

20. Steinberg, D., Budinsky, F., Paternostro, M., Merks, E.: EMF: Eclipse Modeling Framework. Addison-Wesley Professional, Upper Saddle River, NJ (2008)

21. Warmer, J., Kleppe, A.: The Object Constraint Language: Getting Your Models Ready for MDA. Addison-Wesley, Boston, MA, USA (2003)

22. Bettini, L., Efftinge, S.: Implementing domain-specific languages with Xtext and Xtend. Packt Publishing, Birmingham Mumbai (2016)

Employing Multiple Online Translation Services in a Multilingual Database Design Tool

Danijela Banjac[1]([✉]), Milica Matic[2], Nedeljko Cvijanovic[3], Drazen Brdjanin[1],
Goran Banjac[1], and Djordje Stojisavljevic[4]

[1] Faculty of Electrical Engineering, University of Banja Luka Patre 5,
78000 Banja Luka, Bosnia and Herzegovina
{danijela.banjac,drazen.brdjanin,goran.banjac}@etf.unibl.org
[2] Codaxy Ltd, Bulevar srpske vojske 17, 78000 Banja Luka, Bosnia and Herzegovina
milicama.995@gmail.com
[3] Ongulus Software Development, Svetozara Markovica 5, 78000 Banja Luka,
Bosnia and Herzegovina
nedeljko97.cvijanovic@hotmail.com
[4] University Computing Center, University of Banja Luka Bulevar Vojvode Petra
Bojovica 1A, 78000 Banja Luka, Bosnia and Herzegovina
djordje.stojisavljevic@unibl.org

Abstract. TexToData is an online tool for automated database design
with multilingual support. It enables the automated derivation of an
initial conceptual database model from textual specifications in different natural languages, whereby more than a hundred natural languages
are supported. The multilingualism is enabled by the translation of a
source textual specification into the English language and vice versa.
For instance, TexToData first translates the source text into English,
then processes the English text, generates the corresponding conceptual
model in English, and finally translates the model back into the source
natural language. In this paper we analyze the suitability of employing
different online translation services in this process.

Keywords: Automatic conceptual database design · Multilingual
tool · Online translation service · TexToData

1 Introduction

The database design process undergoes several typical steps [9], whereby the
first and the most important step is conceptual design. Conceptual design
results in the *conceptual database model* (CDM) – a platform-independent model
that provides data descriptions at a high level of abstraction. The result of
the subsequent steps can be automatically derived starting from the CDM by
applying straightforward transformation rules. Ultimately, almost everything up
to the database specifications for the specific target *database management system*
(DBMS) can be generated starting from the CDM. Therefore, many research

J. Tekli et al. (Eds.): ADBIS 2024, CCIS 2186, pp. 238–249, 2025.
https://doi.org/10.1007/978-3-031-70421-5_20

publications focus on conceptual design, with a particular emphasis on automatic CDM creation.

The existing approaches typically take textual specifications [21] or models [2] as a source for the automatic CDM derivation. According to [13,15], about 90% of all the requirements in industrial practices are written in *natural language* (NL). Considering all of this, there is a significant potential demand for the tool that can automatically transform NL text into the CDM. Research in *natural language processing* (NLP) started in the 1940 s, but in the last decade it has gained popularity. The first rules for translation of English sentences into E-R diagrams were established in the 1980 s [6], and since then a lot of research [21] has been done in the field of NLP to extract knowledge from requirements specifications and automate CDM design.

Although there has been ongoing research for years, most text-based tools typically support one single source NL (mainly English) and do not provide multilingual support. Only the TexToData tool [5] enables automatic CDM derivation from textual specifications in different source NLs, with more than a hundred supported NLs. TexToData analyzes the source text and detects the source NL, and in case the source NL is not English, forwards the text to an external translation service. The TexToData tool, presented in [5], uses one online translation service (Yandex), which represents a bottleneck of the entire process.

In this paper, we present the most recent improvements of TexToData that enable automated CDM derivation with reduced dependence on a particular external translation service. We have added support for five of the most commonly used online translation services and evaluated their impact on the effectiveness of CDM derivation.

The paper is structured as follows. After this introduction, the second section presents the related work. An overview of the pre-existing TexToData tool is presented in the third section, while the most recent improvements are presented in the fourth section. The fifth section presents the evaluation. The final section concludes the paper.

2   Related Work

In this section, we provide an overview of the (semi-)automatic CDM design approaches, with a particular focus on the text-based approaches.

The existing approaches to (semi-)automatic CDM design can be classified as (Fig. 1): (1) *text-based*, (2) *model-based*, (3) *form-based*, and (4) *speech-based*.

Text-based approaches derive CDMs from textual specifications that are typically unstructured and represented in some NL. NL can be ambiguous – sentences and phrases that potentially have two or more possible interpretations, therefore an effective and accurate analysis is difficult. Furthermore, the same words and phrases can have different meanings according to the context of a sentence. Nevertheless, text-based approaches are the oldest and most dominant category of the (semi-)automatic CDM design approaches. They can be further classified (as suggested in [25]) as: (1) *linguistics-based*, (2) *pattern-based*, (3) *case-based*, (4) *ontology-based*, and (5) *multiple approaches*.

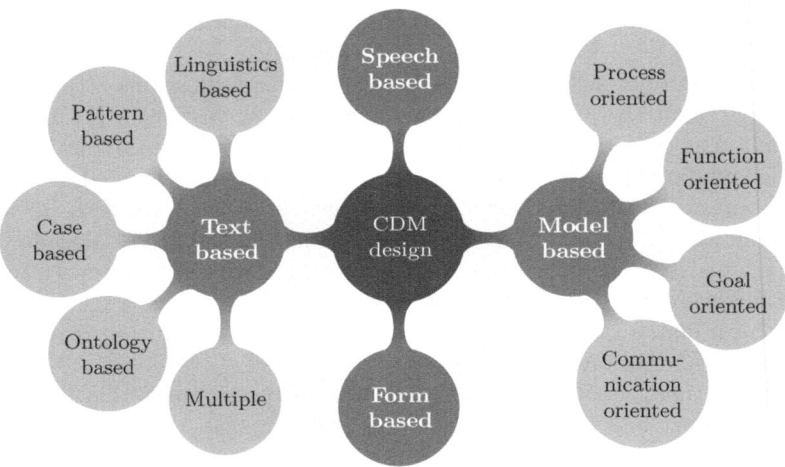

Fig. 1. Taxonomy of approaches to (semi-)automatic CDM design

The majority of text-based approaches and tools are linguistics-based. NLP techniques are used to convert NL text into the CDM. The most important linguistics-based tools are: ER-Converter [17], CM-Builder [10], and LIDA [18]. Our TextToData tool also belongs to this category. The main representatives of other categories are: pattern-based APSARA [19], case-based CABSYDD [8], ontology-based OMDDE [23], and HBT [25] belonging to the multiple group.

Model-based approaches appeared as an alternative to the text-based approaches, in order to avoid their shortcomings mainly related to he modest effectiveness for languages with complex morphology. The source models in the existing approaches are represented by more than 20 different graphical notations, which can be classified as: (1) *process-oriented* (e.g. BPMN), (2) *function-oriented* (e.g. Data Flow Diagram), (3) *communication-oriented* (e.g. Sequence Diagram), and (4) *goal-oriented* (e.g. TROPOS). There are no tools enabling automatic synthesis of the complete target CDM from a source model regardless of the notation. The vast majority of papers present guidelines and informal rules that do not enable automatic CDM derivation. Only a small number of papers present a set of formal rules for automatic CDM derivation (e.g. [1,24]). Most of the proposed tools are actually transformation programs (e.g. [12,20]) specified in some model-to-model transformation language (e.g. ATL [11]), while only a small number of papers present real CASE tools for automatic model-driven CDM synthesis (e.g. [16]). Here we mention the AMADEOS tool [3,22] since it shares some services with TextToData – it is the first online model-driven tool enabling the automatic CDM derivation from a set of business process models, as well as the complete forward database engineering process.

Form-based approaches take collections of forms as the source for CDM derivation. The most important tools are EDDS [7] and IIS*Case [14].

Speech-based approaches take speech as the source for CDM derivation. There is only one paper [4] presenting the SpeeD tool for speech-based database design.

3   TexToData

This section briefly[1] presents the TexToData[2] tool. TexToData is the first online web-based tool for automated database design with multilingual support. It enables automatic derivation of an initial CDM from textual specifications in different NLs, whereby more than a hundred NLs are supported. The entire process of CDM derivation is implemented as an orchestration of a set of internal and external web services. The external services are employed to carry out the core functionalities, such as text translation from the source NL into English and vice versa (to enable multilingual support).

Apart from the automatic CDM derivation, TexToData also enables other phases of the forward database engineering process, i.e. automatic CDM conversion to the corresponding relational model followed by the automatic generation of the DDL[3] script and creation of the physical database schema in the selected DBMS. Figure 2 shows a part of the system's architecture that enables automatic CDM derivation.

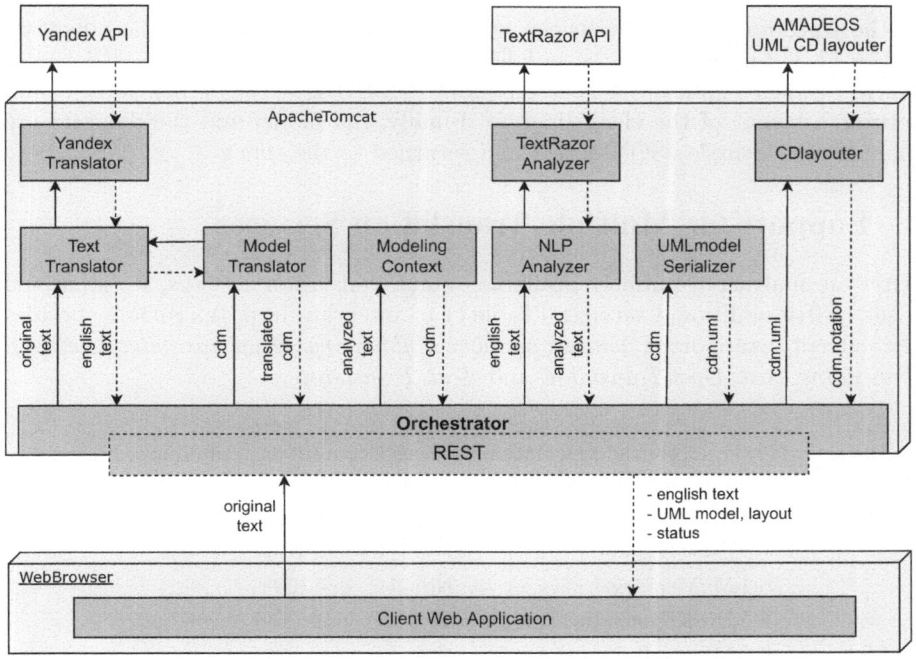

Fig. 2. A part of TexToData that enables automatic CDM derivation [5]

---

[1] For a detailed description we refer the readers to [5].

[2] http://m-lab.etf.unibl.org:8080/Textodata.

[3] Data Definition Language.

The *client web application* allows users to upload a source text and start the CDM derivation process. When the entire process is finished, the client application receives the JSON[4] response and visualizes the generated CDM in the browser.

The *server-side* is implemented as a set of web services. The *Orchestrator* service orchestrates the whole process, whereby each activity is implemented by the corresponding service. In a positive usage scenario, the orchestrator receives a source text and returns the automatically generated CDM. The source text is firstly sent to the *TextTranslator* service which detects the source NL and, in case the source NL is not English, forwards the text to the external translation service (the pre-existing tool was employing the *Yandex* service). The orchestrator further sends the English text to the *NLPAnalyzer* service responsible for NLP, which employs an external NLP service via the corresponding adapter – currently the *TextRazor* service is employed. After NLP is finished, the analyzed text is sent to the *ModelingContext* service which generates an internal CDM representation. If the source NL is not English, then the CDM is sent to the *ModelTranslator* service, which further employs the *TextTranslator* service to translate each model element back into the source NL. The CDM is further sent to the *UMLmodelSerializer* service which serializes the generated class diagram in the XMI[5] format. After the serialization, the model is sent to the *CDlayouter* service, which employs the corresponding *AMADEOS layouting service* and returns a layout of the class diagram. Finally, the model and the diagram are merged into a single JSON object and returned to the client.

4   Support for Multiple Translation Services

After an analysis of publicly available online translation services, we identified a set of five additional services (Table 1) to integrate into TexToData, besides the currently supported *Yandex* service: *MyMemory, DeepTranslate, Microsoft Translator Text, OpenTranslator*, and *Text Translator*.

Table 1. Newly supported translation services (from https://rapidapi.com/hub)

Online translation service		Automatic	Number of
ID	Service name	NL detection	supported NLs
S-1	MyMemory	No	100+
S-2	DeepTranslate	Yes	~120
S-3	Microsoft Translator Text	Yes	~130
S-4	OpenTranslator	Yes	~105
S-5	Text Translator	Yes	100+

---

[4] JavaScript Object Notation.
[5] XML MetaData Interchange.

All analyzed online services expose REST API[6] and return the result as a JSON object. In case of successful processing, the returned object contains the translated text. Otherwise, an error message with the corresponding status code is received. Although all analyzed services provide very similar functionalities, the obtained results are not the same. Considering the precision and quality of the translated text, there are differences, primarily in the translation style. Some services offer a more literal translation of the source text than others. All analyzed services offer different subscription plans, limiting the number of requests/characters sent to the services on a monthly basis. In this paper we analyze only free (basic) subscription plans for all services.

Figure 3 shows the extended architecture of TexToData. Only parts relevant to the support of multiple translation services are shown, while other system components are not modified. The improved tool contains additional implementations of the *TextTranslator* interface, whereby each implementation communicates with the corresponding external service to translate text from the source NL to English, and after the CDM is created, to translate it from English back to the source NL. In the pre-existing tool, there was only one implementation of this interface – *YandexTranslatorAdapter*, which communicates with the external *Yandex* service. Each implementation of the *TextTranslator* interface provides methods for the text translating and detection of the input NL, whose implementation depends on the corresponding external service.

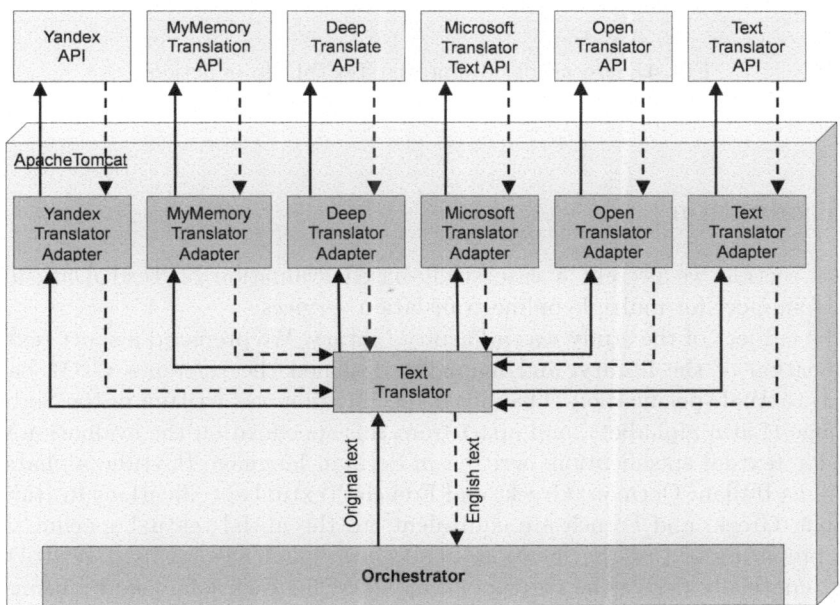

Fig. 3. A part of the TexToData extended architecture for multilingual support

---

[6] REST API (also called RESTful API) is an API that conforms to the design principles of the *representational state transfer* (REST) architectural style.

The client application (Fig. 4) also has been altered to enable the user to select one of the supported translation services. If the selected service does not support automatic NL detection, the user can select a source NL from a list of supported languages for the given service.

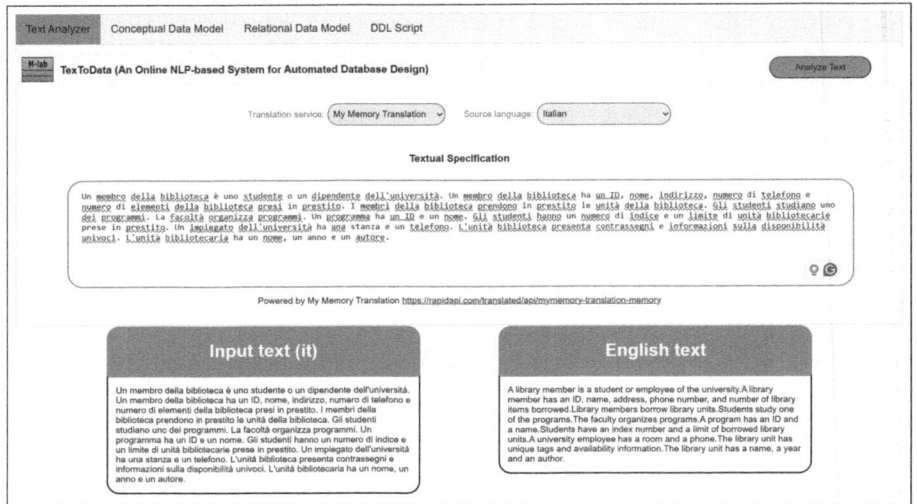

Fig. 4. Screenshot of improved TexToData in action

5   Evaluation

In this section we present a case study-based evaluation of TexToData after adding support for multiple online translation services.

The subject of the study was a Faculty Library. We prepared a short textual specification of the library, and manually designed the *reference CDM* based on this textual specification. The initial specification was written in the Serbian language (Latin alphabet), and apart from this specification the evaluation was done for textual specifications written in Serbian language (Cyrillic alphabet), as well as Italian, German, Greek, and French. Textual specifications in Italian, German, Greek, and French are equivalent[7] to the initial textual specification. After preparing textual specifications in all source languages, we used TexToData to automatically derive the corresponding CDM for each language by using all of the supported translation services, including *Yandex*.

---

[7] Textual specifications are translated by the Google Translate service (https://translate.google.com). There may be some minor differences in the generated CDMs, due to some translation inconsistencies.

For the sake of illustration, Fig. 5 presents the input textual specification in Italian and the corresponding CDMs derived automatically from this specification by using the *MyMemory* and *OpenTranslator* services.

The evaluation was independently performed by each author, whereby each generated CDM was manually evaluated against the reference CDM (36 CDMs in total, i.e. six specifications × six services).

We used the following metrics for the quantitative evaluation of the automatically generated CDMs:

- N_g – number of generated concepts[8] in the generated CDM,
- N_c – number of correctly generated concepts in the generated CDM,
- N_w – number of incorrectly generated concepts in the generated CDM, and
- N_m – number of missing concepts in the generated CDM.

Based on the manually determined metrics, we calculated *recall, precision,* and *F-score* as measures for the quantitative evaluation of the automatically generated CDM.

Recall (R) constitutes a measure of the completeness of the generated CDM, and it is defined as:

$$R = \frac{N_c}{N_c + N_m}.$$ (1)

Precision (P) constitutes a measure of the correctness of the generated CDM, and it is defined as:

$$P = \frac{N_c}{N_c + N_w},$$ (2)

The **effectiveness**, named *F-score* or *F-measure* (F), is defined as a harmonic mean of precision and recall, i.e.

$$F = \frac{2PR}{P + R}.$$ (3)

Table 2 contains the results of the evaluation of the CDMs derived from textual specifications in different languages by employing all supported translation services in the TexToData tool. It contains calculated values for R, P, and *F-score*, where each row contains values obtained by employing corresponding external translation service (S-1, ... , S-5, and Yandex), for specifications in different languages. The last row (Mean per language) contains the average values for all measures for each language, and the last column (Mean per service) contains the average values for all measures for each external translation service.

Table 2 contains summary results for all types of concepts in CDM, i.e. we didn't separate results for each concept type. Therefore, there may be some differences in results depending on the concept type. The results show that each translation service enables generation of the CDMs for all languages used in the evaluation, but none of the generated CDMs is 100% complete nor 100% correct.

---

[8] The "concept" term includes classes, attributes, generalizations, associations, and association end multiplicities.

Un membro della biblioteca è uno studente o un dipendente dell'università. Un membro della biblioteca ha un ID, nome, indirizzo, numero di telefono e numero di elementi della biblioteca presi in prestito. I membri della biblioteca prendono in prestito le unità della biblioteca. Gli studenti studiano uno dei programmi. La facoltà organizza programmi. Un programma ha un ID e un nome. Gli studenti hanno un numero di indice e un limite di unità bibliotecarie prese in prestito. Un impiegato dell'università ha una stanza e un telefono. L'unità biblioteca presenta contrassegni e informazioni sulla disponibilità univoci. L'unità bibliotecaria ha un nome, un anno e un autore.

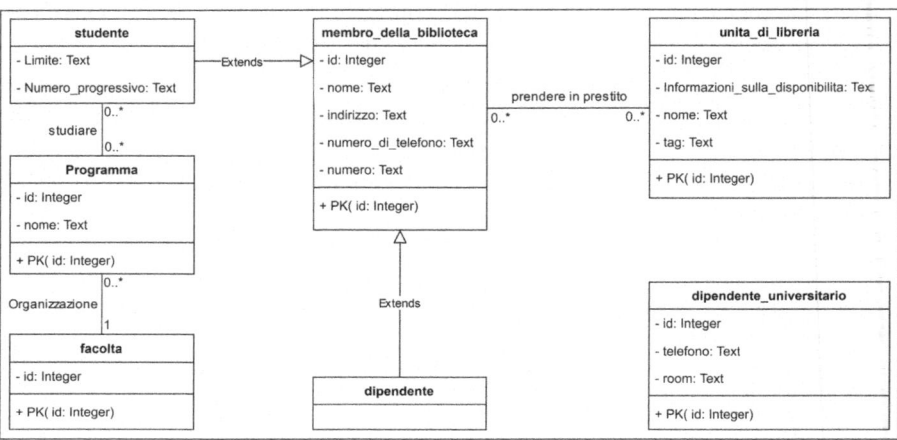

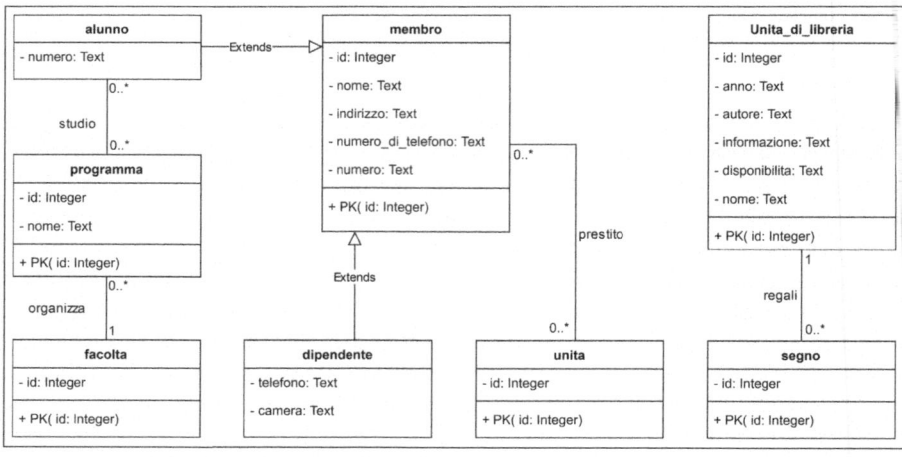

Fig. 5. Textual specification in Italian (top) and the corresponding CDMs generated by using *MyMemory* (middle) and *OpenTranslator* (bottom)

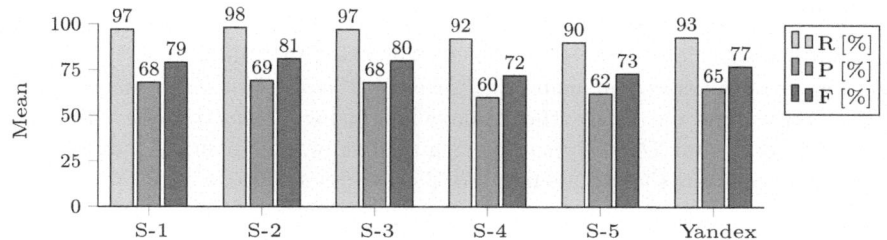

Fig. 6. Comparison of quantitative measures for generated CDMs per service

Table 2. Summary evaluation results

Service		Serbian (Latin)	Serbian (Cyrillic)	Italian	German	Greek	French	Mean per service
S-1	R	0.97	0.97	0.94	0.94	0.97	1.00	0.97
	P	0.71	0.71	0.75	0.68	0.66	0.54	0.68
	F	0.82	0.82	0.83	0.79	0.78	0.70	0.79
S-2	R	0.97	0.94	0.97	1.00	1.00	0.97	**0.98**
	P	0.69	0.76	0.72	0.65	0.71	0.63	**0.69**
	F	0.80	0.84	0.82	0.79	0.83	0.77	**0.81**
S-3	R	0.93	0.93	0.97	1.00	1.00	0.97	0.97
	P	0.60	0.65	0.78	0.77	0.69	0.59	0.68
	F	0.73	0.77	0.86	0.87	0.81	0.73	0.80
S-4	R	0.96	0.90	0.97	0.97	0.83	0.90	0.92
	P	0.68	0.69	0.62	0.72	0.38	0.49	0.60
	F	0.79	0.78	0.75	0.83	0.52	0.64	0.72
S-5	R	0.87	0.90	0.94	0.97	0.83	0.90	0.90
	P	0.76	0.69	0.66	0.72	0.38	0.49	0.62
	F	0.81	0.78	0.77	0.83	0.52	0.64	0.73
Yandex	R	0.97	0.97	0.92	0.91	0.92	0.90	0.93
	P	0.60	0.67	0.60	0.71	0.79	0.55	0.65
	F	0.74	0.79	0.73	0.80	0.85	0.68	0.77
Mean	R	0.95	0.94	0.95	**0.97**	0.93	0.94	
per	P	0.67	0.70	0.69	**0.71**	0.60	0.55	
language	F	0.78	0.80	0.79	**0.82**	0.72	0.69	

The recall for all languages varies between 0.93 (Greek) and 0.97 (German), the precision varies between 0.55 (French) and 0.71 (German), while *F-score* goes from 0.69 (French) to 0.82 (German). The most effective CDM derivation ($R = 0.98$, $P = 0.69$, $F = 0.81$) was obtained by using the *DeepTranslate* service.

The results also show that employing different translation services affects generated CDMs. For example, for the specification in Italian (Fig. 5), the best precision is obtained when using the *Microsoft Translator Text* service, while results concerning recall are the same when using *DeepTranslate*, *Microsoft Translator Text*, and *OpenTranslator*. Figure 6 shows the average values of the quantitative measures for the generated CDMs, per service.

6   Conclusion

In this paper we presented an improvement of the TexToData tool for automated database design with multilingual support. In the process of automatic CDM derivation, if the source language is not English, an external translation service is employed to translate the textual specification into English and to translate the generated model back into the source natural language. The pre-existing TexToData used only one external translation service. Because of the strong dependency on a single service, this was a perceived bottleneck in the overall process.

By adding support for five additional translation services, we achieved automatic CDM derivation that does not depend only on one translation service, and therefore, we made a more robust system. Since we are using only free subscription plans that are limited by the number of requests, in case the free quota on one external service is exhausted, another service can be used.

We also evaluated the impact of added services on the effectiveness of CDM derivation in a case study. The evaluation proved that employing multiple online translation services enables effective automatic CDM derivation.

Our future work will focus on further improvements in the entire approach to the automatic CDM derivation from textual specifications, as well as more extensive evaluation.

References

1. Brdjanin, D., Maric, S.: An approach to automated conceptual database design based on the UML activity diagram. Comput. Sci. Inf. Syst. **9**(1), 249–283 (2012)
2. Brdjanin, D., Maric, S.: Model-driven techniques for data model synthesis. Electronics **17**(2), 130–136 (2013)
3. Brdjanin, D., Vukotic, A., Banjac, D., Banjac, G., Maric, S.: Automatic derivation of the initial conceptual database model from a set of business process models. Comput. Sci. Inf. Syst. **19**(1), 455–493 (2022)
4. Brdjanin, D., Banjac, G., Babic, N., Golubovic, N.: Towards the speech-driven database design. In: Proceedings of TELFOR 2022, pp. 1–4. IEEE (2022)
5. Brdjanin, D., et al.: Towards an online multilingual tool for automated conceptual database design. In: Braubach, L., Jander, K., Bădică, C. (eds.) Intelligent Distributed Computing XV, pp. 144–153. Springer International Publishing, Cham (2023). https://doi.org/10.1007/978-3-031-29104-3_16
6. Chen, P.P.-S.: English sentence structure and entity-relationship diagrams. Inf. Sci. **29**(2–3), 127–149 (1983). https://doi.org/10.1016/0020-0255(83)90014-2

7. Choobineh, J., Mannino, M.V., Nunamaker, J.F., Konsynski, B.R.: An expert database design system based on analysis of forms. IEEE Trans. Softw. Eng. **14**(2), 242–253 (1988). https://doi.org/10.1109/32.4641
8. CHOOBINEH, J.O.O.B.I.N., LO, A.M.B.E.R.W.: CABSYDD: case-based system for database design. J. Manage. Inf. Syst. **21**(3), 281–314 (2004). https://doi.org/10.1080/07421222.2004.11045813
9. Date, C.: An Introduction to Database Systems, 8th edn. Addison-Wesley (2003)
10. Harmain, H., Gaizauskas, R.: CM-Builder: a natural language-based CASE tool for object-oriented analysis. Autom. Softw. Eng. **10**(2), 157–181 (2003)
11. Jouault, F., Allilaire, F., Bezivin, J., Kurtev, I.: ATL: a model transformation tool. Sci. Comput. Program. **72**(1–2), 31–39 (2008)
12. Kriouile, A., Addamssiri, N., Gadi, T.: An MDA method for automatic transformation of models from CIM to PIM. Am. J. Softw. Eng. Appl. **4**(1), 1–14 (2015)
13. Luisa, M., Mariangela, F., Pierluigi, N.: Market research for requirements analysis using linguistic tools. Requirements Eng. **9**, 40–56 (2004)
14. Lukovic, I., Mogin, P., Pavicevic, J., Ristic, S.: An approach to developing complex database schemas using form types. Softw.: Pract. Exper. **37**(15), 1621–1656 (2007)
15. Neill, C., Laplante, P.: Requirements engineering: the state of the practice. IEEE Softw. **20**(6), 40–45 (2003)
16. Nikiforova, O., Gusarovs, K., Gorbiks, O., Pavlova, N.: BrainTool: a tool for generation of the UML class diagrams. In: Proceedings of ICSEA 2012, pp. 60–69. IARIA (2012)
17. Omar, N., Hanna, P., McKevitt, P.: Heuristics-based entity-relationship modelling through natural language processing. In: Proceedings of AICS 2004, pp. 302–313 (2004)
18. Overmyer, S.P., Benoit, L., Owen, R.: Conceptual modeling through linguistic analysis using LIDA. In: Proceedings of ICSE 2001, pp. 401–410. IEEE (2001)
19. Purao, S.: APSARA: a tool to automate system design via intelligent pattern retrieval and synthesis. SIGMIS Database **29**(4), 45–57 (1998)
20. Rodriguez, A., Garcia-Rodriguez de Guzman, I., Fernandez-Medina, E., Piattini, M.: Semi-formal transformation of secure business processes into analysis class and use case models: an MDA approach. Inf. Softw. Technol. **52**(9), 945–971 (2010)
21. Song, I.-Y., Zhu, Y., Ceong, H., Thonggoom, O.: Methodologies for semi-automated conceptual data modeling from requirements. In: Johannesson, P., Lee, M.L., Liddle, S.W., Opdahl, A.L., Pastor López, Ó. (eds.) Conceptual Modeling, pp. 18–31. Springer International Publishing, Cham (2015). https://doi.org/10.1007/978-3-319-25264-3_2
22. Spasic, Z., Vukotic, A., Brdjanin, D., Banjac, D., Banjac, G.: UML-based forward database engineering. In: Proceedings of INFOTEH 2023, pp. 1–6. IEEE (2023)
23. Sugumaran, V., Storey, V.C.: Ontologies for conceptual modeling: their creation, use, and management. Data Knowl. Eng. **42**(3), 251–271 (2002)
24. Tan, H.B.K., Yang, Y., Blan, L.: Systematic transformation of functional analysis model in object oriented design and implementation. IEEE Trans. on Soft. Eng. **32**(2), 111–135 (2006)
25. Thonggoom, O.: Semi-automatic conceptual data modelling using entity and relationship instance repositories. PhD Thesis, Drexel University (2011)

PERS 2024: 3rd Workshop on Personalization and Recommender Systems (PERS)

Development of Collaborative Business Intelligence Framework for Tourism Domain Analysis

Olga Cherednichenko[1]([✉]) [ID] and Oleksandr Sutiahin[2] [ID]

[1] Univ Lyon, UR ERIC, Univ_Lyon 2, 5 Avenue Mendès France, 69676 Bron Cedex, France
`olga.cherednichenko@univ-lyon2.fr`
[2] National Technical University KhPI, Kyrpychova Street, Kharkiv 61002, Ukraine
`sutiahin.oleksandr@cs.khpi.edu.ua`

Abstract. Business Intelligence (BI) is a widely recognized term encompassing analytical applications and IT systems supporting them. Collaborative Business Intelligence (CBI) represents a refinement of traditional BI, emphasizing the cooperative execution of BI tasks. The tourism industry is experiencing rapid growth, leading to heightened competition among businesses. Collaborative Business Intelligence transcends organizational boundaries, creating a virtual space where individuals contribute insights to improve decision-making. In this study, our focus is on developing an application tailored for tourists, empowering them to make well-informed decisions about destinations and activities. Our approach involves aggregating, cleansing, and structuring tourism data to offer a user-friendly interface equipped with collaborative analysis tools. This necessitates gathering data from disparate sources and consolidating it into a unified format. CubeJS serves as a tool in creating an additional layer between the database and the main application, facilitating the generation of an OLAP cube and an API for generating necessary cube slices. MongoDB, chosen as the NOSQL database, is selected for its user-friendly interface, compatibility with CubeJS, and performance capabilities. To populate the database, we utilize data from Google Maps obtained through the corresponding API. Through our experimentation with tourism data, we introduce the CBI framework tailored for the tourism domain. This framework comprises three primary components: 1) Data Exploration; 2) Shared Analysis and Insights; 3) Real-time Decision-making. By integrating these three key components, our CBI framework aims to enhance decision-making processes, foster collaboration among stakeholders, and ultimately improve the overall effectiveness and efficiency of decision making.

Keywords: Collaborative Business Intelligence · Framework · Data Mining · Virtual Assistance · Software Prototype · Tourism

1 Introduction

In today's fast-paced tourism industry, staying ahead requires more than just intuition - it demands data-driven insights and strategic decision-making. Business Intelligence (BI) is a widely recognized term encompassing analytical applications and IT systems

J. Tekli et al. (Eds.): ADBIS 2024, CCIS 2186, pp. 253–262, 2025.
https://doi.org/10.1007/978-3-031-70421-5_21

supporting them. Collaborative Business Intelligence (CBI) represents a refinement of traditional BI, emphasizing the cooperative execution of BI tasks. By using the power of Collaborative Business Intelligence, we aim to suggest a platform where users can visualize, analyze, and compare tourism data in real-time. Collaborative Business Intelligence transcends organizational boundaries, creating a virtual space where individuals contribute insights to improve decision-making.

The BI4people project [1] aims to analyze data and making business decisions collaboratively by leveraging Online Analytical Processing (OLAP) to provide interactive analysis and data visualization. It introduces Collaborative Business Analysis (CBA) as a process of analyzing data and making business decisions collaboratively. Under the idea of developing the software-as-a-service model the BI4people project tries to bring people together in a virtual space and encourage them to share their opinions and comments to collectively solve problems. The concept of CBI involves using social networks, quizzes, brainstorming sessions, and even simple chats [2]. Additionally, the reuse of other collaborators' comments or results is also considered part of CBI, which leads to a more comprehensive approach to BI.

When creating a virtual space or forum for users to share their problems, comments, and solutions, it is essential to gather and analyze their data to unveil the intricate relationships between different pieces of information. It is crucial for the user to provide their feedback on how they utilized the analysis carried out that can make a background for recommendations and reusing.

The main research question is What is the basic scenery for a collaborative tool on BI platform? To answer this question, we suggest the crowdsourcing approach [3] to collect and observe the data of user's behavior. Our project aims to create a playground for Collaborative Business Intelligence (CBI) within the tourism domain. This playground will serve as a dynamic platform for organizing collaborative BI processes, collecting valuable data, and exploring the possibilities of providing personalized recommendations.

BI4Tourism is a web-based application designed to empower users with insights into tourism data, thereby facilitating decision-making. At its core, BI4Tourism enables users to visualize and compare diverse tourism data, unraveling dependencies crucial for making strategic decisions.

Our principal aim in this study is to reveal patterns in creating a more knowledgeable and experience led tourism sector. By addressing understanding of the roles BI and CBI, we aim to develop a structured framework for future research, strengthening the practical impact of BI in tourism domain. Accordingly, our research questions can be formulated as follows: 1) How to help users analyze data, annotate and share analysis results, and to construct knowledge? 2) What are the most important ways of achieving CBI? To answer these questions, we develop the CBI framework which comprises three primary components: 1) Data Exploration; 2) Shared Analysis and Insights; 3) Real-time Decision-making. Data Exploration unit involves the tools for the comprehensive exploration of tourism data, encompassing various sources and datasets. In the component of Shared Analysis and Insights it is presupposed that users can keep, share and explore their insights from the tourism data. And, finally, the Real-time Decision-making provides the tools for collaboration like chat or forum. In order to evaluate the proposed framework in the current research we focus on prototyping the BI4Tourism web application.

The rest of the paper is structured as follows. The next section depicts the Business intelligence in modern tourism and reasons to use collaborative analysis for it. The state-of-the-art is presented. The third section describes a brief summary of the methods. And the fourth section represents the CBI reference model for tourism BI. The next section provides a discussion and, finally, we conclude our results.

2 Background

Business intelligence in modern tourism is quite a popular and deeply investigated topic. We can divide tourism BI sources into problems that sources' authors tried to solve: enhance private tourism experience, improve decision making process for tourism businesses, and gather information for the government's use.

For first category we can refer BITOUR platform [4] that enables data analysis and visualization to answer questions like the most frequented places by tourists, the average stay length, or the view of visitors of some particular destination, it combines different open data sources like Twitter, Openstreetmap, TripAdvisor, and Airbnb to show on map most visited and most rated places. Also recommended systems help users to make their tourism experience much better. Implantation of one of those systems is presented in the paper [5]. The main idea is to extract personal preferences of the user and combine with analyzed before big data to produce personal recommendations. They use the Bayesian network to generate a user model that shows the probability of recommending a certain scenic spot to a user, that is, the user's scenic spot recommendation degree [5].

Another category provides business with graphical or numerical information of existing business data for analyzing and making right decisions. The emotional analysis of open data source could dramatically boost review processing that helps with understanding of tourists' needs [6]. The authors of the article implemented a method of extracting expressions of reviews and assigned to these expression tokens that describes the emotional weight of the review. Twitter, TripAdvisor were used as datasource for the sentimental study. The paper [7] uses adaptive neural network (ANN) technology to conduct tourism demand forecast analysis in order to improve the effect of forecasting tourism demand that makes tourism business more predictable. The system's implementation can assist visitors in making more informed decisions about where to visit, increase their understanding of the area's comfort and temperature, and assist businesses in managing and recommending tourism attractions to visitors.

Assembling and combining large amounts of different data, we can even improve tourism policies in government [8]. Using GPS coordinates of taxi trajectory, this study determined main reasons for passengers' trips and grouped them by different activities. The probability of points of interest to be visited is modeled by Bayes' rules, which take both spatial and temporal constraints into consideration [8]. Eventually such a system can help investigate the flow of tourists in specific points of interests and their typical routes through the city to make more comfortable certain points of interests. But showing a simple dashboard with graphics sometimes could be more helpful than complicated analysis systems. The paper [9] investigates The Data Analytics Framework (DAF), that is used in Public Administrations of Italy, and describes enhancement of the project by introducing a case study created by the author, concerning tourism of Sardinia. DAF gathers information from government datasets of Sardinian Region and provides tool for data exploration.

BI could be enhanced with people collaboration that leads to the combination of social software with business intelligence (collaborative BI) can dramatically improve the quality of decision making. In this point of view among all reviewed articles we can highlight the [4] and [5]. The platform BITOUR proposed in [4] and recommender system suggested by [5] have some collaboration features because of using social networks and tourism forum. But the main disadvantage of those systems is that their users can't directly improve their analysis by making BI collaboratevly. Theoretically they can add some more reviews on TripAdvisor or add some angry tweets on Twitter but generally it doesn't change anything. So we need some system with direct interaction between users that will support BI for tourism.

3   Materials and Methods

Data exploration is an important component of the BI process, which involves collecting, identifying, and analyzing data to discover meaningful insights and patterns. The main goal of data exploration is to identify key business opportunities and challenges that can drive decision-making and improve business performance. However, most tourists do not possess expertise in data analysis or the ability to make rational decisions. Therefore, they require a user-friendly tool that incorporates data visualization capabilities, offering potential solutions and ideas in a comprehensible manner. Chatbots play a vital role in supporting users by providing information, searching data, and executing routine tasks.

In order to investigate the effectiveness of integrating a collaborative tool, we have chosen to utilize a scenario focused on the analysis of tourist data. This scenario requires a set of statistical commands as well as capabilities for communication. The data exploration process entails a systematic examination and analysis of datasets with the goal of gaining insights, identifying patterns, and uncovering valuable information. This process typically involves several key steps, including data collection, data cleaning and pre-processing, descriptive statistics, data visualization, exploratory data analysis (EDA), hypothesis testing, and iterative refinement. In this research, our focus lies in the task of tourist data exploration, where we aim to identify and generalize queries and commands as items. Our ultimate goal is to provide recommendations for non-expert users in the field of data exploration specifically in the tourist domain. As our research is in its early stages, we begin by investigating fundamental aspects to lay the groundwork for developing a recommender assistant framework.

When creating a virtual space or forum for users to share their problems, comments, and solutions, it is essential to gather and analyze their data to unveil the intricate relationships between different pieces of information. Visualizing this data through captivating and informative charts can assist in illuminating these dependencies. Additionally, it is crucial for the user to provide their feedback on how they utilized the analysis carried out. The data for this analysis is sourced from Google Maps reviews and descriptions of tourism destinations. To enhance the user experience, a chatbot employing a pre-trained

LLM model, engages in conversations with users, suggesting existing solutions to their queries as prompted. Upon creating a request, the chatbot deeply analyzes the text, extracting commands and transforming them into Cube JS HTTP queries to gather the necessary data for generating the charts. In cases where the user's text lacks sufficient information to formulate a Cube JS query, the chatbot kindly requests clarification for specific aspects of the text, such as missing filter details. This dialogue aims to achieve a more natural and engaging conversation between the chatbot and the user, utilizing reinforcement learning algorithms and a model based on previously gathered data and manually curated Q&A sets.

CubeJS serves as a tool in creating an additional layer between the database and the main application, facilitating the generation of an OLAP cube and an API for generating necessary cube slices. MongoDB, chosen as the NOSQL database, is selected for its user-friendly interface, compatibility with CubeJS, and performance capabilities. To populate the database, we utilize data from Google Maps obtained through the corresponding API.

4   Results

Usually people can't use a lot amount of open source data to help them with making decision in tourism activities due to lack of experience or knowledge. So the system is web based application, similar to the forum, that helps users visualise and compare tourism data due to resolve specific users' tourism needs, for example user can ask question whether it could be best attraction place to visit in Paris but the visit should be in most uncrowded time of the day and chat bot of the system should ask for more details or show the chart. The BI4Tourism should save answers and questions of the user during his session. Eventually the system should ask the user to leave comment what kind of problem he solved by comparing data on the chart. Also the system contains all question and charts that user asked before in form of forum's topics. So new user could check already created questions and charts before ask new request to the system. Also the chat bot can use pretrained chat GPT to make conversation with user more human. The summary of user's needs is represented in the table 1.

The web system compiles and runs on the some application server or even cloud. The BI4Tourism is written in React, Cube JS and Java, and is source-code portable. The user's environment is any browser on any device that supports browsing. Hardware specs don't matter because of web application. To create the chart that represents dependency of some tourist data, the tourist user needs a few minutes to create topic in application with concrete request, where he must mention which data should be shown on the chart. Also user can check already answered requests of another users that are similar to his question to the chatbot. The last described operation takes a few moments.

The app consists of three large parts: frontend application, backend application, Cube js instance (Fig. 1). Frontend part will be implemented with React JS framework because of simple connecting to Cube js instance. In this app user can request some chart that help him perform BI analysis in tourism area, moreover the system is able to investigate whether the request has some similar questions from previous users. Cube js has API to create appropriate result set of data to create charts in React web app. Source of the data for chart is MongoDB. MongoDB will be filled by open source data gathered by

Table 1. Summary of Key Stakeholder Needs

Need	Concerns	Proposed solution
User wants to plan trip and check when it's better to visit tourism attractions, cafes. Main questions that should be answered: which hour the place is not crowded, during which time the place has best reviews, how the weather influence on occupancy of the place	User can check visualization of tourism data dependency to make decision	Use Cube JS to store raw data. The frontend app should retrieve request from the user and call Cube JS to get formatted for chart
The businessman wants to see when clients leave best reviews and what they leave at the comments and what was occupancy of this place in time when the reviews were left	User can check visualization of tourism data dependency to make decision	Use Cube JS to store raw data. The frontend app should retrieve request from the user and call Cube JS to get formatted for chart
Check the charts that were	It saves time when the user can check already created chart that he/she needs	The backend app should save all requests and charts in the database
Check the purpose of generating charts by another users	After the chart was created and conversation with chat bot finished user has to leave comment that contains the conclusion to which user came after checking the chart. The user can check the conclusion of another user to save his time	The backend app should save all last comments of users in the database
Help with updating chart or generating new chart	After creating chart, user can ask the chat bot to change something or regenerate to make his request more accurate for his needs	The fronted app should contain chat bot that helps user to work with charts and based on pre trained chat GPT model

open data preprocessor (Google maps, trip adviser, weather forecast websites). Also the user has to leave comment after investigation of the chart, the comment should include information how the chart helps user in his study. So the comment, the request and chart data set will be sent to Java backend REST API and saved in Postgres DB. This saved use case will be proposed for next user with similar request. All parts will be dockerized due to easy deployment and CI/CD implementing.

We combine NOSQL databases with OLAP cubes using machine learning, a method of naive Bayesian classifier. But this method is quite complicated and requires full independent implementation, which entails the search for experienced developers and thorough testing of the final algorithm. Therefore, it was decided to look for a ready-made, open-source solution that can be configured to suit your needs.

CubeJS is a program that creates a certain additional layer between the database and the main application, generating an OLAP cube from the database and an API that makes it easy to generate the necessary cube slices, which in turn can be obtained in JSON format for further graphing. CubeJS also has built-in functionality with a web interface that allows you to select measurements and dimensions for further graphing there. MongoDB was chosen as the NOSQL database because of its ease of use, connection to CubeJS, and performance. To fill the database, we chose data from Google Maps, which we previously received from the corresponding API.

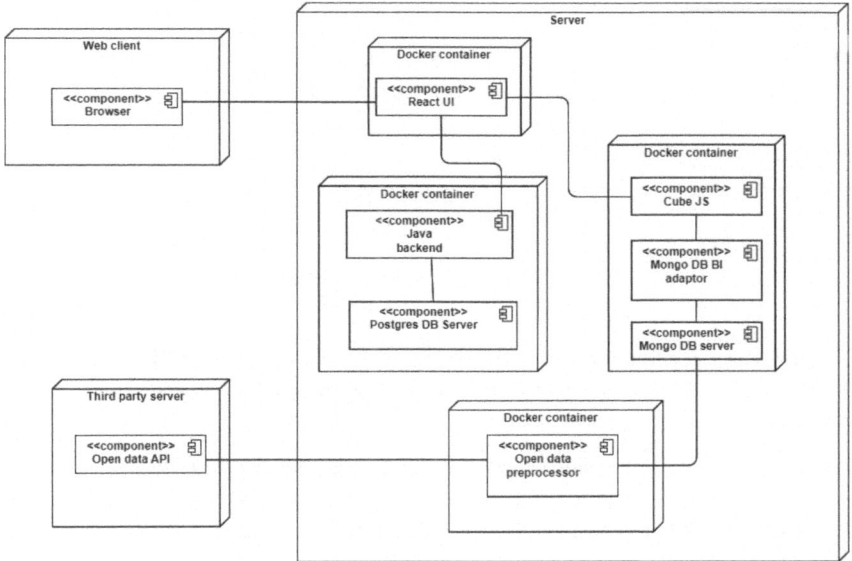

Fig. 1. The system architecture [3]

As a result, using the extensive functionality of CubeJS and MongoDB, an OLAP cube was created from data in JSON format and cube slices were created. Through this endeavor, we aim to streamline the decision-making process for tourists, enhancing their overall experience.

To analyze tourism information, preliminary data preparation is required. We use data from different sources, clean and combine them to build a structure suitable for analysis. The tourism information data model used in our application is shown in Fig. 2. We identify three key concepts for describing and presenting tourism data for analysis. This is a tourist attracttion including location, weather conditions and tourist reviews. The collected data is a set of snapshots of data over a certain period and reflects changes in amount of visitors, sentiment of reviews, ambient temperature depending on the date and time of day.

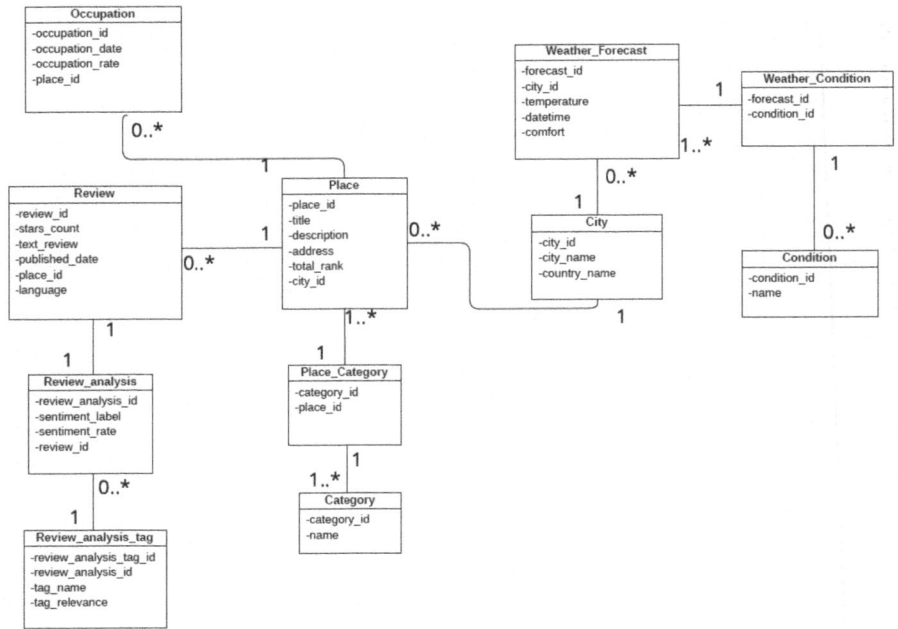

Fig. 2. The data model (fragment)

Through our experimentation with tourism data, we develop a CBI framework tailored specifically for the tourism domain. This framework comprises three primary components:

- Data Exploration: This phase involves the comprehensive exploration of tourism data, encompassing various sources and datasets. It facilitates a deep understanding of the available data landscape, allowing stakeholders to identify trends, patterns, and potential insights.
- Shared Analysis and Insights: In this component, stakeholders engage in collaborative analysis and knowledge sharing. Through shared access to data and analytical tools, participants collaboratively explore, interpret, and derive insights from the tourism data. This collaborative environment fosters a synergistic approach to understanding and addressing challenges within the tourism domain.
- Real-time Decision-making: The final component emphasizes the integration of insights derived from collaborative analysis into real-time decision-making processes. By leveraging the CBI framework, stakeholders can make informed decisions promptly, responding swiftly to changing circumstances and optimizing tourism-related strategies and initiatives.

By integrating these three key components, our CBI framework for the tourism domain aims to enhance decision-making processes, foster collaboration among stakeholders, and ultimately improve the overall effectiveness and efficiency of tourism initiatives.

5 Discussion and Conclusion

This study investigated how collaborative business intelligence could be implemented using prepared chat bot and CubeJS. While earlier studies have focused only on recommendations, business analysis or gathering info from open sources, our application combines all those points and adds some artificial intelligence for better communication with the user. So we investigated that graphics can help users to find some dependency between different types of tourism data and make concrete conclusions that will be used as prepared results for similar requests of other users.

Although other solutions as BITOUR [4] is using a kind of collaborative business analysis, it takes data only from open sources (Twitter, TripAdvisor) and the users can't directly participate in analysis and leave some eventual comments. Morevore our app will contain chatbot assistance and will help the user to adjust the result of graphics according to their needs. The last function is quite rare among all similar applications that we checked.

Our study has not still covered gaps about how to help users to add their own data to our OLAP cubes or create new ones and how to map this new data to CubeJS. Further investigation may explore how to map heterogeneous data to OLAP cube. Our approach involves aggregating, cleansing, and structuring tourism data to offer a user-friendly interface equipped with collaborative analysis tools. Through our experimentation with tourism data, we introduce a Collaborative Business Intelligence (CBI) framework tailored for the tourism domain. This framework comprises three primary components: 1) Data Exploration; 2) Shared Analysis and Insights; 3) Real-time Decision-making. By integrating these three key components, our CBI framework aims to enhance decision-making processes, foster collaboration among stakeholders, and ultimately improve the overall effectiveness and efficiency of decision making.

Acknowledgments. The research study depicted in this paper is funded by the French National Research Agency (ANR), project ANR-19-CE23–0005 BI4people (Business intelligence for the people).

Disclosure of Interests. The authors have no competing interests to declare that are relevant to the content of this article.

References

1. Business intelligence for the people. https://eric.univ-lyon2.fr/bi4people/index-en.html
2. Muhammad, F., Darmont, J., Favre, C.: The Collaborative Business Intelligence Ontology (CBIOnt). 18e journées Business Intelligence et Big Data (EDA-22), B-18. Clermont-Ferrand, Octobre 2022, RNTI (2022)
3. Cherednichenko, O., Vovk, M., Sutiahin, O.: Towards collaborative business intelligent framework: crowdsourcing approach. In: 42nd International Conference on Organizational Science Development, Portoroz, Slovenia, pp.197–207 (2023) https://doi.org/10.18690/um.fov.3.2023.16
4. Bustamante, A., Sebastia, L., Onaindia, E.: BITOUR: a business intelligence platform for tourism analysis. ISPRS Int. J. Geo Inf. **9**(11), 671 (2020). https://doi.org/10.3390/ijgi9110671

5. Yun, L., Jiao, H., Lu, K.: Tourist attraction recommendation method based on megadata and artificial intelligence algorithm. Wireless Commun. Mob. Comput. 2022, 4461165 (2022). https://doi.org/10.1155/2022/4461165
6. Koduru, S.K.R.: Nostalgic analysis of big data in tourism by business intelligence. Am. J. Comput. Res. Repository **7**(1), 1–6 (2022). https://doi.org/10.12691/ajcrr-7-1-1
7. Wang, L.: Tourism demand forecast based on adaptive neural network technology in business intelligence. Comput. Intell. Neurosci. **2022**, 1–14 (2022). https://doi.org/10.1155/2022/337 6296
8. Gong, L., Liu, X., Wu, L., Liu, Y.: Inferring trip purposes and uncovering travel patterns from taxi trajectory data. Cartogr. Geogr. Inf. Sci. **43**, 1–12 (2015). https://doi.org/10.1080/ 15230406.2015.1014424
9. Michele, P., Fallucchi, F., De Luca, E.: Create dashboards and data story with the data & analytics frameworks, 272–283 (2019).https://doi.org/10.1007/978-3-030-36599-8_24

Session-Based Recommendation with Graph Neural Networks with an Examination of the Impact of Local and Global Vectors

Justyna Głogowska[✉], Dariusz Kobiela, and Szymon Mielewczyk

Gdańsk University of Technology, Faculty of Electronics, Telecommunications and Informatics, Department of Multimedia Systems and Department of Software Engineering, Gabriela Narutowicza 11/12, 80-233 Gdańsk, Poland
s197121@student.pg.edu.pl

Abstract. This study investigates the application of graph neural networks (GNN) in session-based recommendation systems (SR), focusing on a key modification involving the use of a global vector. Session-based recommendation systems often face challenges in accurately capturing user behavior due to the limited data available within individual sessions. The SR-GNN model, originally designed for automatic feature extraction from session graphs by leveraging rich connections between nodes, addresses these challenges effectively. In our experiments, we replaced the local vector with a global vector representing the entire session sequence, not just the last element. Our results show that both local and global vectors perform comparably, suggesting that the global vector is sufficient to capture the session context. Additionally, our study indicate that the SR-GNN algorithm maintains consistent performance across various datasets, with minor fluctuations depending on the dataset characteristics. The conducted experiments highlight the resilience and adaptability of the SR-GNN model in diverse scenarios, demonstrating its potential for use in session-based recommendation systems.

Keywords: Recommendation system · session-based · graph neural networks · SR-GNN · local vectors · global vectors · embeddings

1 Introduction

The rapid growth of online information has made recommendation systems necessary for coping with information overload and guiding user choices across diverse web applications, from search engines and e-commerce platforms to multimedia portals [15]. While many recommendation systems rely on persistent user profiles and past behavior [3,6], numerous services operate without user identification, relying solely on data collected during the ongoing session [6,8]. This necessitates the development of effective models that can analyze limited,

J. Tekli et al. (Eds.): ADBIS 2024, CCIS 2186, pp. 263–272, 2025.
https://doi.org/10.1007/978-3-031-70421-5_22

session-specific behavior to generate timely recommendations. The increasing importance of this real-time personalization has spurred significant research in session-based recommendations (SR) [19].

A key challenge in SR lies in accurately capturing the user's intent within the constraints of a single session. This task is complicated by the ephemeral nature of session data and the absence of long-term user profiles. To address this, recent advancements have focused on leveraging graph neural networks (GNN) to model the complex relationships and transitions between items within a session. The SR-GNN model stands out for its ability to extract rich features from session graphs, providing robust recommendations based on the intricate connections between session elements [18].

The main research question addressed in this work is whether the introduction of a global vector, representing the entire session sequence, can enhance the performance of the SR-GNN model compared to the traditional local vector approach, which only considers the last element of the session. This question is essential because it explores the potential for a more holistic understanding of session data, potentially leading to more accurate and contextually relevant recommendations.

This problem is significant due to the increasing reliance on real-time, session-based interactions in various online platforms where user anonymity is preserved. Enhancing the accuracy of session-based recommendations can directly improve user satisfaction and engagement, leading to better service outcomes and higher user retention rates.

Previous studies have demonstrated the efficacy of SR-GNN models in capturing session context through local vectors. [20] However, these approaches may not fully leverage the entirety of session data, potentially missing out on broader contextual patterns. By introducing a global vector, our approach aims to address this gap, offering a comprehensive perspective on session-based interactions.

Our proposed solution involves modifying the SR-GNN model to incorporate a global vector that encapsulates the entire session sequence. We hypothesize that this modification will not only maintain the model's robustness but also enhance its adaptability across different datasets, offering consistent performance.

In our experiments, we observed that the global vector approach performed comparably to the local vector, demonstrating that it is sufficient to capture the session context. This finding indicates that our proposed modification retains the effectiveness of the original model while providing a new perspective on session data utilization.

The remainder of this paper is organized as follows: Sect. 2 reviews related work on session-based recommendation systems and graph neural networks. Section 3 details the methodology of our proposed approach, including the modifications to the SR-GNN model. Section 4 presents the experimental setup and results, highlighting the performance of the global vector compared to the local vector. Section 5 discusses the implications of our findings and potential future work. Finally, Sect. 5 concludes the paper, summarizing the key contributions and outcomes of our research .

2 Background and Related Work

2.1 Graph Neural Networks (GNN)

Graph neural networks (GNNs) are becoming increasingly popular for generating representations of graph-structured data, such as social networks and knowledge bases. These networks excel in capturing the dependencies and interactions between nodes in a graph, making them suitable for a wide range of applications. Classic neural network architectures, such as convolutional neural networks (CNNs) and recurrent neural networks (RNNs), have also been adapted for graph-structured data. For instance, Duvenaud et al. introduced a convolutional neural network that operates directly on graphs of any size and shape [2]. Kipf and Welling proposed a scalable approach to convolutional architectures by locally approximating spectral graph convolutions, enabling efficient operations on graphs [7].

While the above methods are usually designed for undirected graphs, GNNs were proposed for directed graphs using recurrent neural networks [4,17]. The Gated Graph Neural Network (Gated GNN), as a modification of GNN, contains recurrent units with a gate and uses backpropagation through time (BPTT) to compute the gradient [9].

GNNs are designed to generate graph representations and have recently been applied to model structural dependencies in various applications, such as event prediction in scripts [12], situation recognition [11], and image classification [14]. In session-based recommendation systems, directed graphs are constructed based on historical session sequences. Using the session graph, GNN can capture transitions between elements and generate accurate embedding vectors for elements that are difficult to discover by conventional sequential methods such as Markov Chains (MC) and RNN-based methods.

2.2 Session-Based Graph Neural Networks (SB-GNN)

In session-based recommendation systems, modeling user behavior within a single session is crucial due to the lack of long-term user profiles. Previous studies have shown that transition patterns between elements are important and can be used as a local factor in session-based recommendation [10,13]. However, these methods often model one-way transitions between successive elements and omit transitions between contexts, i.e., other elements in the session. Complex transitions between distant elements are often overlooked by these methods.

Session-based GNN (SB-GNN) approaches address this issue by constructing directed graphs from historical session sequences and capturing transitions between elements within these graphs. This allows for the generation of accurate embedding vectors for elements that might be missed by traditional sequential methods.

2.3   DeepFM

In the context of session-based recommendation systems with graph neural networks (SR-GNN), it is essential to understand the roles of local and global vectors. Initially, all session sequences are modeled as directed session graphs, where each session sequence can be treated as a subgraph. Each session graph is processed sequentially, and embedding vectors for all nodes involved in each graph can be obtained using Gated GNN.

The operational schematic of the SR-GNN method is illustrated in Fig. 1 (adapted from Wu et al. [16]). Each session is represented as a composition of global preferences and current user interests within that session. These global and local session embedding vectors are formed by the node embedding vectors. Finally, for each session, the probability that a particular item will be the next click is predicted.

Local and global vectors play key roles in modeling and predicting user behaviors in session-based recommendation systems. The local vector is typically used to represent the last item in a session sequence, which is crucial in session-based recommendations where user behavior is confined to a single session and there is no access to long-term user preferences. The local vector helps capture the user's latest interests, which are often key to generating effective recommendations.

On the other hand, the global vector is used to represent the entire session sequence. This could encompass all items that a user clicked on during a session, or it could be more complex representations, such as those generated by graph neural networks in the SR-GNN method. The global vector is significant as it allows for considering a broader session context, not just the last click. This can lead to more balanced and consistent recommendations that take into account the entirety of a user's behavior in a given session.

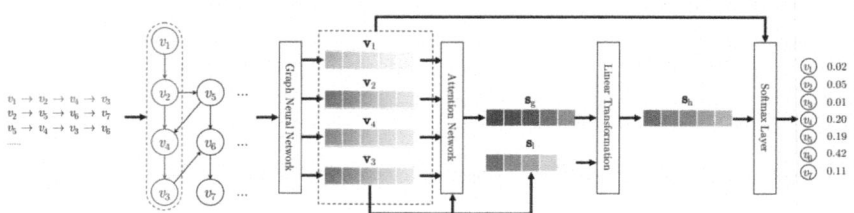

Fig. 1. The SR-GNN architecture [20]

3   Challenging the Use of both Local and Global Vectors in Session-Based Recommendations

To evaluate the effectiveness of using both local and global vectors in session-based recommendation systems, we conducted two consecutive experiments. Both experiments aimed to replicate the work of Wu et al. (2019) with modifications to test different aspects of session embeddings.

3.1   Experiment 1: Using Local Vectors

In the first experiment, we replicated the experiment from Wu et al. (2019) using a new dataset and a local vector. Initially, individual session sequences were modeled as structural graph data, and graph neural networks were used to capture complex transitions between elements. The session-based recommendations were generated using session embeddings, which were derived from the embedding vectors of elements involved in each individual session, without relying on long-term user representations.

3.2   Experiment 2: Using Global Vectors

The second experiment also replicated the experiment from Wu et al. (2019) but with the use of a global vector on the same new dataset. As with the first experiment, session sequences were modeled as directed graphs, and graph neural networks captured the transitions between elements. However, this time, we utilized global vectors to represent the entirety of a session sequence, aiming to understand how global preferences influence recommendation performance.

3.3   Datasets

We evaluate the proposed method on three basic real representative datasets, i.e., Yoochoose[1], Diginetica[2] and eCommerce[3] (see Table 1). The Yoochoose dataset comes from the RecSys Challenge 2015, which contains a stream of user clicks on an e-commerce site over 6 months. The Diginetica dataset [1] comes from the CIKM Cup 2016, where only transactional data are used. The remaining 7,981,580 sessions and 37,483 items constitute the Yoochoose dataset [16], while 204,771 sessions and 43,097 items make up the Diginetica dataset. The new added dataset is the eCommerce behavior data from a multi-category store. The dataset contains 285 million user events from the eCommerce site [5]. These events include various types such as browsing products, adding to the cart, and also finalizing the purchase. In our analysis, we focused on data from October, which contains 4,2448,764 events.

1. RecSys Challenge 2015: https://www.kaggle.com/datasets/chadgostopp/recsys-challenge-2015
2. CIKM Cup 2016: https://competitions.codalab.org/competitions/11161
3. eCommerce behavior data: https://www.kaggle.com/datasets/mkechinov/ecommerce-behavior-data-from-multi-category-store?

Table 1. Statistics of basic datasets used in the experiments

Statistics	Yoochoose 1/64	Diginetica	eCommerce
# of clicks	557,248	982,961	1,235,381
# of training sessions	5,917,745	719,470	215,300
# of test sessions	55,898	60,858	53,825
# of items	16,766	43,097	81,137
Average session length	6,16	5,12	4,59
Size [GB]	1,38	0,04	5,28

3.4 Evaluation Metrics

The modified algorithm's performance was evaluated using the same metrics as Wu et al. (2019): Precision at 20 (P@20) and Mean Reciprocal Rank at 20 (MRR@20).

– **P@20 (Precision)**: The proportion of correctly recommended items among the top 20 items.
– **MRR@20 (Mean Reciprocal Rank)**: The average of the reciprocal ranks of correctly recommended items. The reciprocal rank is set to 0 when the rank exceeds 20, and a high MRR value indicates that correct recommendations appear at the top of the rank list.

4   Results and Discussion

4.1 Experiment 1: Using Local Vectors

To measure the performance of the Wu et al. [20] algorithm, we utilized an additional dataset, eCommerce [5], which has a similar structure based on user sessions. The performance metrics P@20 and MRR@20 achieved significantly higher values compared to Yoochoose [16] and Diginetica [1]. The obtained results are presented in Table 2.

The eCommerce dataset achieved higher performance due to its large number of clicks and unique elements, which provided the model with more information for learning. The shorter average session length (see Table 1) suggests more frequent user interactions, which could also have contributed to the improved results.

4.2 Experiment 2: Using Global Vectors

The purpose of this experiment was to evaluate the impact of changing from local to global vectors on the same datasets. The results indicated no significant differences in performance when using global vectors instead of local ones. Both P@20 and MRR@20 metrics were at similar levels to those reported by Wu et al.

Table 2. Comparison of the results obtained by Wu et al. [20] on the baseline algorithms, on SR-GNN with a local vector, and on SR-GNN with a global vector after adding another dataset

	Yoochoose 1/64		Diginetica		eCommerce	
	P@20	MRR@20	P@20	MRR@20	P@20	MRR@20
POP	6.71%	1.64%	0.89%	0.20%	–	–
S-POP	30.44%	18.32%	21.06%	13.68%	–	–
Item-KNN	51.60%	21.81%	35.75%	11.57%	–	–
BPR-MF	31.31%	12.08%	5.24%	1.98%	–	–
FPMC	45.62%	15.01%	26.53%	6.95%	–	–
GRU4REC	60.64%	22.89%	29.45%	16.17%	–	–
NARM	68.32%	28.63%	49.70%	14.32%	–	–
STAMP	68.74%	29.67%	45.64%	17.59%	–	–
SR-GNN (Wu et al.)	70.57%	30.94%	50.73%	17.59%	–	–
SR-GNN (replication)	70.43%	30.62%	51.63%	17.88%	100%	95.22%
SR-GNN-Global Vector	70.54%	30.05%	51.52%	17.91%	100%	95.28%

Table 3. Comparison of the time obtained by Wu et al. [20] on the SR-GNN with a local vector, and on SR-GNN (time per training for 15 epochs)

	Yoochoose 1/64	Diginetica	eCommerce
	Time	Time	Time
SR-GNN (replication)	1,38 h	1 h	1 h
SR-GNN-Global Vector	1,5 h	1 h	1,13 h

(2019). The eCommerce dataset showed no visible changes in performance after switching the vector type, supporting the hypothesis that global vectors do not significantly impact the prediction accuracy of the SR-GNN model.

Table 3 shows the execution times for training the model with both local and global vectors across different datasets. The training time did not increase significantly when switching from local to global vectors, indicating that the computational cost remains manageable.

4.3   Discussion

Experiment 2 verified that the global vector does not have a direct impact on the results. The training time did not increase significantly after changing from local to global vector. During the replication of the experiment with algorithm by Wu et al. [20] we obtained results at the comparable level as with SR-GNN with the usage of local vector. The obtained results did not significantly differ from the results obtained by Wu et al. [20]; depending on the dataset, they were slightly higher or lower. One of the possible reasons for such results is the minimal impact of the local vector on the performance of the algorithm. The described data splitting method may not be suitable for the studied dataset due to the occurrence of the same item identifier in a given session.

5   Conclusions and Future Work

The algorithm demonstrated effective prediction of the next elements in the sequence of user actions using both global and local vectors. This indicates the potential for employing only the global vector created for a given user. The experiments showed that the performance of the SR-GNN algorithm, whether using local or global vectors, remains consistent, which suggests that the choice between local and global vectors may depend more on computational efficiency and dataset characteristics rather than predictive accuracy alone.

Additionally, it is crucial to apply the algorithm to appropriate data, consisting of sequences of different elements, to fully leverage its capabilities. The performance on the eCommerce dataset highlighted the importance of dataset characteristics, such as the number of clicks and unique elements, which provide the model with richer information for learning.

Future research will focus on testing the algorithm on new datasets that are not based on user sessions. This will provide insights into how the algorithm behaves in unconventional environments. Another research direction may include exploring a hybrid local-global vector approach to further enhance the algorithm's capabilities. This hybrid approach could potentially combine the strengths of both local and global vectors, leading to improved performance across various datasets and scenarios.

Source code. Source code for this paper with detailed instruction of performed experiments is available at: https://github.com/justyn1303/SR-GNN-global-vector-modification.

References

1. CodaLab. CIKM Cup 2016 Track 2: Personalized E-Commerce Search Challenge (2016). https://competitions.codalab.org/competitions/11161. Accessed 01 May 2024
2. Duvenaud, D., et al.: Convolutional networks on graphs for learning molecular fingerprints. In: Advances in Neural Information Processing Systems. NIPS '15. Montreal, Canada: Neural Information Processing Systems, pp. 2224–2232 (2015). ISBN: 9781450349185. https://doi.org/10.48550/arXiv.1509.09292
3. Gałka, A., Grubba, J., Walentukiewicz, K.: Performance and reproducibility of BERT4Rec. In: Abelló, A., et al. (eds.) New Trends in Database and Information Systems: ADBIS 2023 Short Papers, Doctoral Consortium and Workshops: AIDMA, DOING, K-Gals, MADEISD, PeRS, Barcelona, Spain, September 4–7, 2023, Proceedings, pp. 620–628. Springer Nature Switzerland, Cham (2023). https://doi.org/10.1007/978-3-031-42941-5_55

4. Gori, M., Monfardini, G., Scarselli, F.: A new model for learning in graph domains. In: Proceedings of the International Joint Conference on Neural Networks, vol. 2. IJCNN '05. Montreal, Canada: International Joint Conference on Neural Networks, pp. 729–734 (2005). ISBN: 9781450349185. https://doi.org/10.1109/IJCNN.2005.1555942.

5. Kaggle. eCommerce behavior data from multi category store (2019). https://www.kaggle.com/datasets/mkechinov/ecommerce-behavior-datafrom-multi-category-store. Accessed 01 May 2024

6. Karpus, A., Raczyńska, M., Przybylek, A.: Things you might not know about the k-nearest neighbors algorithm. In: Proceedings of the 11th International Joint Conference on Knowledge Discovery, Knowledge Engineering and Knowledge Management (IC3K 2019) - Volume 1: KDIR. INSTICC. SciTePress, pp. 539–547 (2019). ISBN: 978-989-758-382-7. https://doi.org/10.5220/0008365005390547

7. Kipf, T.N., Welling, M.: Semi-supervised classification with graph convolutional networks. In: Proceedings of the International Conference on Learning Representations. ICLR '16. San Juan, Puerto Rico: International Conference on Learning Representations, pp. 1–14 (2016). ISBN: 9781450349185. https://doi.org/10.48550/arXiv.1609.02907.

8. Leszczełowska, P., Bollin, M., Grabski, M.: Systematic literature review on click through rate prediction. In: Abelló, A., et al. (eds.) New Trends in Database and Information Systems: ADBIS 2023 Short Papers, Doctoral Consortium and Workshops: AIDMA, DOING, K-Gals, MADEISD, PeRS, Barcelona, Spain, September 4–7, 2023, Proceedings, pp. 583–590. Springer Nature Switzerland, Cham (2023). https://doi.org/10.1007/978-3-031-42941-5_51

9. Li, J., et al.: Neural attentive session-based recommendation. In: Proceedings of the 2017 ACM on Conference on Information and Knowledge Management. CIKM '17. Singapore, Singapore: Association for Computing Machinery, pp. 1419–1428 (2017). ISBN: 9781450349185. https://doi.org/10.1145/3132847.3132926

10. Li, J., et al.: Neural attentive session-based recommendation. In: Proceedings of the 2017 ACM on Conference on Information and Knowledge Management. CIKM '17. Singapore, Singapore: Association for Computing Machinery, pp. 1419–1428 (2017). ISBN: 9781450349185. https://doi.org/10.1145/3132847.3132926

11. Li, R., et al.: Situation recognition with graph neural networks. In: Proceedings of the IEEE International Conference on Computer Vision, ICCV '17. Venice, Italy, pp. 4183–4192. IEEE (2017). ISBN: 978-1-4503-5552-0. https://doi.org/10.1109/ICCV.2017.447

12. Li, Z., Ding, X., Liu, T.: Constructing narrative event evolutionary graph for script event prediction. In: Proceedings of the 24th ACM SIGKDD International Conference on Knowledge Discovery and Data Mining. KDD '18. London, United Kingdom: Association for Computing Machinery, pp. 1831–1839 (2018). ISBN: 9781450349185. https://doi.org/10.48550/arXiv.1805.05081

13. Liu, Q., et al.: STAMP: short-term attention/memory priority model for session-based recommendation. In: Proceedings of the 24th ACM SIGKDD International Conference on Knowledge Discovery and Data Mining. KDD '18. London, United Kingdom: Association for Computing Machinery, pp. 1831–1839 (2018). ISBN: 978-1-4503-5552-0. https://doi.org/10.1145/3219819.3219950

14. Marino, K., Salakhutdinov, R., Gupta, A.: The more you know: using knowledge graphs for image classification. In: Proceedings of the IEEE Conference on Computer Vision and Pattern Recognition. CVPR '17. Honolulu, HI, USA, pp. 2673–2681. IEEE (2017). ISBN: 978-1-4503-5552-0. https://doi.org/10.48550/arXiv.1612.04844.

15. Przybyłek, A., et al.: Databases and information systems: contributions from ADBIS 2023 workshops and doctoral consortium. In: Abelló, A., et al. (eds.) New Trends in Database and Information Systems, pp. 293–311. Springer Nature Switzerland, Cham (2023)

16. RecSys Challenge 2015 (2015). https://www.kaggle.com/datasets/chadgostopp/recsys-challenge-2015. Accessed 01 May 2024

17. Scarselli, F., et al.: The graph neural network model. In: IEEE Transactions on Neural Networks, vol. 20, no 1, pp. 61–80 (2009). https://doi.org/10.1109/TNN.2008.2005605

18. Wang, S., et al.: A survey on session-based recommender systems. In: ACM Computing Surveys. IF 16.6. ACM (2019). https://doi.org/10.1145/3465401

19. Wang, Z., et al.: Global context enhanced graph neural networks for session-based recommendation. In: Proceedings of the IEEE International Conference on Computer Vision (ICCV). ICCV '17. Venice, Italy, pp. 4183–4192. IEEE (2021). https://doi.org/10.1109/ICCV.2017.447

20. Wu, S., et al.: Session-based recommendation with graph neural networks. In: Proceedings of the AAAI Conference on Artificial Intelligence. AAAI '19. Honolulu, HI, USA: Association for the Advancement of Artificial Intelligence, pp. 346–353 (2019). ISBN: 9781450349185. https://doi.org/10.1609/aaai.v33i01.3301346

Senselife: Service Recommendation and Frailty Prevention Through Knowledge Models

Ghassen Frikha[1](✉), Xavier Lorca[1], Hervé Pingaud[2], Adel Taweel[3],
Christophe Bortolaso[4], Katarzyna Borgiel[4], and Elyes Lamine[1,5]

[1] University of Toulouse, IMT Mines Albi, Industrial Engineering Center,
Route de Teillet, 81013 Albi Cedex 9, France
`ghassen.frikha@mines-albi.fr`
[2] CNRS-LGC, Champollion National University Institute, University of Toulouse,
Albi, France
[3] Faculty of Engineering and Technology, Birzeit University, Birzeit, Palestine
[4] Research and Innovation Division, Berger-Levrault, Labège, France
[5] University of Toulouse, ISIS, Champollion National University Institute,
Rue Firmin-Oulès, 81104 Castres, France

Abstract. The aging global population presents unique challenges, particularly in managing frailty—a condition defined by declines in physical, cognitive, and social capacities. This paper introduces Senselife, a recommender system tailored for frailty management in elderly individuals. Senselife leverages hypergraph-based knowledge models to intelligently recommend personalized services aimed at mitigating frailty and enhancing life quality. Our methodology integrates diverse data types through Heterogeneous Information Networks (HINs), allowing for nuanced user-service interactions that significantly improve recommendation accuracy and relevance. This paper details the development of these models, emphasizing the transition from conventional data handling to advanced, knowledge-driven approaches that consider both user and service complexities. By incorporating these sophisticated models, Senselife aims to provide a scalable solution for frailty prevention, offering a significant contribution to personalized elderly care.

Keywords: Frailty · Elderly · Knowledge Models · Hypergraphs · Heterogeneous Information Networks · Service Recommendation

1 Introduction

The process of ageing is frequently accompanied by a gradual decline in various functional domains, including physical abilities, cognitive capacities, and social engagement. This decline, often compounded by the cumulative loss of resources, can precipitate a state of frailty, characterized by an increased vulnerability and a decline in overall well-being. As the elderly population expands, particularly

J. Tekli et al. (Eds.): ADBIS 2024, CCIS 2186, pp. 273–285, 2025.
https://doi.org/10.1007/978-3-031-70421-5_23

noted within developed countries (around 25% of the population of France in 2040) [2], the repercussions of frailty manifest not only in personal health trajectories but also in broader socioeconomic terms, necessitating an urgent response.

Effective strategies to mitigate the detrimental effects of frailty should emphasize enhancing individuals' abilities to achieve their own sense of well-being. Investing in this capacity necessitates a proactive approach that includes adopting adaptive strategies, life-management practices, and self-management techniques, with focusing on gaining new advantages or improvements instead of just trying to fix problems or make up for things they have lost [16].

Tackling frailty is a complex yet promising challenge. Various strategies, including physical activity, proper nutrition, and active social participation, have proven to be effective in forestalling frailty [1]. Nonetheless, these strategies can pose implementation challenges and typically necessitate a cooperative effort among health practitioners, caregivers, and the individuals at risk.

With the emergence of digital innovations, there is an increasing focus on leveraging technology to strengthen frailty prevention efforts. One innovative and promising technological solution is the application of recommender systems [6]. These are sophisticated algorithms that analyze users' historical data and preferences to suggest relevant products, services, or activities, potentially offering a customized approach to prevent frailty [14].

In this context, we propose Senselife: a recommender system that addresses the challenge of managing frailty in older adults by utilizing self-evaluations to provide personalized recommendations tailored to their unique needs and preferences. The system recommends services aimed at enhancing the functional abilities of older adults and matches the supply and demand of services available in the elderly environment.

During the development of this system, we encountered challenges, primarily related to (i) finding suitable data that accurately represents the frailty characteristics of elderly individuals in their profiles, and (ii) determining the techniques and methods for generating recommendations within the system.

In a previous work [6], we explained how these challenges were partially addressed and how we constructed the Senselife framework, which comprises three main components: data collection, recommendation generation, and service consumption and usage.

The paper proposes knowledge models to improve the accuracy of recommendation results. It begins by presenting a comprehensive overview of traditional recommendation techniques commonly employed in various contexts. Furthermore, we provide a concise explanation of two important concepts: Heterogeneous Information Network and hypergraph. In the subsequent section, we delve into our proposed knowledge model, explaining the different element of our Heterogeneous Information Networks and their extension to incorporate hypergraphs. Additionally, we describe the details of the algorithm used for service identification and explore the use of the french National Directory of Health and Social Care Services and Support[1]. In th fourth section, we present a use case

---

[1] https://esante.gouv.fr/produits-services/repertoire-ror.

to illustrate the effectiveness of this method. Lastly, we conclude our paper by summarizing our key findings and offering potential directions for future research in this field.

2 Background

2.1 Recommender Systems

The development of recommender systems (RS) can be categorized into three main generations, each incorporating advancements in technology and methodology. The first generation (1995–2005) employs three primary approaches: content-based filtering (CBF), collaborative filtering (CF), and hybrid methods. Content-based filtering (CBF) recommends items that are similar to those a user has previously liked by analyzing the content features of the items [5,18]. Collaborative filtering (CF), on the other hand, recommends items based on the preferences and interactions of other users who have similar tastes. Hybrid methods combine both CBF and CF to leverage their respective strengths and mitigate their weaknesses, providing more accurate and diverse recommendations [18].

The second generation of recommender systems (2003–2014) introduces contextual factors into the recommendation process, enhancing the relevance of recommendations by considering situational contexts such as time, location, and user group ratings. Context-Aware Recommender Systems (CARS) have become the leading approach for integrating context information, operating on the premise that context can be defined by a predefined set of stable, observable attributes [5]. Recently, recommender systems have expanded to include factors such as geometric information and the presence of others (e.g., friends, romantic partners, relatives, or colleagues). By incorporating these contextual elements, recommender systems gain access to richer data, significantly enhancing their ability to provide relevant and personalized recommendations, especially in scenarios where user and item information alone is insufficient [12].

Research on this second generation of RS is ongoing, but the third-generation is gaining increasing interest. These RSs concentrate on semantic models of representation and the utilization of all knowledge components involved in the recommendation process [15].

Knowledge-Based Filtering operates by suggesting items to users based on domain knowledge regarding how well the items align with the user's preferences [4]. According to [18], these systems should incorporate three types of knowledge: knowledge about the users, knowledge about the items, and knowledge about the relationship between the item and the user's needs. Knowledge graphs offer additional information that can address the challenges faced by collaborative filtering and content-based filtering approaches [22], as their recommendations are not solely reliant on ratings but rather leverage domain knowledge. However, one significant drawback of knowledge-based recommenders is that their development requires expertise in knowledge engineering [3]. On the other hand, the semantic relationships present in a knowledge graph can be utilized by the system to enhance accuracy and diversify the recommended items.

In order to enhance the effectiveness and efficiency of knowledge-based systems, it is crucial to adopt more advanced models that can capture and process the intricate interconnections present in real-world data. In this context, the utilization of Heterogeneous Information Networks (HINs) emerges as a compelling approach (see Sect. 2.2). HINs enable the integration of diverse data types and relationships, thereby providing a more comprehensive semantic context. To fully leverage the potential of HINs, we propose representing these networks in the form of hypergraphs (see Sect. 2.3). This innovative representation enhances the ability to model multi-way relationships that are often oversimplified in traditional graph-based approaches.

2.2 Heterogeneous Information Networks

Heterogeneous Information Networks (HINs) provide a sophisticated framework for modeling diverse and complex data in recommender systems (RS) [17]. Essentially, HINs are networks that consist of different types of nodes and edges, which represent various objects and their interrelationships, respectively. This structure enables the integration of multiple data types and sources, enhancing the contextual richness available for recommendations.

HINs facilitate a nuanced representation of relationships between users, items, and other auxiliary entities like tags or categories, thus improving recommendation accuracy and relevance [17]. The exploration of HINs can be done through different methods. One of these methods is the use of meta-paths. It helps capture semantic relationships between entities and provides more personalized and accurate recommendations by exploring different paths through the network [19].

By incorporating various types of information, HINs address several limitations of traditional RS, such as sparsity and the cold-start problem [21]. HINs enhance the system's ability to infer user preferences and item attributes from connected but previously underutilized auxiliary information.

In summary, HINs significantly enrich recommender systems by providing a robust structure to incorporate heterogeneous data and complex relationships, thereby improving the personalization and accuracy of recommendations [17].

2.3 Hypergraphs

Hypergraphs, unlike traditional graphs, are advanced mathematical structures where hyperedges can connect more than two vertices. This unique characteristic allows for the modeling of complex and higher-order interactions, making them valuable in reducing modeling complexity within recommender systems.

The application of hypergraphs in recommender systems offers several benefits. Firstly, they excel in representing multifaceted relationships, enabling the capture of intricate user-item interactions and significantly improving recommendation accuracy [8]. Hypergraphs also address issues like popularity bias, promoting fairness in recommendations by providing a more equitable distribution across different stakeholders [7]. Furthermore, in [9] work, the hypergraph model represents complex interactions between users and songs, leading to improved recommendation performance, especially in scenarios with the cold-start problem.

Hypergraph-based recommender systems (RSs) have been applied in e-commerce [13,20]. For example, a multipartite hypergraph models relationships between users, restaurants, and attributes in a multi-objective framework [13], effectively representing item attributes and user-item interactions. Hypergraphs naturally model relationships between various stakeholders, making them suitable for multi-stakeholder RSs [7]. They also help break the filter bubble by generating diverse recommendation lists based on user history [10].

In settings where the order of interactions matters, hypergraphs prove particularly effective in sequential recommendation scenarios. Methods such as Hyperbolic Hypergraphs leverage a hyperbolic space to manage sparsity and the dynamic nature of user-item interactions over time, resulting in more contextually relevant recommendations [11].

Overall, hypergraphs provide a robust and versatile framework for enhancing recommender systems across various domains, reducing modeling complexity and enabling more accurate and context-aware recommendations.

3   Our Proposed Knowledge Model

During the development of our recommendation platform Senselife, due to insufficient data for conventional recommendation techniques, we were obliged to use methods based on knowledge models. These models address distinct objectives: initially, we concentrate on (i) understanding each user's unique profile, gathering relevant information such as characteristics, preferences, demographics, and contextual data. Subsequently, we (ii) identify users' specific needs based on their features, identified in the previous step from the data collected from the survey [6]. Finally, we proceed to (iii) recommend the most suitable services aligning with users' needs and preferences.

To facilitate these recommendations, we devised a process integrating the aforementioned knowledge models. These models, representing various stages of the recommendation pipeline, offer valuable insights and aid in decision-making.

Acknowledging overlaps between these models, they can be effectively fused into a unified general knowledge model. This general model, presented in this paper, captures our data's complexity and justifies employing Heterogeneous Information Networks. Harnessing this approach's power, we aim to lay robust groundwork for an efficient recommendation system upon acquiring sufficient data.

3.1   Senselife Network Schema

Figure 1 presents a network schema that provides an overview of Senselife's Heterogeneous Information Network (HIN).

We introduce a formal mathematical model to represent the Senselife Heterogeneous Information Network. Let's consider Senselife (HIN) as a graph $G = (V, E)$ where V represents the set of nodes, each node denoting an entity within the network, and E denotes the set of edges, each edge representing a relationship between the entities.

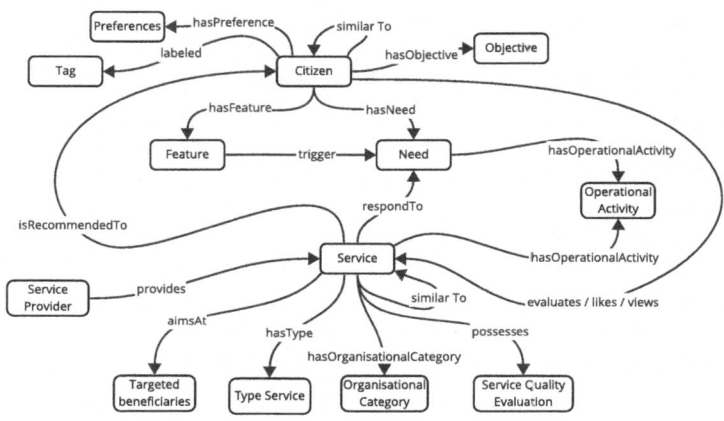

Fig. 1. Senselife network schema.

Entity and Relationship Types. The heterogeneity of the network arises from the assortment of entities and relationships, which we denote as A and R respectively. The function $\phi : V \rightarrow A$ maps each node to one of the entity types in A, and the function $\psi : E \rightarrow R$ maps each edge to one of the relationship types in R, allowing us to formalize the semantic meaning of each node and edge in the network.

The recommendation flow within our platform proceeds as follows: Firstly, (i) we begin by identifying features through survey responses while simultaneously understanding citizens' preferences and objectives. These aspects are depicted in the upper portion of the Senselife network schema (see Fig 1). Next, (ii) these features initiate the creation of needs following the predefined rules in our system. This process is illustrated by the "trigger" relationship in Fig 1. Finally, (iii) service identification is conducted by identifying Triplets in the knowledge graph (OperationalActivities-Service-OrganisationalActivity), taking into account citizens' preferences and situations.

In fact, the list of recommended services is sourced from the French National Directory of Health and Social Care Services and Support. This directory allows us to categorize the services based on their organizational structure and understand the operational activities associated with each service. Additionally, based on our mapping, we are able to identify the specific operational activities required to fulfill each identified need.

3.2   Service Selection Using the Knowledge Model

During the service selection phase in the recommendation process, the algorithm utilizes a previously created profile that includes the citizens' features and needs. Additionally, the algorithm has access to two crucial datasets: the list of all operational activities and the list of all organizational categories considered by the system.

The Algorithm 1 initiates by identifying the needed operational activities based on the citizens' needs and updating the profile accordingly. Subsequently, all potential services that could be recommended are identified. To achieve this, We use Cypher which is a declarative query language for the Neo4j graph database, as shown in Listing 1.1. The query incorporates various parameters to personalize the selection; in this example, we are using only the geographical parameter (city_name) to simplify the query.

Algorithm 1 Service Selection Based on Operational Activities.

Require:

    Input: $profile$: the profile of the user containing the features and needs,

        $operationalActivities$: list of all operational activities considered by the system,

        $organisationalCategories$: list of all organisational categories considered by the system

Ensure:

    Output: $recommendation$: package of services to be recommended

 1: UpdateProfile($profile$,$operationalActivities$)

 2: $listOfServices \leftarrow$ IdentifyTriplets($operationalActivities$,$organisationalCategories$)

 3: $connectedOAForEachService \leftarrow$ GetAllOA($listOfServices$)

 4: $recommendation$.Initialise()

 5: **while** (not all OAs in $profile$ are satisfied) AND VerifyServiceExistence() **do**

 6:      $coverageList \leftarrow$ CalculateCoverage($connectedOAForEachService$,$profile$)

 7:      $bestService \leftarrow$ CompareServices(coverageList) // Get the service that satisfies most of needed OAs

 8:      $recommendation$.Update($bestService$) // Get the service that satisfies most needed OAs

 9:      $connectedOAForEachService$.Update($bestService$) // Update list of needed OAs and services based on selected_service

10: **end while**

11: $profile \leftarrow$ GetRecommendation($recommendation$) // Update the profile and associate the possible recommendation to it.

12: **End**

Once the services are identified, they are compared based on their ability to address the operational activities. This is ensured by the method *CalculateCoverage(connectedOAForEachService,profile)*. This methods calculate the coverage of each service. For example if a service satisfies two OA and the profile contains 10 needed OA then the coverage of the service will be 20%.

Afterwards, the algorithm selects the best service with the highest coverage percentage and repeats these two steps and eliminating the already satisfied OA from the profile. This process continues until a package of different services is constructed.

This package comprises services that align with the person's needs and the required operational activities. Each package and recommendation is associated with a total coverage percentage, indicating the extent to which the person's needs are met.

Listing 1.1. Cypher Query used to identify triplets in the knowledge base.

```
MATCH (cat:OrganisationalCategory)
<-[:HAS_OrganisationalCategory]-
(n:UniteElementaire {city_name:$name})
-[:HAS_OperationalActivity]->
(act:OperationalActivity)
WHERE cat.displayName IN $categories
AND act.displayName IN $activities
RETURN COLLECT(n.entity_id) AS nodes
```

3.3   Hypergraph Extension of the Knowledge Model

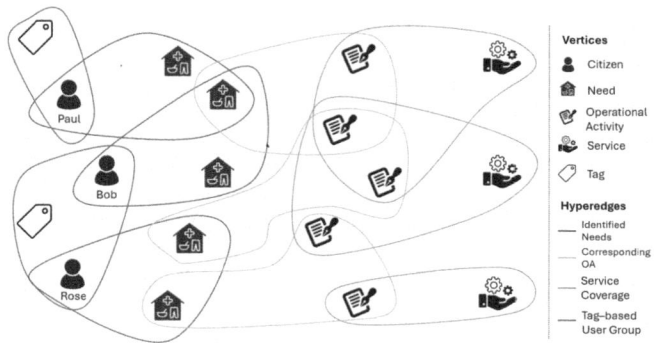

Fig. 2. Representation of Senselife complex associations.

To address complex associations, we extend the traditional graph model to a hypergraph $H = (V, E')$, where each hyperedge in E' can connect to any number of vertices in V. This extension is pivotal for representing scenarios where services are, related to multiple operational activities or multiple services, recommended to a citizen with many possibilities.

Figure 2 illustrates a subgraph of H, which represents a subset of V. The vertices in the figure are denoted as:

$$V' = \{\text{Citizen, Need, Operational Activity (OA), Service, Tag}\} \subseteq V$$

By utilizing hyperedges, we visually capture the different associations within our knowledge model.

For instance, the light blue hyperedge represents the service coverage in relation to operational activities. Similarly, the light green hyperedge represents the connections between each need and multiple operational activities. By considering these two hyperedges, we can infer the coverage of services in terms of operational activities for different needs.

Figure 3 illustrates how a package of services addresses citizen needs in terms of OAs. For example, considering citizen Rose, based on her profile and service coverage information, a recommendation of two services highlighted in blue represents the optimal combination to fulfill her needs. On the other hand, a recommendation comprising only one service suffices to meet Bob's needs, as depicted in Fig. 3.

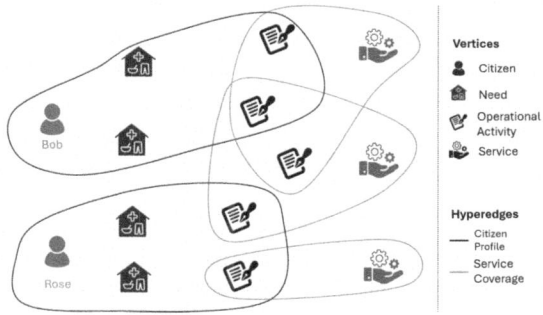

Fig. 3. Service Coverage - Citizen Profile representation.

4   Use Case

The service selection, which is the final phase of Senselife recommendation process, is illustrated through the following use case. Mr. Bob, a 75-year-old retiree, answered the survey that identified 6 features. Based on these features, four specific needs were identified: help with food preparation, help with transportation, dental care, and doctor evaluation.

The first step is to identify the operational activities that can help with Mr. Bob's needs. Table 1 provides the list of operational activities identified for Mr. Bob's profile.

Table 1. Identified operational activities.

List of Needs	Identified operational activities
Help with food preparation	Catering service
Help with transportation	Support for activities of daily living
Dental care	Organization of personal transportation
Doctor evaluation	Support for autonomy in mobility and travel
	Dentistry
	Stomatology

Afterwards, the algorithm continues by identifying potential services from our knowledge base, which is enriched with data from the French National Directory of Health and Social Care Services and Support. This directory encompasses 558 types of operational activities and 137 organizational categories. For our specific context - providing services to the elderly for frailty prevention from home - we focus on 40 operational activities and 39 organizational activities. Information such as instance structure types, legal status types, or establishment categories are available in the directory but currently not relevant for our framework.

Within this directory, services are defined using all available types of information, but we select only those services defined by the 40 operational activities and 39 organizational activities pertinent to frailty prevention. This ensures that the chosen services effectively support our goal of preventing frailty. Leveraging this extensive dataset, we employ queries like Listing 1.1 to identify services tailored to our users' needs. This process involves pinpointing relevant triplets within the dataset to ensure a precise match between user requirements and available services.

Table 2 presents an example of operational activities and organizational categories associated with the services. This approach allows us to filter out services that are not relevant to specific demographics such as disabled individuals or economically disadvantaged families. By incorporating both operational activities and organizational categories in our queries, we can narrow our focus to better cater to the needs of the elderly population.

Table 2. Example of operational activities and organisational categories.

Operational activities	Organisational categories
Support for living in a house	Home Nursing Care Services
Support for carrying out domestic activities	Home Help and Support Service
Specialized Geriatric Physiotherapy	Medical and Rehabilitation Care Services

Table 3 displays the list of services identified for Mr. Bob, with each service addressing specific operational activities (OA). In Mr. Bob's case, the first identified service is "Home care provider for Elderly people," which offers comprehensive home care and covers 66% of the required OAs. This service is deemed a good match. The algorithm then proceeds to search for services that can fulfill the remaining OAs.

In our example, we have two remaining OAs related to dentistry and stomatology. The service "Stomatology consultation with Dr. GERALD ATLAN" is identified, which covers only one of the two OAs. These are the only two services recommended to Mr. Bob, providing a total coverage of 83%.

The absence of a service that satisfies the "Dentistry" OA can be attributed to its incompatibility with Mr. Bob's preferences or contextual situation. Factors such as location, insurance coverage, or specific requirements may contribute to the unavailability of some services for Mr. Bob.

Lastly, it is worth mentioning that the service "Stomatology consultation with Dr. GERALD ATLAN" may be sufficient for Mr. Bob's case due to the similarity between dentistry and stomatology. Given the overlap in these fields, the identified service may effectively address Mr. Bob's dental care needs, despite not specifically falling under the category of "Dentistry."

Table 3. Identified Services and their coverage.

Service Name	Operational activities	Service Coverage
Home care provider for Elderly people	Catering service Support for activities of daily living Organization of personal transportation Support for autonomy in mobility and travel	66%
Stomatology consultation with Dr. GERALD ATLAN	Stomatology	16%

5 Conclusion

The paper introduces a formal representation of knowledge models within the Senselife framework, utilizing Heterogeneous Information Networks and extending it with a hypergraph structure. The process of identifying and recommending services is thoroughly explained, including the details of the algorithm employed. To evaluate the effectiveness of the proposed method we created different scenarios featuring distinct profiles. We then evaluated the coverage of Operational Activities (OAs) within the recommendation package and verified the compatibility of the recommendation with contextual factors and citizen preferences. Finally, we present a use case to contextualize these findings.

Considering future perspectives, there are several aspects to consider. Firstly, the addition of other datasets, such as the directory of French associations, that could enhance the recommendations by diversifying the available services and potentially enriching the list of operational activities. Additionally, graph-based recommendation strategies based on meta-paths or weighted meta-paths could be introduced, providing further optimization and personalization in the recommendation process.

References

1. Apóstolo, J., et al.: Effectiveness of interventions to prevent pre-frailty and frailty progression in older adults: a systematic review. JBI Evid. Synth. **16**(1), 140–232 (2018)
2. Blanpain, N., Buisson, G.: Projections de population à l'horizon 2070, les personnes de plus de 75 ans deux fois plus nombreuses qu'en 2013. Insee Première (2016)
3. Burke, R.: Hybrid web recommender systems. In: Brusilovsky, P., Kobsa, A., Nejdl, W. (eds.) The Adaptive Web: Methods and Strategies of Web Personalization. LNCS, vol. 4321, pp. 377–408. Springer, Heidelberg (2007). https://doi.org/10.1007/978-3-540-72079-9_12
4. Colombo-Mendoza, L.O., Valencia-García, R., Rodríguez-González, A., Alor-Hernández, G., Samper-Zapater, J.J.: Recommetz: a context-aware knowledge-based mobile recommender system for movie showtimes. Expert Syst. Appl. **42**(3), 1202–1222 (2015)
5. Desrosiers, C., Karypis, G., Ricci, F., Rokach, L., Shapira, B., Kantor, P.: Recommender Systems Handbook. Springer, New York (2011). https://doi.org/10.1007/978-0-387-85820-3
6. Frikha, G., Lorca, X., Pingaud, H., Taweel, A., Bortolaso, C., Borgiel, K., Lamine, E.: A recommendation system for personalized daily life services to promote frailty prevention. In: Abelló, A., et al. (eds.) ADBIS 2023. CCIS, vol. 1850, pp. 563–574. Springer, Cham (2023). https://doi.org/10.1007/978-3-031-42941-5_49
7. Gharahighehi, A., Vens, C., Pliakos, K.: Fair multi-stakeholder news recommender system with hypergraph ranking. Inf. Process. Manag. **58**(5), 102663 (2021)
8. Gharahighehi, A., Vens, C., Pliakos, K.: Hypers: building a hypergraph-driven ensemble recommender system. arXiv preprint arXiv:2306.12800 (2023)
9. La Gatta, V., Moscato, V., Pennone, M., Postiglione, M., Sperlí, G.: Music recommendation via hypergraph embedding. IEEE Trans. Neural Netw. Learn. Syst. (2022)
10. Li, L., Li, T.: News recommendation via hypergraph learning: encapsulation of user behavior and news content. In: Proceedings of the sixth ACM International Conference on Web Search and Data Mining, pp. 305–314 (2013)
11. Li, Y., et al.: Hyperbolic hypergraphs for sequential recommendation. In: Proceedings of the 30th ACM International Conference on Information and Knowledge Management, pp. 988–997 (2021)
12. Lu, J., Wu, D., Mao, M., Wang, W., Zhang, G.: Recommender system application developments: a survey. Decis. Support Syst. **74**, 12–32 (2015)
13. Mao, M., Lu, J., Han, J., Zhang, G.: Multiobjective e-commerce recommendations based on hypergraph ranking. Inf. Sci. **471**, 269–287 (2019)
14. Marcucci, M., et al.: Interventions to prevent, delay or reverse frailty in older people: a journey towards clinical guidelines. BMC Med. **17**(1), 1–11 (2019)
15. Rizun, M.: Concept of recommender system for building an individual educational profile. In: BIR Workshops, pp. 165–176 (2019)

16. Schuurmans, J.E.H.M.: Promoting well-being in frail elderly people: theory and intervention (2004)
17. Shi, C., Philip, S.Y.: Heterogeneous Information Network Analysis and Applications. Springer, Cham (2017). https://doi.org/10.1007/978-3-319-56212-4
18. Tarus, J.K., Niu, Z., Mustafa, G.: Knowledge-based recommendation: a review of ontology-based recommender systems for e-learning. Artif. Intell. Rev. **50**, 21–48 (2018)
19. Thomas, T., Mathew, B., Manoharan, A., Joseph, N., Binny, B.: An overview of heterogeneous information networks based on recommendation system. Int. J. Eng. Technol. Manag. Sci. (2022). https://doi.org/10.46647/ijetms.2022.v06i05.096
20. Wang, J., Ding, K., Hong, L., Liu, H., Caverlee, J.: Next-item recommendation with sequential hypergraphs. In: Proceedings of the 43rd International ACM SIGIR Conference on Research and Development in Information Retrieval, pp. 1101–1110 (2020)
21. Zhao, H., Yao, Q., Li, J., Song, Y., Lee, D.L.: Meta-graph based recommendation fusion over heterogeneous information networks. In: Proceedings of the 23rd ACM SIGKDD International Conference on Knowledge Discovery and Data Mining, pp. 635–644 (2017)
22. Zou, X.: A survey on application of knowledge graph. J. Phys. Conf. Seri. **1487**, 012016 (2020). IOP Publishing

Evaluating Diversity in Sequential Group Recommendations

Haider Zulfiqar[1], Emilia Lenzi[2], and Kostas Stefanidis[1(✉)]

[1] Tampere University, Tampere, Finland
{haider.zulfiqar,konstantinos.stefanidis}@tuni.fi
[2] Politecnico di Milano, Milan, Italy
emilia.lenzi@polimi.it

Abstract. The increasing complexity of user preferences in group activities, such as movie watching, necessitates more dynamic recommendation systems. SQUIRREL, utilizing Reinforcement Learning, adapts to the changing preferences of group members through sequential group recommendations. This paper introduces an enhancement to SQUIRREL by permitting the re-recommendation of items, reflecting more realistic scenarios where users may revisit previously enjoyed content. We implemented and evaluated three distinct reward functions—focusing on maximizing group satisfaction, balancing satisfaction with diversity, and maximizing diversity—to investigate their impact on recommendation diversity and group satisfaction. Experimental results, using the 20M MovieLens dataset, demonstrate that while diversity-focused rewards enhance recommendation variety, it requires careful balance with user satisfaction to avoid diminishing user engagement.

Keywords: Group Recommendations · Diversity

1 Introduction

The SQUIRREL [6,14] framework utilizes Reinforcement Learning (RL) to refine group recommendation algorithms through successive interactions. This model recognizes and adapts to the dynamic nature of user preferences within a group, diverging from traditional recommendation systems that do not consider the sequence of interactions in their decision-making processes. SQUIRREL specifically integrates this sequence, capturing the evolving preferences of group members effectively [11].

Group recommendation systems are essential for providing personalized suggestions to a group of users with potentially diverse preferences. Traditional recommendation systems often focus on individual users, but as group activities such as watching movies or dining out become more prevalent, there is a growing need for effective group recommendation systems. Balancing the diverse preferences of group members is challenging, and achieving a high level of satisfaction for all members often requires sophisticated techniques.

J. Tekli et al. (Eds.): ADBIS 2024, CCIS 2186, pp. 286–298, 2025.
https://doi.org/10.1007/978-3-031-70421-5_24

This work delves into the SQUIRREL framework's capacity to handle diverse and dynamic group preferences through its innovative use of RL. Specifically, SQUIRREL defines states based on the current preferences within the group, identifies actions as potential recommendation strategies, and measures rewards by the level of group satisfaction. The framework's ability to learn and adapt over time makes it uniquely capable of handling the dynamic and often conflicting preferences within group settings.

In our study, we introduce enhancements to the traditional SQUIRREL model by allowing re-recommendation of items, reflecting realistic scenarios where users may repeatedly enjoy or revisit specific content. This adaptation, alongside the implementation of diverse reward functions, aims to balance individual and collective satisfaction while promoting a broader exploration of content, addressing the core challenges of diversity in group recommendations. Our experimental analysis, using the 20M MovieLens dataset, evaluates the effectiveness of these adaptations in improving the diversity and satisfaction metrics of group recommendations, offering insights into their practical applications and implications for real-world systems. By focusing on the dynamic interplay between user satisfaction and recommendation diversity, this paper contributes to the ongoing discourse on optimizing group recommender systems, highlighting the potential of RL in enhancing the adaptability and fairness of these systems. Future directions will explore more granular adaptations of reward functions and the integration of fairness-oriented metrics to further refine the balance between diversity and user satisfaction in group settings.

2   Related Work

Recommender systems have evolved to tackle the complexities of group decision-making, recognizing the social dynamics in collaborative settings. Transitioning traditional systems tailored for individual users to group contexts presents challenges like aggregating diverse preferences and balancing group member influence, as discussed by [10]. While traditional techniques, such as Collaborative Filtering (CF), have advanced with techniques like neural networks [9], they often overlook the intricacies of group recommendations, emphasizing the importance of balancing satisfaction among all members. Early group recommendation methods relied on simplistic strategies, like averaging preferences [11]. However, more sophisticated models that acknowledge individual influence and group dynamics have emerged [7].

The shift towards sequential group recommendations, driven by the need to analyze group dynamics and evolving user preferences over time [8], has led to the adoption of techniques, such as RNNs and CNNs [15]. However, these approaches may overlook fairness and diversity considerations. The recent development of SQUIRREL represents a significant step in this direction, leveraging reinforcement learning to optimize group satisfaction and ensure fair treatment of members [14]. Fairness in recommender systems extends beyond accuracy metrics, encompassing equitable treatment for users and items. Multisided fairness frameworks, introduced by [1], emphasize balancing user satisfaction and item visibility to prevent favoritism towards popular items or user groups, while according to [3], fairness often involves addressing diversity to mitigate biases.

Diversity plays a crucial role in ensuring fairness and increasing user satisfaction over time [5]. Efforts to boost diversity have introduced metrics like the Gini index and item coverage, offering insights into recommendation richness and coverage [2]. Integration of diversity into sequential recommendation systems presents unique opportunities, explored e.g., in [8,12]. In the context of sequential group recommendations, SQUIRREL exemplifies adaptive approaches, dynamically adjusting recommendations based on group dynamics and diversity needs [14]. The research underscores the intersection of fairness and diversity, requiring sophisticated models to adapt to changing group dynamics [4]. Managing disagreement within groups is also vital, as highlighted by [13]. Despite advancements, gaps remain in capturing nuanced user engagement with diverse content and predicting diversity's long-term impact on satisfaction.

3   The SQUIRREL Framework

SQUIRREL is formulated as a Markov Decision Process (MDP), where the RL agent iteratively improves its action selection policy based on observed rewards, aiming to maximize the cumulative group satisfaction over time. The process is defined by states (S), actions (A), state transitions (P), and rewards (R). **State (S):** Represents the collective satisfaction of the group, integrating each member's response to previous recommendations. It encapsulates the complexity of group preferences over time, providing a holistic view of the group's historical interaction data. **Actions (A):** These are the various recommendation strategies that SQUIRREL can employ. The choice of action at any given time is crucial as it directly influences the subsequent state of the system by attempting to optimize the satisfaction derived from group recommendations. **Rewards (R):** Rewards in SQUIRREL are determined by the satisfaction levels of group members with the recommendations provided. It aims to quantify the effectiveness of the actions taken by measuring aspects such as user engagement and satisfaction metrics like click-through rates or ratings. **Transition Probabilities (P):** $P(s, s')$ represents the probability of moving from the current state s to a new state s' under the action a. This transition encapsulates the likelihood of evolving group preferences following a specific recommendation strategy. **Reward Function:** Defined as $R(s, s')$, this function quantifies the immediate feedback (reward) received after transitioning from one state to another due to an action. It is a crucial component that guides the learning process of the RL agent by reinforcing strategies that yield higher satisfaction.

The RL component of SQUIRREL is vital because it allows the system to learn from the group's feedback loop continuously, thereby refining its recommendation strategies to better suit the group's changing preferences. This adaptive approach is designed to enhance user satisfaction by aligning recommendations more closely with user expectations and preferences as they evolve.

Each round of recommendations considers the current state of group satisfaction, chooses an action that is expected to maximize future rewards and updates the system's understanding based on the outcome. For each user u in the group G, the satisfaction $\text{sat}(u, GL_j^G)$ with a group list GL_j^G at any round j is calculated by comparing the relevance of recommended items to the user's preferences, as defined in [14]. Mathematically, it is expressed as: $\text{sat}(u, GL_j^G) = \frac{\sum_{d \in GL_j^G} P_j(u,d)}{\sum_{d \in B(u,k)^j} P_j(u,d)}$, where $P_j(u, d)$ is the predicted preference score of item d for user u at round j. The overall state of the group can then be determined by calculating the average satisfaction of all its members up to the present round of recommendations. In turn, the satisfaction of group G for a specific group recommendation list GL_j^G is the mean of the satisfaction scores across all group members, while the overall satisfaction of the group $\text{groupSat}(GL_j^G)$ is the average of individual satisfactions, providing a measure that reflects the collective contentment with the recommendations. This overall satisfaction can serve as a measure of the reward generated by action a in a particular recommendation round j, represented as $R_s(GR^j) = \text{groupSatO}(GR^j)$, where GR^j encompasses all rounds up to the jth one.

To consider disagreements among users, we calculate the harmonic mean (FScore) between the overall group satisfaction, groupSatO, and the overall group disagreement, groupDisO, where $\text{groupDisO}(GR)$ is defined as the difference between the highest and lowest satisfaction levels within the group members, i.e., $\max_{u \in G} \text{satO}(u, GR) - \min_{u \in G} \text{satO}(u, GR)$.

4   Diversity in SQUIRREL

In SQUIRREL, each recommendation round is designed to suggest new and unique data items, aligning well with the framework's goal of exploring a variety of options. However, this approach does not always reflect realistic scenarios where users may want to revisit previously recommended items, such as movies they particularly enjoyed or might have missed in earlier sessions. To address this and better simulate real-life group dynamics, we introduce an adaptation that permits items to be recommended more than once across different recommendation rounds. This change is crucial for studying the impact of diversity because it introduces the possibility of repeated exposures to items, mirroring how real groups might choose to consume a favorite item multiple times.

With the possibility of re-recommendation, diversity is not only about the breadth of unique items suggested but also about how the model balances new and old recommendations to satisfy group preferences over time. By examining particular metrics over multiple rounds with and without the possibility of item re-recommendation, we observe notable differences in the framework's behavior:

Overall Satisfaction: Trends show that allowing re-recommendations often leads to higher overall satisfaction, particularly in later rounds (X-axis corresponds to rounds of recommendations - Fig. 1). This indicates that as the model adapts and learns from past interactions, it becomes more efficient in cycling back to highly regarded content that aligns well with the group's consolidated preferences.

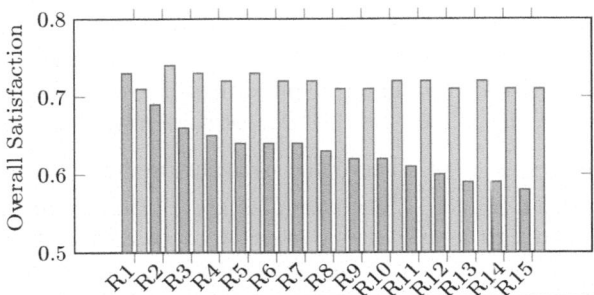

Fig. 1. Overall Satisfaction

MaxMin (Disagreement): The difference between the most and least satisfied group members tends to be smaller when re-recommendation is allowed (Fig. 2). This suggests a potential reduction in group disagreement, possibly because the re-recommended content has already been vetted and accepted by the group in previous interactions.

NDCG: The Normalized Discounted Cumulative Gain (NDCG) metric is used to evaluate the ranking quality of recommended items. It measures how well the recommended items are ranked according to their relevance to the user. The metric experiences more consistency and occasionally higher values with re-recommendations, hinting at a more predictable and satisfying sequence of recommendations (Fig. 3).

FScore: Generally shows improvement when re-recommendation is allowed due to the reduced disagreement and the positive reinforcement of content that meets the group's approval (Fig. 4).

Our approach, including re-recommendation, enhances user satisfaction by aligning recommendations more closely with both individual preferences and observed group behaviors. While the original SQUIRREL model prioritizes novelty, the adapted approach allows for a nuanced strategy where diversity encompasses both new discoveries and revisits to items with significant group appeal. This strategy is particularly relevant for scenarios mimicking real-life social interactions, such as groups of friends planning movie nights over time, enhancing the system's ability to facilitate decision-making dynamics and improve user engagement with the platform.

4.1 Reward Functions

To further enhance the capability of SQUIRREL in handling diverse group preferences, we developed new reward functions, focusing on promoting diversity while maintaining satisfaction. These adaptations aim to provide a more balanced and nuanced approach to recommendation that reflects both individual and collective preferences.

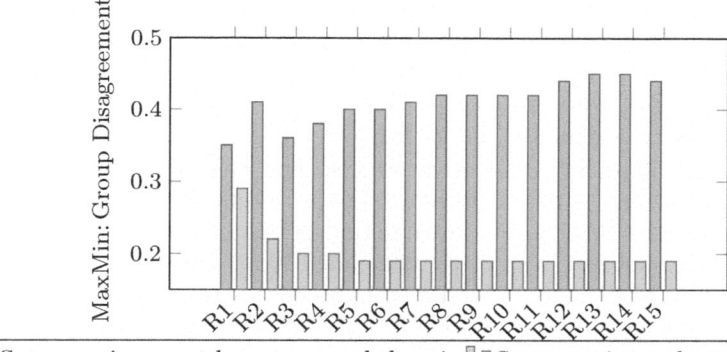

Fig. 2. MaxMin: Group Disagreement

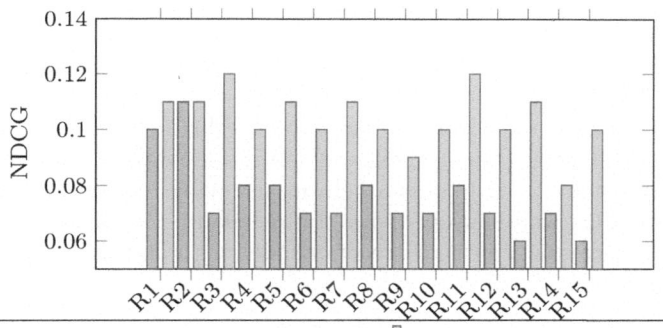

Fig. 3. NDCG Scores Over Rounds: Comparing the quality of ranking for recommended items across different reward functions.

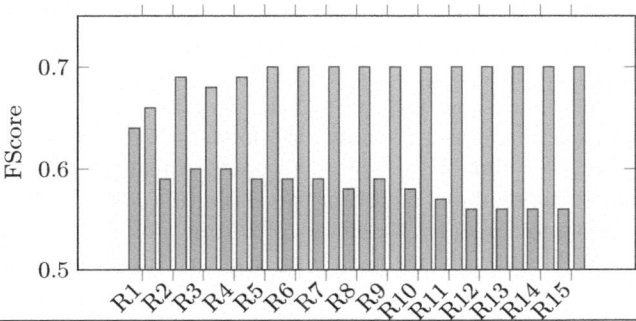

Fig. 4. FScore Over Rounds: Evaluating the balance between relevance and diversity in recommendations.

The foundational reward function in SQUIRREL [14], primarily focuses on maximizing group satisfaction calculated as the average satisfaction of all group members concerning the recommended items. To address the need for diversity in recommendations, we introduce the Diversity Reward, which incorporates the Jaccard Index, calculated as: $J(A, B) = \frac{|A \cap B|}{|A \cup B|}$, where A and B are sets of recommended items in consecutive rounds. The diversity score, which we name $JaccardDistance$, is then defined as: Jaccard Distance $= 1 - J(A, B)$. This measure guarantees that the reward function promotes diversity by minimizing the similarity between sets of recommended items from one round to the next. By integrating the Jaccard Index, this function ensures that the system continuously provides a range of diverse options. thereby enhancing user engagement and preventing the stagnation of content. However, focusing too much on diversity might lead to less satisfactory recommendations if diverse items are not aligned with user preferences. To balance between user satisfaction and diversity, the $BalancedReward$ function integrates both aspects: BalancedReward $= \alpha \cdot$ GroupSatisfaction $+ (1 - \alpha) \cdot$ Jaccard Distance, where α is a tuning parameter that balances the importance of satisfaction versus diversity. The parameter alpha in the BalancedReward function was selected through cross-validation. We experimented with different values to find the optimal balance between satisfaction and diversity. The chosen value reflects a compromise that maximizes overall group engagement without significantly sacrificing either metric. Further fine-tuning of alpha can help in achieving more precise control over this balance. This formulation allows for a balanced approach, ensuring that both new and engaging content is recommended without significantly drifting away from known user preferences.

5   Experimental Setup

In our evaluation, we use the 20M MovieLens dataset as described in [14], considering 20 recommendations per round. We focused on the 4+1 group configuration where one dissimilar member joins four similar members to simulate a realistic scenario where a new member with different preferences joins an established group. This configuration helps to highlight the challenges in balancing individual and group satisfaction. We adopt a user-based Collaborative Filtering (CF) technique for generating single-user recommendations, utilizing the Pearson Correlation for assessing user similarity. The CF system considers users similar if they share a similarity score above a certain threshold (0.8 here) and have rated at least five common movies. The threshold of 0.8 in Collaborative Filtering was selected here based on empirical analysis. Higher threshold ensured that only the most similar users were considered, improving the accuracy of recommendations. This threshold further balances precision and recall, optimizing the recommendation quality for the MovieLens dataset.

Genre Diversity Measures: In the 20M MovieLens dataset, each movie is categorized into one or more of 19 possible genres, represented by a 19-dimensional vector in a genre-specific space using one-hot encoding. To analyze the diversity and similarity of movie recommendations, distances between these vectors are calculated using the Euclidean distance and the Cosine similarity. These metrics help quantify the differences/similarities in genre composition between movie lists across recommendation rounds. Both provide a comprehensive assessment of diversity in movie recommendations. While Euclidean distance quantifies the actual differences in genre composition between movie lists, Cosine similarity measures the orientation or angle between genre vectors, reflecting the similarity in genre distribution. Together, these metrics ensure that the diversity of recommendations is evaluated both in terms of variety (Euclidean distance) and alignment (Cosine similarity), providing a well-rounded view of how diverse the content is across consecutive rounds.

6   Experimental Evaluation

In this section, we delve into the results of our experiments designed to test the effectiveness of the SQUIRREL framework under various reward configurations. The aim is to assess how these configurations influence the diversity and satisfaction of group recommendations for the 4+1 group configuration.

Diversity Analysis. This analysis utilizes the diversity score, derived from the Jaccard Index, to gauge the overlap between consecutive rounds of recommendations. The diversity score is calculated as $1 - $ Jaccard Index, where a higher score indicates fewer overlaps and, thus, a greater exploration within the recommendation space. We compared three reward functions in their ability to promote recommendation diversity: **Group Satisfaction**: Prioritizes the overall satisfaction of the group, potentially at the expense of individual preferences' diversity; **Balanced Reward:** A hybrid approach that balances overall satisfaction with the diversity of recommendations, factoring in both aspects to moderate extents; **Diversity Reward:** Focuses predominantly on maximizing diversity among recommendations to ensure a broad representation of genres and preferences. Each of the 15 rounds of recommendations was analyzed to compute the diversity scores across 20 test groups, tracking the evolution of recommendation diversity across different reward configurations.

Table 1 shows a clear trend where diversity increases as the reward functions shift from focusing solely on satisfaction to integrating and then prioritizing diversity. While maintaining user satisfaction, group satisfaction shows limited diversity, which could lead to content stagnation over time. Balanced Reward signifies an improvement in achieving higher diversity scores, suggesting that integrating satisfaction and diversity leads to more effective recommendations.

The highest diversity scores were observed with the Diversity Reward function, affirming its efficacy in expanding the recommendation space, which is crucial in diverse group settings to sustain engagement. However, this heightened diversity may potentially compromise satisfaction. This analysis underpins the importance of adaptive reward functions in enhancing group recommender

Table 1. Diversity scores by reward function

Reward Function	Avg Diversity Score	Std. Deviation
Group Satisfaction	0.5336	0.3280
Balanced Reward	0.6607	0.3254
Diversity Reward	0.7238	0.2861

Table 2. Impact of reward functions on genre diversity

Reward Function	Avg Euclidean Distance	Avg Cosine Similarity	Std. Deviation (ED)	Std. Deviation (CS)
Group Satisfaction	1.841	0.298	0.114	0.048
Balanced Reward	1.838	0.300	0.137	0.057
Diversity Reward	1.815	0.304	0.151	0.068

systems' effectiveness. By transitioning to diversity-aware configurations, recommender systems can significantly improve user engagement, ensuring relevance and appeal over successive interactions.

Genre Diversity Assessment. Genre diversity is crucial as it directly affects user engagement and satisfaction. A recommender system that effectively diversifies its genre offerings can cater to a broader array of preferences, potentially enhancing user satisfaction and reducing fatigue, which is especially important in maintaining long-term user engagement in diverse group settings. The focus here is to scrutinize how the integration of diversity-aware metrics into reward functions affects the variation in genre representation across consecutive recommendation rounds, using the Euclidean distance and the Cosine similarity.

The results in Table 2 illustrate how each reward function impacts genre diversity. Overall, the slight decrease in the average Euclidean Distance from Group Satisfaction to the Diversity reward function suggests a minor reduction in absolute genre diversity. Conversely, an increase in average Cosine similarity indicates a better alignment in genre distributions across recommendations. Both metrics exhibit a broader range under the Balanced and Diversity reward functions, indicating greater variability in genre diversity. This suggests that more extreme cases of high and low genre diversity are more frequent under these models. Finally, an increase in standard deviation for both metrics across the reward functions indicates more fluctuation in genre diversity, which could be due to the complex dynamics introduced by prioritizing diversity.

While the Diversity and Balanced rewards foster exploration across a wider range of genres, they do not consistently enhance genre diversity compared to the Group Satisfaction reward. This suggests that while the framework is effective in introducing variability, the depth of genre diversification may need further enhancement to better meet diverse group preferences. This underscores the complex interplay between designed reward functions and actual outcomes in

Table 3. Top Recommended Movies Across Reward Functions

Rank	Movie ID (Sat)	Freq (Sat)	Movie ID (Balanced)	Freq (Balanced)	Movie ID (Diversity)	Freq (Diversity)
1	733	14 (4.91%)	733	12 (4.00%)	733	9 (3.00%)
2	858	13 (4.56%)	4886	8 (2.67%)	4993	8 (2.67%)
3	1148	13 (4.56%)	2858	7 (2.33%)	858	8 (2.67%)
4	750	13 (4.56%)	858	7 (2.33%)	5952	8 (2.67%)
5	1193	12 (4.21%)	1148	7 (2.33%)	750	8 (2.67%)

Table 4. Popularity Metric Across Reward Functions

Rank	Movie ID (Sat)	Freq (Sat)	Movie ID (Balanced)	Freq (Balanced)	Movie ID (Diversity)	Freq (Diversity)
1	912	161 (2.68%)	1193	135 (2.25%)	912	106 (1.77%)
2	1193	140 (2.33%)	858	124 (2.07%)	193	90 (1.50%)
3	858	131 (2.18%)	50	124 (2.07%)	858	77 (1.28%)
4	4993	124 (2.07%)	1198	105 (1.75%)	50	68 (1.33%)
5	1198	112 (1.87%)	4993	102 (1.70%)	750	68 (2.67%)

genre diversity. Moving forward, future research should consider more targeted approaches that explicitly enhance genre diversity. Integrating genre-specific reward mechanisms or adjusting existing rewards to emphasize genre diversity could provide deeper insights into optimizing group recommendation systems.

Union and Frequency Analysis. By exploring both the union of recommended items across iterations and their overall popularity, we assess how different reward functions influence the recommendation patterns. This section aims to determine if the recommender system disproportionately favors certain items, potentially leading to stagnation in user experience and diminishing the system's capability to expose viewers to a broader spectrum of content. To do so, we aggregate all recommended items across multiple iterations to observe each movie's appearance frequency. This provides a granular view of which movies are persistently recommended, helping to understand the system's inclination towards specific movies and whether it adequately explores the diverse tastes within a group. We also examine how frequently different movies are recommended across various reward functions, providing insights into the system's tendency to favor certain movies. A summary of the results appears in Table 3.

Popularity Metric. Popularity Metric quantifies how frequently each movie is recommended across all iterations, providing a broader view of its appeal within the recommendation system. It is instrumental in detecting biases toward popular movies and evaluating the system's effectiveness in balancing the representation of diverse movie genres and choices. Table 4 presents an overview of the most popular movies and their frequencies across different reward functions.

Overall, Group Satisfaction tends to repeat certain movies more frequently than the Balanced and Diversity rewards, suggesting a potential bias towards

Table 5. Average Overall Satisfaction, Average MaxMin, Average NDCG, Average FScore

Reward Function	Avg Overall Satisfaction	Avg MaxMin	Avg NDCG	Avg Fscore
Group Satisfaction	0.717	0.196	0.103	0.616
Balanced Reward	0.661	0.228	0.105	0.643
Diversity Reward	0.614	0.310	0.110	0.570

popular content that could limit diversity. In contrast, both Balanced and Diversity demonstrate a broader spread in movie recommendations, which could enhance user exposure to varied content and improve overall satisfaction. Although there is some overlap in the top movies recommended across each reward function, the decrease in frequency from Group Satisfaction to Balanced and Diversity rewards suggests that incorporating diversity-oriented rewards can effectively promote long-tail movies. This shift in recommendation patterns across different reward functions underscores the critical need for systems that balance popularity with diversity. This balance is essential for maintaining user engagement and satisfaction over time, particularly in diverse group settings.

Implications of Reward Functions. Overall Satisfaction reflects the user's contentment with the recommendations. Our findings (Table 5) indicate that while the introduction of the Balanced and Diversity rewards enriches content diversity, they compromise overall user satisfaction, meaning that enhancing diversity may sometimes occur at the expense of reducing individual satisfaction. **MaxMin**, which measures the disparity in satisfaction within the group, sheds light on fairness. Lower MaxMin scores indicate less variance in satisfaction, suggesting a more equitable distribution of content quality across the group. The Diversity reward, with the highest MaxMin score, suggests more pronounced disparities in satisfaction, potentially reflecting its focus on diversity at the expense of uniform satisfaction. The increase in MaxMin scores with the introduction of Diversity illustrates a positive shift toward fairness in recommendations. In turn, **NDCG** assesses the quality of the ranking of recommended items, with higher scores indicating that items of interest are presented earlier in the list. The stability of NDCG scores across all settings implies that despite the strategic shifts towards diversity, the quality of the recommendations remains consistently high. **FScore** reflects the balance of relevance and diversity in the recommendations. An increase in FScore from Group Satisfaction to Diversity reward suggests that despite a slight compromise in user satisfaction, the reward functions targeting diversity improve the balance between relevant and diverse recommendations.

Overall, the results illustrate a complex interplay between user satisfaction, fairness, and diversity. **Trade-off Between Satisfaction and Diversity:** Although overall satisfaction decreases with diversity-focused rewards, improvements in MaxMin and FScore indicate a more equitable distribution of satisfaction across the group. **Fairness and Diversity Synergy:** The increase in fairness metrics with diversity-focused rewards suggests these rewards can reduce biases, promoting a more inclusive recommendation approach. **Sustained Recommendation Quality:** The consistent NDCG scores affirm that enhancing diversity does not detract from the fundamental quality of recommendations.

7 Summary

This study has demonstrated the capability of SQUIRREL to adaptively manage group recommendations by incorporating a re-recommendation feature and testing different reward functions for sequential group settings. Our findings indicate that while diversity-oriented reward functions enrich the recommendation landscape by offering a broader array of choices, they may occasionally reduce overall satisfaction. These insights underscore the delicate balance required in real-world recommendation systems to cater simultaneously to group preferences and individual diversity needs.

References

1. Abdollahpouri, H., Mansoury, M., Burke, R., Mobasher, B.: The impact of popularity bias on fairness and calibration in recommendation. CoRR abs/1910.05755 (2019)
2. Adomavicius, G., Kwon, Y.: Improving aggregate recommendation diversity using ranking-based techniques. IEEE TKDE **24**(5), 896–911 (2012)
3. Biswas, A., Patro, G., Ganguly, N., Gummadi, K., Chakraborty, A.: Toward fair recommendation in two-sided platforms. ACM Trans. Web **16**(2) (2022)
4. Burke, R.: Multisided fairness for recommendation. CoRR abs/1707.00093 (2017)
5. Castells, P., Hurley, N.J., Vargas, S.: Novelty and diversity in recommender systems. In: Recommender Systems Handbook (2015)
6. Hasan, M.M., Pervez, S., Stratigi, M., Stefanidis, K.: SQUIRREL 2.0: fairness & explanations for sequential group recommendations. In: DOLAP (2024)
7. Jameson, A., Smyth, B.: Recommendation to Groups, pp. 596–627. Springer, Heidelberg (2007). https://doi.org/10.1007/978-3-540-72079-9_20
8. Koren, Y.: Collaborative filtering with temporal dynamics. Knowl. Discov. Data Min. (2009)
9. Li, Y., Liu, K., Satapathy, R., Wang, S., Cambria, E.: Recent developments in recommender systems: a survey. CoRR abs/2306.12680 (2023)
10. Masthoff, J.: Group recommender systems: combining individual models. In: Ricci, F., Rokach, L., Shapira, B., Kantor, P.B. (eds.) Recommender Systems Handbook, pp. 677–702 (2011)
11. Ntoutsi, E., Stefanidis, K., Nørvåg, K., Kriegel, H.: Fast group recommendations by applying user clustering. In: Conceptual Modeling - ER (2012)
12. Quadrana, M., Cremonesi, P., Jannach, D.: Sequence-aware recommender systems. ACM Comput. Surv. **51**(4) (2018)
13. Stratigi, M., Pitoura, E., Nummenmaa, J., Stefanidis, K.: Sequential group recommendations based on satisfaction and disagreement scores. J. Intell. Inf. Syst. **58**(2), 227–254 (2022)

14. Stratigi, M., Pitoura, E., Stefanidis, K.: SQUIRREL: a framework for sequential group recommendations through reinforcement learning. Inf. Syst. **112** (2023)
15. Zhang, S., Yao, L., Sun, A., Tay, Y.: Deep learning based recommender system: a survey and new perspectives. ACM Comput. Surv. **52**(1) (2019)

ADBIS DC 2024: Doctoral Consortium

Using Graph Theory for Clinical Data Management

Ilaria Lazzaro[1,2]([envelope]) [iD]

[1] Department of Medical and Surgical Sciences, University "Magna Græcia",
88100 Catanzaro, Italy
[2] Data Analytics Research Center, University "Magna Græcia", 88100 Catanzaro,
Italy
ilaria.lazzaro@unicz.it

Abstract. The increasing complexity of these data and their related interactions (e.g., interactions between biological systems, genes, proteins, and neurons) requires an appropriate approach, which is realized through network models capable of systematically analysing the relationships between entities within a multi-level interconnected system. Multilayer networks best represent this approach. The workflow begins with exploring the structural and functional aspects of a multilayer network, shedding light on advanced techniques for model analysis, management, and visualization. Although various approaches have been implemented, the study of these complex structures presents several open challenges. Through a synthesis of theoretical insights and empirical observations, a comprehensive understanding of how multilayer networks model and respond to various stimuli is provided; however, several limitations related to data accessibility and computational costs are highlighted. It is also considered, the application of multilayer networks in the field of neuroimaging, aiming to leverage the available data from neuroimaging studies to understand brain connectivity and function better. The final objective involves designing and developing methodologies that enable the graphical representation of multilayer networks, the research and analysis of the topological metrics that comprise the network, to facilitate the evolution of medicine in the realm of data-driven healthcare.

Keywords: Network Sciences · Complex Networks · Brain Networks

1 Introduction

The design and development of new methodologies and software environments for the management and analysis of clinical data can be effectively achieved through the application of graph theory [15]. This paradigm provides insight into the dynamics of network models; it is a powerful mathematical framework for analysing and modelling systems ranging from simple graphs to more complex graphs that enable the representation of a wide range of dynamic patterns. However, the increasing complexity of clinical data, the intricate interactions

J. Tekli et al. (Eds.): ADBIS 2024, CCIS 2186, pp. 301–308, 2025.
https://doi.org/10.1007/978-3-031-70421-5_25

between biological systems, genes, proteins, neurons find their place within a multi-layered interconnected system such as multilayer networks.

This paper begins by examining the structural and functional aspects of multilayer networks, describing existing techniques for their analysis, management and visualization. Although several different methodologies have been developed, the study of these complex structures still presents many unsolved challenges. The primary motivation behind this research is to address these challenges by advancing understanding of multilayer network theory and its practical applications. Furthermore, the application of multilayer networks in neuroimaging is examined, with the goal of leveraging neuroimaging data to enhance our understanding of brain connectivity and function, which can lead to significant advancements in diagnosing and treating neurological disorders. Future developments include the design and development of methodologies that enable the graphical representation of multilayer networks, as well as the study and analysis of their topological metrics.

The rest of the document is structured as follows: In Sect. 2, provides an overview of the basic concepts of multilayer networks and the available tools for modeling, analyzing and visualization them. In Sect. 3, the main applications of this approach in the field of brain network and neuroimaging are presented. In Sect. 4, we discussion and conclude the paper.

2   Basic Concepts of Multilayer Networks

The concept of multilayer networks has been extensively discussed [3,6,10], and its introduction enables the study of dynamic processes by considering the complex interactions between entities within the network. The general model of multilayer network is used to specify most systems that include interactions between system actors, where each layer may contain a complete set of nodes or a subset of them. The multilayer network model (m,n) is defined by a collection $G(m,n) = (G1, G2, ..., Gm)$ of m simple graphs $G = ([n], E)$ having a set of vertices in common $[n] = 1, ..., n$, in which the edges E varies from layer to layer [20]. In other words, can be defined through a tuple $M=(A,L,P,M)$ in which the set of nodes A represent the actors in the system, the set of layers L represent the interconnections and relationships in the system, $P \subseteq A \times L$ encodes information on which node participates in a particular type of relationship and defines the node-layer component. A distinction is made between intra-layer and iter-layer edges. The former connect nodes belonging to the same layer $EA = ((u,\alpha),(v,\beta)) \in EM \mid \alpha = \beta$, inter-layer nodes instead connect vertices of two distinct layers: $EC = EM \setminus EA$. In fact, graph entities may exist on different layers and assume different types of relationships. Recent advancements in multilayer network theory have focused on developing mathematical models and algorithms to understand better and utilize these complex structures. To reconcile the diverse terminology employed, various prevailing multilayer network models are delineated, each assuming distinct configurations contingent upon their configuration and implementation attributes. The most significant types are:

- *Multiplex Networks*: These networks involve interconnected nodes across different layers. Each pair of levels maps nodes one by one, with links existing solely between corresponding nodes across different levels. Each layer represents a specific aspect or type of relationship, allowing for a comprehensive understanding of complex systems with diverse interactions.
- *Multilevel Networks*: In multilevel networks, nodes are categorized into different levels based on certain criteria or attributes. Links within and between levels signify relationships, facilitating the analysis of hierarchical structures and interactions across various layers. Unlike multiplex networks, nodes in multilevel networks do not necessarily exist across all layers.
- *Hypergraphs*: Hypergraphs provide a flexible framework for modeling complex relationships by mapping nodes onto a single layer. Unlike traditional graphs, hypergraphs allow for hyperedges, which can connect multiple nodes simultaneously, representing more intricate connections and dependencies.
- *Interconnected Networks*: These networks play pivotal roles in telecommunications, transportation, and other domains where nodes are interconnected through sets of edges. By decomposing nodes into subsets and linking them across different layers or systems, interconnected networks facilitate the analysis interactions in complex systems.

2.1 Multilayer Network Analysis and Visualization: Tools and Methodologies

Although the classification of a multilayer network defines the implementation characteristics, however, topological measurements involve analyzing the structural properties of multilayer networks. These measurements provide a deep understanding of the parts that determine a multilayer network and that are measured to perform a given analysis. Let us denote the multilayer network as M=(A,L,P,M) where A is the set of nodes, L is the set of layers, $P \subseteq A \times L$ is the node-layer participation set, and M represents the set of edges, including both intra-layer and inter-layer connections.

Below are key topological concepts involved in analysis metrics:

- **Degree distribution**: the degree of a node in a multilayer network represents the number of connections it has in all layers. Analyzing the degree distribution helps to understand how nodes are connected between layers. Degree distributions can be calculated for each layer separately and for the aggregated multilayer network. For a node $v \in A$, the degree $d(v)$ across all layers is:

$$d(v) = \sum_{\alpha \in L} d_\alpha(v) + d_{inter}(v) \tag{1}$$

The degree distribution $P(k)$ is given by:

$$P(k) = \frac{|\{v \in A \mid d(v) = k\}|}{|A|} \tag{2}$$

– **Centrality Measures**: centrality measures the importance of nodes in a network [2]. Several concepts of centrality have been defined such as degree of centrality, eigenvector centrality, closeness centrality and betweenness centrality defined according to the structure and dynamics of the multilayer network. These metrics help to identify the key players or hubs in the network.
- *Degree Centrality:*

$$C_D(v) = \frac{d(v)}{|A| - 1} \tag{3}$$

- *Eigenvector Centrality:*

$$C_E(v) \propto \sum_{u \in A} a_{vu} C_E(u) \tag{4}$$

where a_{vu} is the adjacency matrix element;
- *Closeness Centrality:*

$$C_C(v) = \frac{1}{\sum_{u \in A} d(v, u)} \tag{5}$$

where $d(v, u)$ is the shortest path distance between nodes v and u;
Betweenness Centrality:

$$C_B(v) = \sum_{s \neq v \neq t} \frac{\sigma_{st}(v)}{\sigma_{st}} \tag{6}$$

where σ_{st} is the total number of shortest paths from node s to node t, and $\sigma_{st}(v)$ is the number of those paths that pass through v.
– **Clustering Coefficient**: The clustering coefficient defines the extent to which the nodes in a multilayer network tend to cluster together. It quantifies the presence of triangles in the network. Clustering coefficients can be calculated for individual layers and for the entire network. For node v in layer α:

$$C_\alpha(v) = \frac{2T_\alpha(v)}{d_\alpha(v)(d_\alpha(v) - 1)} \tag{7}$$

where $T_\alpha(v)$ is the number of triangles through node v in layer α.
– **Mean global node overlapping**: indicates the extent to which nodes are shared or interconnected at multiple levels. Node overlap provides information about the roles and importance of nodes in a multilayer system.

$$O_N = \frac{1}{|A|} \sum_{v \in A} \frac{|\{\alpha \in L \mid (v, \alpha) \in P\}|}{|L|} \tag{8}$$

– **Mean global edge overlapping**: quantifies the extent to which arcs (connections) exist simultaneously in multiple layers of the network. This value provides information about how edges are shared across layers, which may be indicative of interactions or relationships that extend to various aspects of the multilayer system.

$$O_E = \frac{1}{|M|} \sum_{e \in M} \frac{|\{\alpha \in L \mid e \in E_\alpha\}|}{|L|} \tag{9}$$

where E_α is the set of edges in layer α.

In addition to topological measures, network alignment and community detection emerge as two fundamental methodologies in network analysis.

- **Network alignment** is process that compares two or more networks to identify regions with similar topology or function [19]. Network alignment methods are distinguished as local and global, with the former extracting small regions between the compared networks through many-to-many mapping. Global alignment aims at overlapping the compared networks across large network regions through one-to-one mapping of nodes.
- **Community detection** provides the identification of densely connected nodes within multilayer networks. Different methods of community extraction are proposed in the literature, categorized into flattening methods, aggregation methods, and direct methods. *Flattening methods* convert multilayer networks into single-layer networks for traditional community detection algorithms, but they may overlook layer-specific communities. *Aggregation methods* detect layer-specific communities and aggregate them to derive the final community structure. *Direct methods* calculate community structures directly on the input multilayer network without flattening.

The visualization of multilayer networks allows the graphical representation of different layers and nodes in order to understand complex relationships. Several different network visualization and representation techniques have been developed, such as:

- *Heatmaps* [14] display multilayer networks using color gradients to represent the strength or frequency of connections between nodes across different layers. Darker colors indicate stronger connections, while lighter colors represent weaker connections. Heatmaps provide a visual summary of the network's connectivity patterns and help identify densely connected regions within and across layers.
- *Matrix representation* [7] visualizes multilayer networks as matrices, where rows and columns correspond to nodes, and entries in the matrix indicate the presence or absence of connections between nodes in different layers. This visualization technique simplifies the representation of multilayer networks and allows for easy identification of patterns and relationships between nodes across layers.
- *GUI-based (Graphical User Interface)* and *API-based (Graphical User Interface)* libraries [17]. Each approach offers different advantages and is adapted to the characteristics of the network, particularly the API-based methods are used when considering large networks that need high-level functions.

3   Application in Brain Network and Neuroimaging

Over the years, multilayer networks have found applications in a wide range of fields, from sociology to social networks [8,16], from transport to telecommunications [4]. Recently, their application in the medical field, in particular on brain networks, has allowed us to understand how the brain is structured and how its architecture relates to its function [5]. This approach has opened up new possibilities for understanding neurological processes and disorders, marking a significant advance in medical science.

Beginning with the general idea of a network, specifically a multilayer network, one can represent the human brain as a sophisticated network composed of nodes representing distinct brain regions, and edges symbolizing the structural or functional links between these regions [13]. The modeling approaches of multilayer brain networks have been extensively utilized to gain insights into both the structural and functional mechanisms of the brain, exploring the characteristics of the network. Furthermore, these approaches have been instrumental in advancing diagnosis and treatment strategies for brain-related disorders.

- The *Structural connectivity* [1] of the brain refers to the physical arrangement of neural connections, their integrity, and anatomical organization. Utilizing diffusion tensor imaging (DTI), it is possible to directly map neural fibers in the brain and assess their integrity and organization. This information is represented as multilayer networks, where each layer reflects a different mode of connection, such as connectivity between cortical regions or connections between brain areas and white matter fibers.
- *Functional connectivity* refers to patterns of synchronized neural activity between different brain regions during the execution of cognitive tasks or under resting conditions. This synchronization can be measured using brain imaging techniques such as functional magnetic resonance imaging (fMRI) or electroencephalography (EEG) [9]. Functional networks reveal the neural circuits involved in cognitive, emotional, and behavioral processes [18].

4   Discussion and Conclusion

The study of multilayer networks has become fundamental for analyzing the interactions and relationships between entities within a multi-layered interconnected system. This approach is constantly evolving, offering new perspectives and solutions to address complex problems across various fields. Notably, in medicine, multilayer networks are essential for the manipulation and management of clinical data, enhancing our understanding of complex biological processes and disease mechanisms.

This paper presented an overview of the basic concept of multilayer networks, including the mathematical formulas and properties that characterize different network configurations. Additionally, it described the main visualization methods used to manage and analyze complex network data, emphasizing the importance of effectively interpreting these intricate systems.

Despite significant progress, visualizing large-scale networks remains a considerable challenge. Current methods often fall short in handling the complexity and scale of the data, thereby limiting the understanding and interpretation of the results [11]. Addressing these challenges is crucial for advancing the field and enabling more effective applications of multilayer networks.

Future work aims to refine methodologies for studying multilayer networks. The ultimate goal is to design an effective network analysis tool that improves both graphical representation and topological analysis methodologies. On the other hand, in the context of neuroimaging, we propose the development and implementation of a software pipeline for the construction and analysis of multilayer brain networks [12].

We will explore how community extraction algorithms can improve our understanding of multiple sclerosis (MS) pathogenesis and progression by identifying the brain regions involved. In addition, the future goal will be the integration of alignment algorithms could further refine the analysis, providing a better view of the dynamics of brain networks.

Acknowledgement. This work was funded by the Next Generation EU - Italian NRRP, Mission 4, Component 2, Investment 1.5, call for the creation and strengthening of 'Innovation Ecosystems', building 'Territorial R&D Leaders' (Directorial Decree n. 2021/3277) - project Tech4You - Technologies for climate change adaptation and quality of life improvement, n. ECS0000009. This work reflects only the authors' views and opinions, neither the Ministry for University and Research nor the European Commission can be considered responsible for them.

References

1. Bastiani, M., Roebroeck, A.: Unraveling the multiscale structural organization and connectivity of the human brain: the role of diffusion MRI. Front. Neuroanat. **9**, 77 (2015)
2. Bianconi, G.: 170Centrality Measures. In: Multilayer Networks: Structure and Function. Oxford University Press (2018). https://doi.org/10.1093/oso/9780198753919.003.0009
3. Boccaletti, S., et al.: The structure and dynamics of multilayer networks. Phys. Rep. **544**(1), 1–122 (2014)
4. Cardillo, A., Zanin, M., Gómez-Gardenes, J., Romance, M., García del Amo, A.J., Boccaletti, S.: Modeling the multi-layer nature of the European air transport network: resilience and passengers re-scheduling under random failures. Eur. Phys. J. Spec. Top. **215**, 23–33 (2013)
5. De Domenico, M.: Multilayer modeling and analysis of human brain networks. Giga Sci. **6**(5), gix004 (2017)
6. De Domenico, M., et al.: Mathematical formulation of multilayer networks. Phys. Rev. X **3**(4), 041022 (2013)
7. De Domenico, M., Solé-Ribalta, A., Omodei, E., Gómez, S., Arenas, A.: Ranking in interconnected multilayer networks reveals versatile nodes. Nat. Commun. **6**(1), 1–6 (2015)
8. Dickison, M.E., Magnani, M., Rossi, L.: Multilayer social networks. Cambridge University Press, Cambridge (2016)

9. Fox, M., Greicius, M.: Clinical applications of resting state functional connectivity. Front. Syst. Neurosci. **4**, 19 (2010)

10. Hammoud, Z., Kramer, F.: Multilayer networks: aspects, implementations, and application in biomedicine. Big Data Anal. **5**(1), 2 (2020)

11. Lazzaro, I., Milano, M., Cannataro, M.: Analysis and visualization of multilayer networks. Numer. Comput. Theory Algorithms NUMTA **2023**, 133 (2023)

12. Lazzaro, I., Milano, M., Cannataro, M.: A pipeline for the analysis of multilayer brain networks. In: Franco, L., de Mulatier, C., Paszynski, M., Krzhizhanovskaya, V.V., Dongarra, J.J., Sloot, P.M.A. (eds.) Computational Science - ICCS 2024, pp. 86–98. Springer Nature Switzerland, Cham (2024)

13. Mandke, K., et al.: Comparing multilayer brain networks between groups: Introducing graph metrics and recommendations. NeuroImage **166**, 371–384 (2018). https://doi.org/10.1016/j.neuroimage.2017.11.016, https://www.sciencedirect.com/science/article/pii/S1053811917309230

14. McGee, F., Ghoniem, M., Melançon, G., Otjacques, B., Pinaud, B.: The state of the art in multilayer network visualization. In: Computer Graphics Forum. vol. 38, pp. 125–149. Wiley Online Library (2019)

15. Pavlopoulos, G.A., et al.: Using graph theory to analyze biological networks. BioData Mining **4**, 1–27 (2011)

16. Scott, J.: Network Analysis: A Handbook. Sage Publications (1992)

17. Škrlj, B., Kralj, J., Lavrač, N.: Py3plex: a library for scalable multilayer network analysis and visualization. In: Aiello, L.M., Cherifi, C., Cherifi, H., Lambiotte, R., Lió, P., Rocha, L.M. (eds.) COMPLEX NETWORKS 2018. SCI, vol. 812, pp. 757–768. Springer, Cham (2019). https://doi.org/10.1007/978-3-030-05411-3_60

18. Ting, C.M., Samdin, S.B., Tang, M., Ombao, H.: Detecting dynamic community structure in functional brain networks across individuals: a multilayer approach. IEEE Trans. Med. Imaging **40**(2), 468–480 (2020)

19. Vijayan, V., Critchlow, D., Milenković, T.: Alignment of dynamic networks. Bioinformatics **33**(14), i180–i189 (2017). https://doi.org/10.1093/bioinformatics/btx246, https://doi.org/10.1093/bioinformatics/btx246

20. Wilson, J.D., Palowitch, J., Bhamidi, S., Nobel, A.B.: Community extraction in multilayer networks with heterogeneous community structure. J. Mach. Learn. Res. **18**(1), 5458–5506 (2017)

Advancing Legal NLP: Application of Pre-trained Language Models in the Legal Domain

Candida Maria Greco$^{(\boxtimes)}$ (iD)

University of Calabria, Arcavacata, Italy
candida.greco@dimes.unical.it

Abstract. Pre-trained Language Models (PLMs) are advanced technologies that enable deep-learning-based solutions for natural language processing and understanding (NLP). They show exceptional results on general domain texts, but they have been used in domain-specific fields like legal and health. However, processing texts in these fields is challenging due to syntax, structure, specialized terminology, privacy, and lack of domain-specific data. The research activities during the PhD course by the author have been focused on exploring opportunities coming from PLMs in the legal domain. Specifically, the research activities include the first systematic overview of PLM-based methods for AI-driven problems and tasks in the legal sphere, and the development of PLM-based methods to analyze the texts of European constitutions that share common topics.

Keywords: Transformer · Law · NLP

1 Introduction

Pre-trained language models (PLMs) have demonstrated exceptional performance in Natural Language Processing (NLP) applications since the introduction of BERT [11]. These models can learn a contextual language understanding model, capturing language semantics, non-linear relationships, and complex lexical patterns, resulting in comprehensive local and global feature representations without feature engineering. However, moving from general-domain to domain-specific texts presents a number of issues. Medical and legal fields, for example, use specialist vocabulary and jargon that are uncommon in normal English. Domain-specific texts sometimes consist of complicated sentence patterns, making parsing and generating tasks particularly challenging. Furthermore, since these documents often include sensitive information, managing such data while protecting privacy and security is a considerably difficult. Another crucial component for language models is the necessity for large amount of training data; yet, the availability of such volumes in the specific domains is very limited. On the other hand, the employment of PLMs in the legal NLP sector has resulted in considerable breakthroughs in the application of AI systems

J. Tekli et al. (Eds.): ADBIS 2024, CCIS 2186, pp. 309–317, 2025.
https://doi.org/10.1007/978-3-031-70421-5_26

to assist legal practitioners. A plethora of PLM-based models [5,7,18,21] and datasets [2,10,16,24] have been developed over the years for several applications (e.g. analysis of privacy policies [1], legal judgment prediction [5], GDPR article retrieval [22]), bringing the state-of-the-art to significant improvements. Hence there is an increasing demand for effective AI solutions in the legal sphere. To bridge the gap between the capabilities of PLMs and the demands of legal applications, this research consists of the following activities:

- Provide the first comprehensive review of PLM-based approaches for AI-driven challenges and tasks in the legal domain. This is not only synthesizes the current state of research but also highlights the limitations of existing models when handling legal texts, thereby guiding future improvements.
- Employ PLM-based methods to process portion of European constitutions that share common areas of interest. By focusing on these texts, the research explores the capability of PLMs to handle complex legal language and extract meaningful insights, which are critical for understanding the nuances of legal frameworks across different jurisdictions.

Motivations for the first activity arise from the evidence that PLMs are gradually gaining relevance in the legal sphere and, in the other hand, the development of PLMs for legal tasks in recent years has been remarkably rapid. Hence, it becomes essential to bring order to how PLMs are applied in the legal field, not only to map the current landscape, but also to identify gaps where PLMs could be improved or further developed to meet the specific needs of the legal sector.

One aspect that emerged from such review is that most of the efforts so far have focused on processing documents such as legal cases, contracts and statutes. There are several types of legal resources that are still overlooked, and these include constitutions. A constitution is a foundational document that outlines the fundamental concepts, institutions, functions, and powers of a country's government. A significant amount of the constitution is dedicated to outlining the rights and duties (RD for short) of citizens. These provisions are crucial for establishing and safeguarding the position of citizens both as individuals and as members of society. This led to the second activity of this research, which focuses on the analysis of the range of topics shared in the European constitutions that guarantee rights and duties to citizens. Revealing cross-border constitutional components may help policymakers and legal experts find common ground faster and easier, speeding up decision-making and policymaking, it may enhance legal alignment, efficiency, and transparency, paving the way for a more integrated and cooperative legal environment.

2   Related Works

The use of AI in law has garnered attention for its impact on legal practice and administration, as well as ethical considerations for applying AI to legal data (e.g., [4,23]). Zhong et al. [25] categorize legal AI tasks and applications summarizes the essential aspects of current techniques for legal AI, including early PLMs. [20] provides an overview of legal information retrieval methods, including natural language, ontology, and deep-learning systems. Francesconi et al. [12] presented his vision for the future of AI and law, including the use of machine and deep learning to extract knowledge from legal data, legal knowledge representation, and models for legal reasoning. Locke et al. [17] focuses on querying for ad-hoc case law retrieval, covering boolean and natural language, conceptual search, case-based retrieval, question answering, query expansion, query reduction, search result diversification, citation networks, and text comprehension. However, none of the above offers a systematic analysis of approaches and methods using PLMs for legal problems and tasks, which is the focus of this review activity.

Several works focuses on legal texts from European legal sources, such as European Convention on Human Rights [5,9], European Union Laws [6], statutes and regulations [2,15].

However, the drafting of such review highlights a notable gap in the literature concerning the use of artificial intelligence, specifically PLMs, for the analysis of European constitutions. Bayrak et al. [3] analyze the US constitution in relation to four geographical areas (Africa, Asia, Europe, and the Middle East) to discern disparities and commonalities in terms of citizen rights and governmental institutions. The research employs text mining methodologies, specifically eliminating context-independent approaches. In contrast to [3], the work proposed during the PhD course seeks to exploit PLMs to solve the gap in comprehending constitutional sections pertaining to citizens' rights and duties in European countries. This is achieved through similarity techniques to identify and comprehend shared topics across European constitutions.

3   Outline and Main Findings

The review activity first offered a thorough deep discussion of cutting-edge PLMs, with an emphasis on those employed in the legal field. Then, primary difficulties and related tasks in the legal AI sector that are being handled by PLM-based approaches were been investigated and classified, as well as relevant benchmarks in the legal area. This resulted in a thorough examination of the key current PLM-based approaches. Finally, such study analyzed the key discoveries and limits of present PLM-based methodologies, as well as open issues and potential directions for the next generation of legal AI tools. Specifically, a key factor in the effectiveness of a PLM-based framework w.r.t. other deep learning approaches lies in the training methodology employed, which mainly consists of task-adaptive fine-tuning or domain adaptation approaches. The leading frameworks currently leverage external techniques such as data augmentation, data enrichment, and ensemble strategies to enhance performance. The scarcity of legal data for specific tasks can hinder adequate model training, preventing the model from fully grasping the nuances of legal texts and limiting its generalization ability. The limited data availability becomes more pronounced when transitioning from high-resource languages like English to languages with fewer resources. In contrast to earlier methods constrained the maximum number of token in input, which limited PLMs' ability to process long documents, the modern approaches favor a more conservative model, which handles texts at the paragraph level or processing the entire document directly. Additionally, current techniques increasingly focus on injecting knowledge into models and enhancing their explainability, interpretability, and adherence to ethical standards. More details about such work can be found in the published paper [13].

As regards the analysis of European constitutions, the research consisted of conducting a lexical and semantic similarity analysis of European constitutions utilizing PLMs.

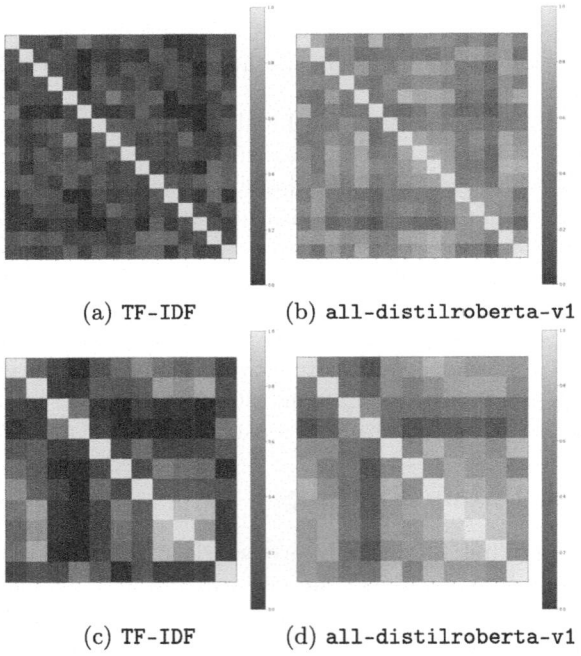

(a) TF-IDF   (b) `all-distilroberta-v1`

(c) TF-IDF   (d) `all-distilroberta-v1`

Fig. 1. Heatmaps corresponding to (a-b) micro-topic "*Limits on employment of children*" and (c-d) micro-topic "*Right to renounce citizenship*".

To this regard, a multi-level similarity analysis was conducted. The first analysis includes a comparison of texts from different countries that share a common topic (*micro-topic level*). The analysis is then extended to texts with different topics but belonging to the same macro-topic (*macro-topic level*), and finally, comparing texts from all macro-topics (*RD-level*). Constitutional texts were embedded using various PLMs, considering word-level encoders (e.g., BERT [11]), sentence-level encoders (i.e., Sentence-Transformers [19]), and models specifically designed for the legal domain (e.g., Legal-BERT [8]).

Table 1. Evaluation of the models. The *micro-topic* column shows the average value (μ) on the mean similarity scores between texts of the same micro-topic. The *macro-topic* column shows the average value (M) on the mean similarity scores for texts within the same macro-topic but different micro-topics. The *RD-level* column represents the average value (M') on the mean similarity scores for texts from different macro-topics and micro-topics. Other columns indicate the differences in values among these categories. The greater the difference, the better the model is at distinguishing texts belonging to the different categories.

model	micro-topic (μ)	macro-topic (M)	RD (M')	$|\mu - M|$	$|M - M'|$	$|\mu - M'|$
bert-base-uncased	0.853	0.799	0.762	0.053	0.037	0.091
legal-bert-uncased	0.905	0.871	0.852	0.033	0.019	0.052
legal-bert-500k	0.853	0.798	0.766	0.055	0.031	0.087
legal-bert-eurlex	0.853	0.796	0.756	0.057	0.039	0.096
legal-bert-echr	0.765	0.687	0.643	0.078	0.044	0.122
all-distilroberta-v1	0.576	0.399	0.313	0.177	0.086	0.263
all-mpnet-base-v1	0.603	0.414	0.323	*0.189*	*0.090*	*0.279*
gtr-t5-large	0.769	0.682	0.643	0.086	0.039	0.126

Experimental results revealed that:

– European constitutions exhibit numerous commonalities in terms of rights and duties;
– Lexical analysis alone does not adequately capture the nuanced similarities between different countries. In Fig. 1 can be noticed that the lexical model shows significantly lower scores than the semantic model. Specifically,

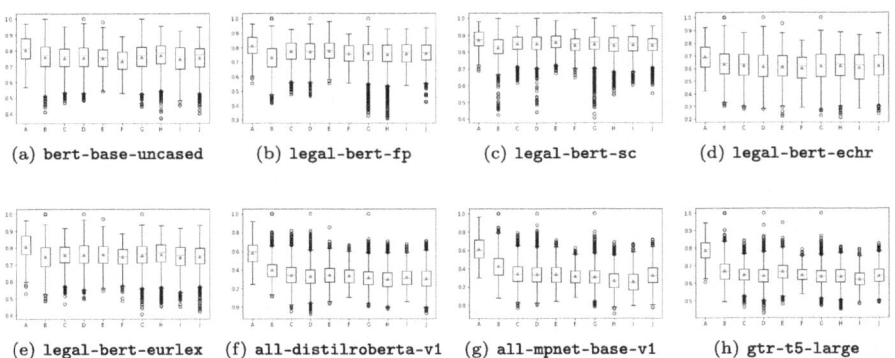

(a) bert-base-uncased (b) legal-bert-fp (c) legal-bert-sc (d) legal-bert-echr

(e) legal-bert-eurlex (f) all-distilroberta-v1 (g) all-mpnet-base-v1 (h) gtr-t5-large

Fig. 2. Boxplots on the micro-topic "*Prohibition of slavery*" vs the texts of constitutions referring to its macro-topic ("*Physical Integrity Rights*") and all the others macro-topics. A - *Prohibition of slavery*; B - *Physical Integrity Rights*; C - *Social Rights*; D - *Economic Rights*; E - *Citizen Duties*; F - *General Duties*; G - *Civil and Political Rights*; H - *Legal Procedural Rights*; I - *Enforcement*; J - *Minority Rights*.

`all-distilroberta-v1` in Fig. 1(b) identifies some similarity matches that are not detected by `TF-IDF` in Fig. 1(a). Moreover, although the heatmaps in Fig. 1(c) and Fig. 1(d) have similar shapes, the semantic model consistently assigns higher scores, capturing strong similarities more effectively and distinguishing texts discussing the same micro-topic, unlike the lexical model which often fails in this regard;

– The considered legal models were not ideally suited for similarity assessments. In contrast, sentence-level encoders yield markedly better performance, even though they are not specifically trained on legal texts. In Table 1 it can be observed that the generic `bert-base-uncased` and the legal models are the least effective in distinguishing between different micro-topics (the first three columns are generally high, while the last three columns are low). Among all the models, `all-mpnet-base-v1`, a Sentence-Transformer, achieves the most favorable results, demonstrating the largest disparity among all scenarios (last three columns) while maintaining an adequate score within all scenarios (first three columns). An illustrative example is provided in Fig. 2, in which the boxplots of the micro-topic "*Prohibition of slavery*" w.r.t. micro-topic level, macro-topic level and RD-level analysis are shown. In the case of `all-mpnet-base-v1`, the first boxplot, which corresponds to the micro-topic level analysis, shows the highest and most consistent values. The second boxplot, representing the macro-topic level analysis, shows values that are significantly lower than the first but higher than those of the RD-level analysis.

For a complete analysis, refer to the published paper [14].

4   Conclusion and Future Directions

This research explores the potential of Pre-trained Language Models (PLMs) in addressing AI-driven problems within the legal domain, particularly analyzing European constitutions. It provides the first systematic overview of PLM-based methods tailored to legal applications and employs PLMs to investigate similarities across European constitutions.

However, when the topics were closely aligned, all models struggled to distinguish subtle differences. Conducting extensive fine-tuning of PLMs on a broader range of constitutions-related texts could improve their ability to detect subtle differences in closely aligned topics. In addition, many different legal models may have been used in the experiments to provide further insight into legal architectures' similarity task performance. Moreover, the similarity task alone is not sufficient to understand in what constitutions are similar or in what they different. More complex tasks, such as summarization, can be useful in providing a clearer and more immediate overview of convergence or divergence between countries. Finally, expanding the analysis to include cross-linguistic and multilingual evaluations of European constitutions could uncover additional layers of similarity and difference, highlighting how language and translation impact legal interpretation and application. These considerations are the basis for future improvements of the work.

References

1. Ahmad, W.U., Chi, J., Le, T., Norton, T., Tian, Y., Chang, K.: Intent classification and slot filling for privacy policies. In: Zong, C., Xia, F., Li, W., Navigli, R. (eds.) Proceedings of the 59th Annual Meeting of the Association for Computational Linguistics and the 11th International Joint Conference on Natural Language Processing, ACL/IJCNLP 2021, (Volume 1: Long Papers), Virtual Event, 1–6 August 2021, pp. 4402–4417. Association for Computational Linguistics (2021). https://doi.org/10.18653/V1/2021.acl-long.340

2. Aumiller, D., Chouhan, A., Gertz, M.: EUR-Lex-Sum: a multi- and cross-lingual dataset for long-form summarization in the legal domain. In: Goldberg, Y., Kozareva, Z., Zhang, Y. (eds.) Proceedings of the 2022 Conference on Empirical Methods in Natural Language Processing, EMNLP 2022, Abu Dhabi, United Arab Emirates, 7–11 December 2022, pp. 7626–7639. Association for Computational Linguistics (2022). https://doi.org/10.18653/v1/2022.emnlp-main.519

3. Bayrak, T.: A comparative analysis of the world's constitutions: a text mining approach. Soc. Netw. Anal. Min. **12**(1), 26 (2022)

4. Callister, P.D.: Law, artificial intelligence, and natural language processing: a funny thing happened on the way to my search results. Law Libr. J. **112**, 161 (2020)

5. Chalkidis, I., Androutsopoulos, I., Aletras, N.: Neural legal judgment prediction in English. In: Korhonen, A., Traum, D.R., Màrquez, L. (eds.) Proceedings of the 57th Conference of the Association for Computational Linguistics, ACL 2019, Florence, Italy, July 28- August 2 2019, Volume 1: Long Papers, pp. 4317–4323. Association for Computational Linguistics (2019). https://doi.org/10.18653/v1/p19-1424

6. Chalkidis, I., Fergadiotis, M., Androutsopoulos, I.: MultiEURLEX - a multi-lingual and multi-label legal document classification dataset for zero-shot cross-lingual transfer. In: Moens, M., Huang, X., Specia, L., Yih, S.W. (eds.) Proceedings of the 2021 Conference on Empirical Methods in Natural Language Processing, EMNLP 2021, Virtual Event / Punta Cana, Dominican Republic, 7–11 November 2021, pp. 6974–6996. Association for Computational Linguistics (2021). https://doi.org/10.18653/v1/2021.emnlp-main.559

7. Chalkidis, I., Fergadiotis, M., Malakasiotis, P., Aletras, N., Androutsopoulos, I.: LEGAL-BERT: the muppets straight out of law school. In: Cohn, T., He, Y., Liu, Y. (eds.) Findings of the Association for Computational Linguistics: EMNLP 2020, pp. 2898–2904. Association for Computational Linguistics, Online, November 2020. https://doi.org/10.18653/v1/2020.findings-emnlp.261. https://aclanthology.org/2020.findings-emnlp.261

8. Chalkidis, I., Fergadiotis, M., Malakasiotis, P., Aletras, N., Androutsopoulos, I.: LEGAL-BERT: the Muppets straight out of law school. arXiv preprint arXiv:2010.02559 (2020)

9. Chalkidis, I., Fergadiotis, M., Tsarapatsanis, D., Aletras, N., Androutsopoulos, I., Malakasiotis, P.: Paragraph-level rationale extraction through regularization: a case study on European court of human rights cases. In: Toutanova, K., et al. (eds.) Proceedings of the 2021 Conference of the North American Chapter of the Association for Computational Linguistics: Human Language Technologies, NAACL-HLT 2021, Online, 6–11 June 2021, pp. 226–241. Association for Computational Linguistics (2021). https://doi.org/10.18653/v1/2021.naacl-main.22

10. Chalkidis, I., et al.: LexGLUE: a benchmark dataset for legal language understanding in English. In: Muresan, S., Nakov, P., Villavicencio, A. (eds.) Proceedings of the 60th Annual Meeting of the Association for Computational Linguistics (Volume 1: Long Papers), ACL 2022, Dublin, Ireland, 22–27 May 2022. pp. 4310–4330. Association for Computational Linguistics (2022). https://doi.org/10.18653/v1/2022.acl-long.297

11. Devlin, J., Chang, M.W., Lee, K., Toutanova, K.: BERT: pre-training of deep bidirectional transformers for language understanding. arXiv preprint arXiv:1810.04805 (2018)

12. Francesconi, E.: The winter, the summer and the summer dream of artificial intelligence in law. Artif. Intell. Law **30**(2), 147–161 (2022). https://doi.org/10.1007/s10506-022-09309-8

13. Greco, C.M., Tagarelli, A.: Bringing order into the realm of transformer-based language models for artificial intelligence and law. Artif. Intell. Law., 1–148 (2023)

14. Greco, C.M., Tagarelli, A.: Topic similarities in rights and duties across European constitutions using transformer-based language models. In: Proceedings of the ACM Hypertext Workshop on Legal Information Retrieval meets Artificial Intelligence (LIRAI) (2023)

15. Klaus, S., Hecke, R.V., Naini, K.D., Altingovde, I.S., Bernabé-Moreno, J., Herrera-Viedma, E.: Summarizing legal regulatory documents using transformers. In: Proceedings of the ACM SIGIR Conference on Research and Development in Information Retrieval (SIGIR), pp. 2426–2430. ACM (2022)

16. Koreeda, Y., Manning, C.D.: ContractNLI: a dataset for document-level natural language inference for contracts. In: Moens, M., Huang, X., Specia, L., Yih, S.W. (eds.) Findings of the Association for Computational Linguistics: EMNLP 2021, Virtual Event / Punta Cana, Dominican Republic, 16–20 November 2021, pp. 1907–1919. Association for Computational Linguistics (2021). https://doi.org/10.18653/v1/2021.findings-emnlp.164

17. Locke, D., Zuccon, G.: Case law retrieval: problems, methods, challenges and evaluations in the last 20 years. arXiv preprint arXiv:2202.07209 (2022)

18. Nguyen, H., Nguyen, L.: Sublanguage: a serious issue affects pretrained models in legal domain. CoRR **abs/2104.07782** (2021). https://arxiv.org/abs/2104.07782

19. Reimers, N., Gurevych, I.: Sentence-BERT: sentence embeddings using Siamese BERT-networks. arXiv preprint arXiv:1908.10084 (2019)

20. Sansone, C., Sperlí, G.: Legal information retrieval systems: state-of-the-art and open issues. Inf. Syst. **106**, 101967 (2022)

21. Shao, Y., et al.: BERT-PLI: modeling paragraph-level interactions for legal case retrieval. In: Bessiere, C. (ed.) Proceedings of the Twenty-Ninth International Joint Conference on Artificial Intelligence, IJCAI 2020, pp. 3501–3507. ijcai.org (2020). https://doi.org/10.24963/ijcai.2020/484

22. Simeri, A., Tagarelli, A.: GDPR article retrieval based on domain-adaptive and task-adaptive legal pre-trained language models. In: Wehnert, S., Fiorelli, M., Picca, D., Luca, E.W.D., Stellato, A. (eds.) Proceedings of the 1st Legal Information Retrieval meets Artificial Intelligence Workshop LIRAI 2023 co-located with the 34th ACM Hypertext Conference HT 2023, Rome, Italy, 04 September 2023. CEUR Workshop Proceedings, vol. 3594, pp. 63–76. CEUR-WS.org (2023). https://ceur-ws.org/Vol-3594/paper5.pdf

23. Surden, H.: Artificial intelligence and law: an overview. Georgia State Univ. Law Rev. **35**, 19–22 (2019)

24. Wang, S.Het al.: MAUD: an expert-annotated legal NLP dataset for merger agreement understanding. In: Bouamor, H., Pino, J., Bali, K. (eds.) Proceedings of the 2023 Conference on Empirical Methods in Natural Language Processing, EMNLP 2023, Singapore, 6–10 December 2023, pp. 16369–16382. Association for Computational Linguistics (2023). https://doi.org/10.18653/V1/2023.EMNLP-MAIN.1019, https://doi.org/10.18653/v1/2023.emnlp-main.1019

25. Zhong, H., Xiao, C., Tu, C., Zhang, T., Liu, Z., Sun, M.: How does NLP benefit legal system: a summary of legal artificial intelligence. arXiv preprint arXiv:2004.12158 (2020)

An Application for Scoliosis Screening and Follow-Up: A First Proposal

Lorella Bottino[1,2(✉)] [iD]

[1] Department of Medical and Surgical Sciences, University Magna Graecia of Catanzaro, Catanzaro, Italy
lorella.bottino@unicz.it
[2] Data Analytics Research Center, University Magna Graecia of Catanzaro, Catanzaro, Italy

Abstract. Adolescent idiopathic scoliosis (AIS) is a spinal deformity that tends to get worse as children grow and affects overall health. Generally, scoliosis is diagnosed by measuring the Cobb angle on radiographs, which represents the gold standard for scoliosis grade quantification. More recently, smartphone applications have been developed to replace the traditional methods of scoliosis management and to be included in clinical practice ensuring good accuracy.

Using a new methodology proposed by us, we evaluated currently available applications based on their main technological and functional characteristics. However current applications, for some aspects, show limitations. I therefore aim to develop a new application for remote screening and monitoring of scoliosis to achieve better management of scoliotic curvature and to simplify the doctor's workload. The application will be developed by taking ideal images of the patient's back and will subsequently be validated on a larger sample.

Keywords: spine · monitoring · data

1 Introduction

In the age of technological innovation, digital health is emerging as a player in the healthcare sector by revolutionizing the way healthcare services are delivered, managed and distributed [1].

In particular, external technological devices such as smartphones, thanks to the numerous sensors (accelerometer, gyroscope, proximity sensor) integrated on them, are increasingly present in the remote patient assessment; simple applications downloadable on these devices can be used to bring the doctor closer to the patient, improve the treatment process and allow access to care to an increasingly larger portion of the population.

In this context of remote healthcare, adolescent idiopathic scoliosis (AIS), a evolutionary pathology which for reasons not yet fully understood ("idiopathic")

J. Tekli et al. (Eds.): ADBIS 2024, CCIS 2186, pp. 318–324, 2025.
https://doi.org/10.1007/978-3-031-70421-5_27

arises generally during puberty and progresses until the end of growth, can be managed effectively.

Scoliosis is a lateral deviation of the spine in the frontal plane with rotation of the vertebrae on the horizontal plane [2].

Considering that during pubertal development (which in girls coincides with the arrival of menarche) there is a greater risk of progression of the scoliotic curve, greater attention have to be paid to the monitoring of adolescent patients precisely in this phase to prevent the progression of the pathology and reduce its impact.

In particular, regular checks carried out independently at home, in combination with periodic visits to the orthopedic doctor, can determine the success of the treatment and avoid the clinical worsening of the scoliotic curve, to the point of avoiding the need for surgical correction as occurs in the most serious cases.

The gold standard for the diagnosis and control of scoliosis is the x-ray which is an internal photograph and is used to measure the Cobb angle [3]. The Cobb angle is measured by determining the most tilted vertebrae on the superior and inferior aspects of the apex. The lines are drawn along the top of the superior tilted vertebra and the bottom of the inferior tilted vertebra. Two additional lines are then drawn perpendicularly to those. The angle of intersection of the perpendicular lines is the Cobb angle expressed in degrees.

The radiographic investigation is accompanied by a clinical evaluation of the patient which consists in the observation by the pediatric orthopedist of some specific symptoms of scoliosis.

In those patients with a greater evolutionary risk, the follow-up times required to monitor the evolution of scoliosis can be very close to each other.

However, go continuously to the clinic to carry out the medical checks in times when families are always very busy it can be quite difficult.

My project proposes an application capable of making alternative measurements to the Cobb angle on x-rays, with a double goal: 1) reduce exposure to ionizing radiation used in radiographic investigations; 2) overcome the current empirical methodology based on the doctor's experience in detecting signs of scoliosis.

The proposed application has a dual functionality:

- screening activities: the system allows to determine scoliosis in the early stages, before the patient has even had a consultation with the doctor. The system stratifies patients based on the presence or absence of scoliosis. If scoliosis is present, the system can suggest the patient to go to the doctor for an visit.
- monitoring activities: for patients who have already had the visit and know they have scoliosis, the system allows to take scoliosis measurements serialized over time and send them to the reference medical center. Thanks to this new monitoring service we aim to reduce exposure to x-rays, costs and waiting times, guaranteeing constant control and at the same time lightening the doctor's workload.

The objective of this work is to develop an application that will be based on the acquisition of images of the patient's spine which will be processed by a computer. Before proceeding with the development phase, it will be necessary to provide a medical/clinical definition of the signs of scoliosis that can be detected on a photo of the patient. The main challenges to be faced will be establishing the positioning of the patient that needs to be acquired and the anonymization of the image to ensure the privacy of the individual.

Furthermore, the application will allow the collection of a significant amount of data relating to scoliosis, which will be used for the implementation of predictive algorithms, which are not intended to replace the doctor but support him in his decisions by adequately motivating their decisions.

The rest of the paper is organized as follows: Sect. 2 presents a summary of the my survey work on apps related to scoliosis, highlighting the limitations of current systems. Section 3 discusses the infrastructure and functions that the application will be able to implement. The concluding remarks are presented in Sect. 4.

2   Background

My research project consists of a first phase in which I make a state of the art of the apps and information systems that can be used in the clinical practice to measure and monitor scoliosis.

More specifically I proposed a classification and evaluation methodology to assess the main technical and functional features of the available scoliosis apps. Furthermore, the study of this first phase is intended to provide a guideline that offers support in the choice of the scoliosis tools that are best suited to specific needs.

The methodology defines 5 aspects, named macro-categories, to assess the apps [4].

These 5 macro-categories are: Availability, Technology, Measurement, Functions, Qualitative evaluation.

Each macro-category includes a group of homogeneous categories used to characterize the apps in more depth.

The Availability macro-category includes that set of conditions, such as reachability and accessibility which make the app ready to be used by the user.

Reachability data represents the source from which the app can be downloaded (e.g. Apple store or Google store); accessibility data provides information on the payment, i.e. if app is obtainable through payment of a fee or is free of charge.

Another feature that is part of this first macro-category informs whether a tutorial or a user guide is provided.

The Technology macro-category includes information on technological aspects of the apps, such as operating system (e.g. iOS or Android) with which the app is compatible.

The Measurement macro-category refers to the tools integrated into the smartphone and used to acquire data relating scoliosis (e.g. gyroscope, accelerometer or the camera) as well as the measured parameters (e.g. Cobb angle).

The Functions macro-category provides information about the functionalities implemented by the app for scoliosis management.

Some relevant Functions data are: the posture monitoring; the presence of a medical interface which allows the doctor to interact with the patient in real time and control the scoliosis data; prescribing exercises and monitoring their benefits; collecting and recording additional information such as height, age and weight to monitor growth and check their impact on scoliosis.

Qualitative evaluation macro-category is a final evaluation that takes into consideration those characteristics that make the app a useful and effective tool in the management of scoliosis.

Apps are rated on their strengths, weaknesses, and user-friendliness.

Below the values of the technological and functional features for some of the most relevant scoliosis apps currently available are made explicit.

Scoliotrack allows the patient to track the progression of the spinal curve by using the smartphone accelerometer. The application is available on Apple Store and Play Store, and requires payment of a fee. Moreover Scoliotrack provides a video tutorial on how to use the app. As for the operating system, the app is compatible with both iOS and Android. It is simple enough for personal use at home and allows to store scoliosis measurements for future monitoring. Scoliotrack measures the patient's angle of trunk rotation and tracks changes in scoliosis over time. The application also stores additional information such as height, weight and age. Among the strengths of the app there are: the app shows data output in graph format; the app tracks changes over time; the app is easy and simple to use.

Scoliometer is available on Apple Store and Play Store. It provides a simple four-step guide to teach how to measure scoliosis. The application, provided at an affordable price, represents a simple and fast way to measure and monitor the angle of trunk rotation. The measurement is performed with a gyroscope/accelerometer. The app is compatible with both iOS and Android. Scoliometer is a simpler and cheaper version of the same app by developer of Scoliotrack.

CobbMeter is used to measure the Cobb angle on radiographs using the angle sensor available on the iPhone (microelectromechanical system accelerometer—MEMS). CobbMeter, freely available, is an app designed for spinal professionals, and it turns the iPhone into a professional measuring tool.

The APECS (Artificial Intelligence Posture Evaluation and Correction System) app performs the posture evaluations using markers that are positioned on the patient body photo, and it uses AI algorithms for performing accurate body symmetry assessment.

However currently available apps analyzed according to the proposed evaluation methodology have the following limitations:

- they are user-dependent: apps such as Scoliotrack and Scoliometer require the user to follow the profile of the spine with the digital goniometer and their measurements are therefore subject to error [4];
- they need x-ray: apps such as Cobbmeter calculate the Cobb angle on radiographs and therefore they are designed to be used by the doctor when the patient goes for a visit and takes an x-ray;
- they do not provide information on how to take the photo: apps like APECS which acquire the patient's image via camera and place markers on it to evaluate the symmetry of the body do not give information on how to take the photo at a geometric level and this can lead to an inaccuracy of the measurement [4].

In a second phase the objective of my research project is to develop a new system that is capable of measuring not the usual values that currently available apps measure but new values on easily transportable data such as images.

3   A Novel Application for Remote Management of Scoliosis Information

The new proposed remote scoliosis diagnosis and monitoring system based on a simple Android/iOS application downloadable on a smartphone is characterized by: 1) the patient's parent's smartphone camera which will be used as a scoliosis data acquisition device, 2) a server that receives the acquired data, 3) a processor which performing standard analyses generates results, 4) the creation of a PDF file containing the results of the processing, which can be exported and shared with the doctor via text message or email.

Thanks to this infrastructure the orthopedist is supported in the diagnosis and monitoring of scoliosis patients: from a single location it is possible to visit and monitor patients, all in real time and guaranteeing greater accuracy.

The patient, on the other side, receives feedback from the doctor which may include having to go to the medical clinic for a check-up, doing some exercises, or simply receiving suggestions (see Fig. 1).

The system offers the opportunity to view the measurement history of scoliosis data, with the possibility of viewing the report of each measurement session.

The system also offers the possibility of viewing the graph of the values of the measurements detected to understand the evolution of scoliosis over time.

Furthermore, the system allows the sharing of data obtained during the diagnosis and follow-up phase through a cloud-based repository.

Scoliosis data and information related patients are collected and properly arranged in databases.

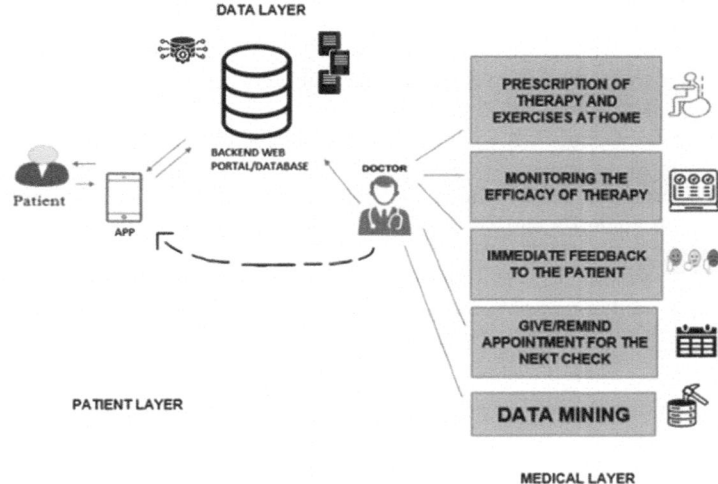

Fig. 1. Novel application workflow for scoliosis monitoring

4   Conclusion and Future Work

The use of apps for scoliosis screening and monitoring spinal curvature progression brings several advantages to both patients and health professionals, in particular to orthopedics specialists. Having the possibility of monitoring scoliosis from home is not only convenient because, through a user friendly app within everyone's reach, it is possible to carry out periodic scoliosis checks without the need to go to the doctor too frequently, but also because such system has a notable impact on pathology management. The doctor can remotely check the evolution of the scoliosis and send the patient immediate feedback. In future work, we plan to develop an app to objectify scoliosis evaluation through the acquisition of photographic images and the subsequent measurement of new values on the photos taken.

We also plan to make a selection from the data collected to model and train AI algorithms to support the orthopedist in the decision-making process.

Acknowledgement. This work was funded by the Next Generation EU - Italian NRRP, Mission 4, Component 2, Investment 1.5, call for the creation and strengthening of "Innovation Ecosystems", building "Territorial R&D Leaders" (Directorial Decree n. 2021/3277) - project Tech4You - Technologies for climate change adaptation and quality of life improvement, n. ECS0000009. This work reflects only the authors' views and opinions, neither the Ministry for University and Research nor the European Commission can be considered responsible for them.

References

1. Shinbane, J.S., Saxon, L.A.: Digital monitoring and care: virtual medicine. Trends Cardiovasc. Med. **26**(8), 722–730 (2016)
2. Weinstein, S.L., Dolan, L.A., Cheng, J.C., Danielsson, A., Morcuende, J.A.: Adolescent idiopathic scoliosis. The Lancet **371**(9623), 1527–1537 (2008)
3. Wang, J., Zhang, J., Xu, R., Chen, T.G., Zhou, K.S., Zhang, H.H.: Measurement of scoliosis Cobb angle by end vertebra tilt angle method. J. Orthop. Surg. Res. **13**, 1–7 (2018)
4. Bottino, L., Settino, M., Promenzio, L., Cannataro, M.: Scoliosis management through apps and software tools. Int. J. Environ. Res. Public Health **20**(8), 5520 (2023)

Negation Detection in Italian: A Key Challenge in Sentiment Analysis

Maria Chiara Martinis[1,2](✉) ⓘ

[1] Department of Medical and Surgical Sciences, University "Magna Græcia", 88100 Catanzaro, Italy
[2] Data Analytics Research Center, University "Magna Græcia", 88100 Catanzaro, Italy
martinis@unicz.it

Abstract. Sentiment analysis, a process aimed at identifying opinions and emotions present in textual data, presents significant challenges despite the progress made in text classification thanks to the new generation of neural language models such as Transformers. These challenges are particularly evident in situations characterized by limited lexical production or specific contexts with specific domains. One of the main challenges in sentiment analysis is the correct identification of negation. Detecting such relationships within Italian texts is crucial. Negation can lead to the opposite meaning of a sentence, significantly influencing overall understanding. In this approach, it is important to follow linguistic cues of negation such as adverbs or complex constructions and can be used in various contexts and languages, making it more flexible. In fields like bioinformatics and medical literature, negation detection plays a crucial role in building connections between genes and diseases, medication prescription history, or the lack of documented complaints by patients. The purpose of this article is to provide a general overview of the search for more effective solutions to detect negation in Italian texts, addressing current limitations such as the scarcity of datasets in the medical domain. Future work will focus on the use of high-level approaches to further improve the performance of negation identification and enhance the stability of sentiment analysis, especially in complex contexts.

Keywords: NLP · Sentiment analysis · Negation detection

1 Introduction

Sentiment analysis, fundamental in the field of Natural Language Processing (NLP), represents a useful tool for understanding the subjective aspects of textual data. Its main objective is to automatically identify the prevailing sentiment within a text, distinguishing whether it is positive, negative, or neutral. This research activity finds extensive applications in various sectors, including social media monitoring, customer feedback analysis, market trend evaluation, and more, but particularly, in this study, the focus will be on the medical

© The Author(s), under exclusive license to Springer Nature Switzerland AG 2025
J. Tekli et al. (Eds.): ADBIS 2024, CCIS 2186, pp. 325–330, 2025.
https://doi.org/10.1007/978-3-031-70421-5_28

field. Through the analysis of sentiments present in textual content, this method enables businesses, scholars, and institutions to extract valuable insights regarding public opinion, user sentiments, and particularly those of patients. A primary obstacle in sentiment analysis lies in the accurate identification and interpretation of negation within texts, especially in languages like Italian. Negation detection assumes critical importance, as it enables a nuanced understanding of expressed sentiments and facilitates adjustments to polarity scores accordingly. The presence of negation substantially alters the meaning of a sentence, profoundly influencing the overall comprehension of the text. Sectors heavily reliant on sentiment analysis include marketing and the medical field, which aid in determining associations between, for example, genes and diseases, evaluating medication prescription statuses and assessing the absence or presence of symptoms in patients' medical records. By delving into more efficient solutions for negation detection, this study aims to overcome current limitations, such as the scarcity of specific datasets in the medical domain. Furthermore, future research directions will focus on adopting advanced methodologies to further refine negation detection and enhance the overall robustness of sentiment analysis, especially in complex linguistic and domain-specific contexts. The rest of the paper is organized as follows: Sect. 2 presents related works, Sect. 3 discusses the crucial importance of sentiment analysis for a wide range of applications while highlighting the complex challenges associated with negation detection, especially in languages other than English. Section 4 addresses the issue of negation detection in Italian texts, highlighting the various linguistic peculiarities that make it a challenging task, but advanced NLP techniques and domain understanding are proposed as solutions to improve the accuracy of sentiment analysis; Sect. 5 discusses the results of paper and finally, Sect. 6 concludes the document.

2 Related Works

Sentiment analysis has made significant strides thanks to the adoption of advanced neural language models like Transformers. However, significant challenges persist in contexts with limited linguistic resources and specific domains. In particular, negation detection remains one of the primary difficulties. Previous studies have shown that negation can significantly alter the meaning of sentences, thereby influencing the accuracy of sentiment analysis results. Despite significant progress in recent years, particularly driven by sophisticated neural language models like Transformers, the application of sentiment analysis faces notable challenges [4]. These challenges are particularly evident in contexts characterized by limited linguistic resources or specific domains. Recent research has focused on using advanced NLP techniques to improve negation detection, especially in languages other than English. For instance, studies in bioinformatics and medicine have demonstrated the importance of sentiment analysis in determining associations between genes and diseases, evaluating medication prescriptions, and analyzing the presence or absence of symptoms in clinical data. However, the scarcity of specific datasets in the Italian medical domain limits the effectiveness of current methodologies. Proposed techniques include using pre-trained language models, manual annotation of specific datasets, and integration of domain knowledge to enhance the precision of sentiment analysis. Omitting negation can lead to misinterpretations and compromise the accuracy of sentiment analysis results [1]. This study explores the complexities of negation detection in Italian texts, particularly in the fields of bioinformatics and medical literature. Future research will concentrate on adopting these techniques to overcome current limitations and improve the overall robustness of sentiment analysis in complex linguistic and domain-specific contexts.

3   Challenges in Sentiment Analysis: a Particular Focus on Negation Detection

The challenges in sentiment analysis, with a particular focus on negation detection, are complex and highlight the difficulties involved in accurately deciphering the sentiments embedded within textual data. While sentiment analysis holds immense importance in various applications, including market research, social media monitoring, and customer feedback analysis, its weak point lies in overcoming hurdles related to negation detection, especially in languages other than English, where discoveries in this field are still limited. Negation detection constitutes a crucial aspect of sentiment analysis, as it directly influences the interpretation of text by reversing or negating the expressed sentiment. Failure to accurately identify negation can lead to distorted analyses. One of the main challenges is to distinguish subtle linguistic cues indicative of negation, such as double negatives or word order variations. These differences are particularly pronounced in languages like Italian, renowned for their rich grammatical structures and nuanced expressions. Additionally, idiomatic expressions further complicate the accurate detection of negation, as their figurative meanings may not align with literal interpretations. Furthermore, contextual factors introduce additional layers of complexity to negation detection. Texts from different domains exhibit varying patterns of negation usage, necessitating domain-specific approaches for robust analysis. For example, medical literature may utilize specialized terminology and negation constructs distinct from those found in general discourse [7]. Addressing these challenges requires the development of advanced Natural Language Processing (NLP) techniques capable of analyzing and interpreting textual data with precision. Advanced machine learning algorithms, along with extensive linguistic resources and domain-specific knowledge, offer a good starting point for improving the accuracy of negation detection. By overcoming these challenges, sentiment analysis can achieve greater reliability and utility across various domains.

4   Negation Detection in Italian Texts

The detection of negation in Italian texts faces multiple linguistic obstacles stemming from the language's morphological complexity, the flexibility in word arrangement, and the extensive use of idiomatic expressions. This context presents a series of challenges and opportunities, which are elaborated upon as follows:

- Syntax of the language: Italian stands out for its considerable syntactic flexibility compared to English. The placement of negative elements, such as "non," can vary about the verb, depending on the emphasis or stylistic preferences of the speaker. This variability poses a challenge for automated negation detection, as its arrangement may evade a predefined model.
- Double Negation: Italian frequently employs double negation, using two or more negative elements to emphasize a concept or for rhetorical purposes.
- Use of Idiomatic Expressions: Like many languages, Italian includes idiomatic expressions where negation may not be explicitly articulated.
- Implications in Sentiment Analysis: Accurate detection of negation holds crucial importance in sentiment analysis, especially in fields such as bioinformatics and medical literature. Misinterpreting negated statements can lead to distorted evaluations of sentiment polarity, with repercussions on subsequent applications such as disease diagnosis or drug efficacy analysis.
- Ambiguity: Resolving ambiguity in negation detection is critical for precise sentiment analysis. Utilizing contextual information, syntactic analyses, and semantic techniques allows for distinguishing negated statements and determining whether the negative sign applies to the entire sentence or only a specific part.

The use of advanced NLP techniques, such as deep learning models, transformative architectures, and pre-trained linguistic representations like BERT, can enhance negation detection in Italian texts [2]. These approaches enable the learning of complex patterns and capture semantic nuances, thereby facilitating the accurate identification of negated statements. Integrating negation detection systems with domain-specific knowledge bases or ontologies enriches the context and improves precision. Linking negated statements to relevant concepts or entities in knowledge bases contributes to more accurate sentiment analysis, especially in specialized domains. In conclusion, despite the significant linguistic challenges in negation detection in Italian texts, advanced NLP techniques and deep domain understanding can foster more precise sentiment analysis and enhance comprehension of textual data in diverse contexts.

5   Discussion and Results

This section aims to explore the outcomes derived from research on negation detection in Italian texts, specifically focusing on bioinformatics and medical literature. It commences by presenting the findings unearthed during the study, shedding light on the pivotal observations gleaned from the analysis of negation within textual data. These findings are subsequently placed within the broader landscape of sentiment analysis and its myriad potential applications. In [6], the authors unveil VADER-IT as an outcome of their research endeavors. Serving as an adaptation of the VADER tool tailored specifically for analyzing polarity within Italian texts, VADER-IT leverages the Italian lexicon and refines five rules inherited from the original VADER framework [3]. Moreover, VADER-IT integrates a novel method capable of predicting polarity categories in addition to polarity scores. The software underwent rigorous evaluation utilizing a dataset sourced from QSalute, a platform where patients articulate their reviews and assessments about health-related subjects. Impressively, the results showcased an F1 score of 85% across a comprehensive array of 5495 reviews, all revolving around gynecology topics. Similarly, in [5], VADER-IT emerged as a tangible product of the research endeavors, undergoing rigorous testing for sentiment extraction. The software was put through its paces utilizing 526 reviews published on the QSalute review platform (https://www.qsalute.it/), originating from patients who had availed themselves of medical care across various hospitals and healthcare facilities throughout Italy. Leveraging VADER-IT, the sentiment polarity was effectively extracted from each Italian review, contributing further to the burgeoning landscape of sentiment analysis research. In conclusion, the implementation and testing of VADER-IT in Italian medical contexts represent significant strides forward in applying sentiment analysis, demonstrating its potential in assessing patient satisfaction and medical experience analysis.

6 Conclusion

Despite significant advances in linguistic classification through advanced neural language models like Transformers, sentiment analysis continues to pose significant challenges. These difficulties are particularly pronounced when there are limited datasets available in the specific language, especially within specific industry domains. It is well-known that sentiment analysis becomes especially complex when dealing with identifying negations in texts, as negations have the power to radically alter the meaning of sentences, heavily influencing their interpretation. Ignoring negations can lead to misunderstandings and compromise the overall accuracy of the analysis. To address these challenges, it is crucial to precisely identify specific features of negations such as adverbs and complex constructions, whose nuances can vary significantly across contexts and languages.

This work aims to make a significant contribution to the field of Natural Language Processing (NLP), focusing primarily on sentiment analysis, which is crucial for understanding the subjective aspects of textual data.

One of the primary challenges in sentiment analysis, especially in languages like Italian, is the accurate identification and interpretation of negations in texts. Properly identifying negations is essential because it allows for a more detailed understanding of the expressed sentiments, directly influencing polarity results.

Future research directions will focus on adopting advanced methodologies to further enhance negation detection, aiming to increase the overall robustness of sentiment analysis, particularly in complex linguistic and sector-specific contexts.

Acknowledgement. This work was funded by the Next Generation EU - Italian NRRP, Mission 4, Component 2, Investment 1.5, call for the creation and strengthening of 'Innovation Ecosystems', building 'Territorial R&D Leaders' (Directorial Decree n. 2021/3277) - project Tech4You - Technologies for climate change adaptation and quality of life improvement, n. ECS0000009. This work reflects only the authors' views and opinions, neither the Ministry for University and Research nor the European Commission can be considered responsible for them.

References

1. Agarwal, A., Xie, B., Vovsha, I., Rambow, O., Passonneau, R.J.: Sentiment analysis of Twitter data. In: Proceedings of the Workshop on Language in Social Media (LSM 2011), pp. 30–38 (2011)
2. Devlin, J., Chang, M.W., Lee, K., Toutanova, K.: BERT: pre-training of deep bidirectional transformers for language understanding. arXiv preprint arXiv:1810.04805 (2018)
3. Hutto, C., Vader, G.E.: A parsimonious rule-based model for sentiment analysis of social media text. In: Proceedings of the Eighth International AAAI Conference on Weblogs and Social Media, pp. 1–4 (2014)
4. Kiritchenko, S., Zhu, X., Mohammad, S.M.: Sentiment analysis of short informal texts. J. Artif. Intell. Res. **50**, 723–762 (2014)
5. Martinis, M.C., Zucco, C.: Evaluation of healthcare structures in Italy through sentiment analysis. Numer. Comput. Theory Algorithms NUMTA **2023**, 147 (2023)
6. Martinis, M.C., Zucco, C., Cannataro, M.: An Italian lexicon-based sentiment analysis approach for medical applications. In: Proceedings of the 13th ACM International Conference on Bioinformatics, Computational Biology and Health Informatics, pp. 1–4 (2022)
7. Morante, R., Blanco, E.: Recent advances in processing negation. Nat. Lang. Eng. **27**(2), 121–130 (2021). https://doi.org/10.1017/S1351324920000534

Optimizing Federated Learning and Increasing Efficiency

Mihailo Ilić$^{(\boxtimes)}$ (iD)

University of Novi Sad, Faculty of Sciences, Trg Dositeja Obradovića 3,
Novi Sad 21000, Republic of Serbia
milic@dmi.uns.ac.rs
https://www.pmf.uns.ac.rs/en/

Abstract. Federated learning (FL) is a distributed machine learning (ML) paradigm which has lately enjoyed the attention of various different domains. Since it brings added privacy, security, and computability advantages, it is a natural fit for many different distributed ML environments. As a consequence of it being a paradigm with such broad applications, many different algorithmic, privacy, personalization, communication, and training efficiency challenges arise, just to name a few. This work focuses on optimizing the FL process in terms of strategy selection, communication, and computational efficiency, for both cross-silo and cross-device settings. This can be achieved during the setup phase by analysing the FL environment, and by selecting optimal network architectures.

Keywords: Federated Learning · Edge Computing · Model Sparsification · Optimisation

1 Introduction

Federated learning is a type of distributed learning setup where data collection and storage is carried out at the edge, and where data is never shared among the participants of the learning process. This approach brings numerous security, privacy, communication, and computational benefits. As such, it enjoys a wide range of applications which can be categorized by various criteria [12].

Classification into *cross-silo* and *cross-device* federated learning is based on the number of participants (edge nodes), the former having a handful of clients, while the latter can have even hundreds of thousands. Different challenges are present in each setup. However, for critical systems [7, 10, 25], it is vital to obtain reliable models as soon as possible.

A wide variety of research directions can be identified, some of the more prominent challenges being privacy, algorithms, communication efficiency, and model personalization. Considerable effort has already been devoted to these challenges. This work, however, focuses on organizational aspects of FL from 2

J. Tekli et al. (Eds.): ADBIS 2024, CCIS 2186, pp. 331–336, 2025.
https://doi.org/10.1007/978-3-031-70421-5_29

perspectives, particularly in environments with data availability and computational power constraints. The research question can be summarized as – *How can the FL process be optimized during the setup phase to ensure optimal model performance and minimal resource usage?* Two stages make up the research plan:

- First, an analysis of learning orchestration techniques is carried out to determine the FL strategy which yields the best models in cross-silo environments, tackling the cold start problem, i.e. situations where it is crucial to obtain well performing models as soon as possible.
- The second part of this work explores the use of Watts-Strogatz [24] network priors in FL as an alternative to network pruning. This technique aims to reduce computational load and communication overhead when training models at the edge, which poses a challenge in IoT environments with limited computational power.

The methods outlined in this research plan would allow for more efficient use of computational resources during FL through smart preparation during the setup phase, prior to learning. The first point in the research plan would secure a faster way of obtaining optimal models in environments with less data available. The second point would bring value in massively distributed environments where efficiency is key, by selecting optimal architectures during setup.

The rest of the paper is structured as follows. Section 2 briefly covers related work. Section 3 covers the impact of different model updating strategies on the performance of FL models. Section 4 explains the use of sparsification techniques in deep learning and the application of Watts-Strogatz models to FL environments. Concluding remarks are given in Sect. 5.

2   Related Work

The development of different FL aggregation techniques like FedAvg, FedProx, and FedHybrid cover *algorithmic aspects* of research [15–17]. This work focuses on algorithmic optimizations of FL, while the research outlined in this paper focuses on structural and organizational optimization. *Model personalization* tackles balancing between models' generalization ability and optimal performance at each individual edge node [3,8,9,14,20,23,27]. *Communication efficiency* solutions propose message quantization and optimized user sampling [1,5,6,13,18,19]. Contrary to these proposals, this work aims to optimize message size indirectly, by first optimizing FL model architectures. Comparisons of concurrent and incremental model updating strategies have been made in [21,22], focusing mainly on medical imaging data. The work described in this paper extends this to tabular data and different tasks like regression, binary and multiclass classification.

3   Optimal Selection of FL Strategies

Algorithmic aspects of FL are prevalent in related work. The main takeaway from this research is that there exists no clear best choice for all different possible FL configurations. Consequently, my research is focused on exploring efficient organization of the FL process through extensive experimentation. The robustness of 2 concurrent and 2 incremental strategies in cross-silo environments was tested on 15 different datasets [11], including extremely non-IID data from the LEAF collection of datasets [4]. Input consisted of both tabular and image data, for both classification and regression problems.

The main focus was on the relationship between data availability and the number of participating edge nodes. The FL strategies tested were: *incremental, cyclic-incremental, concurrent, semi-concurrent*. Various data distributions between edge nodes were explored. A general trend was observed where incremental methods provided more reliable models in situations where data was spread thin between edge nodes. In other words, incremental methods proved to be more robust in cross-silo settings with less data available. Concurrent methods caught up to incremental ones once more data became available in the federated environment.

These findings point to the importance of assessing the state of the distributed system, particularly the data availability, as training orchestration can have a significant impact on the performance of models in early stages of data collection in federated environments. Critical systems based on cross-silo FL which need to have access to reliable models as soon as possible may benefit from initially using incremental methods, and later switching to concurrent methods of learning, once more data becomes available.

4   FL Model Sparsification Using Network Priors

Computational costs and communication overhead present challenges in massively distributed environments, such as cross-device FL. Having potentially hundreds of thousands of participants communicating with the FL server in centralized FL exposes this communication bottleneck. Proposed methods aim to resolve this issue either through reducing the size of the messages being communicated, or by optimally selecting a subset of the edge nodes for training.

Since neural networks are commonly used in FL settings, network pruning can be utilized to reduce model complexity and size, which in turn reduces message size when communicating model updates. This way, more sparse neural network architectures are obtained. In order to determine which connections to remove in a trained neural network, some sort of importance metric needs to be introduced like weight magnitude or activation importance. This process is carried out during training. In order to truly reduce communication and computational costs, this would ideally be conducted prior to training, during the initial setup of the FL environment. By utilizing knowledge gained from network science, graphs called *"small worlds"* with certain properties like short average path lengths and high clustering can be used as a baseline for hidden layer architectures.

The *Watts-Strogatz* model is an approach to constructing random graphs with *"small world"* properties. WS graphs can be generated by first creating a ring lattice where each of the N nodes is connected to k of its closest neighbours. After this, each of the existing edges has a chance p of being randomly rewired. The graph prior can then be transformed to a hidden layer architecture [2,24] by transforming the undirected graph to a directed one (Fig. 1a). Nodes with an in-degree of 0 represent the input layer, while nodes with an out-degree equal to 0 are placed in the last hidden layer (Fig. 1b).

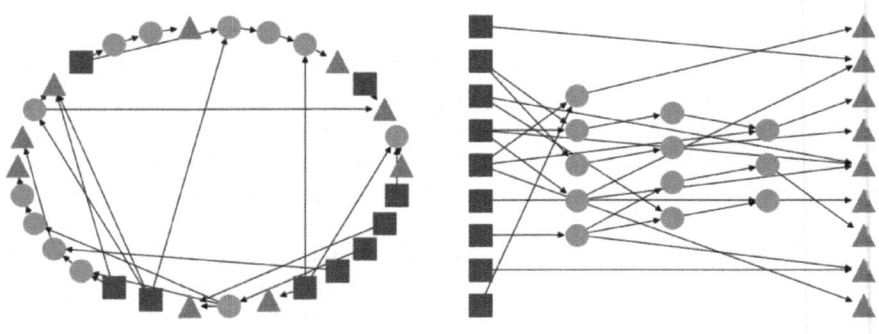

(a) Directed WS graph.   (b) WS-based hidden layer architecture.

Fig. 1. The transformation of a Watts-Strogatz graph into a hidden layer architecture. The input parameters for the WS model are $N = 30, k = 2, p = 0.5$. Blue squares have an in-degree equal to 0 and are placed in the first layer, red triangles have an out-degree of 0 and are placed in the last layer, while all other nodes (green circles) are in various layers in between. (Color figure online)

These kind of hidden layer architectures have been shown to produce quality models in centralized learning scenarios [26], while significantly reducing the cost of training. My work aims to optimize the federated learning process in terms of communication overhead and computational efficiency by introducing network priors to FL as a baseline for neural network architectures.

The first stage of research will validate the impact of WS hyperparameters (N, k, and p) on the performance of models based on network priors. Following this, the best performing network architectures will be compared to regular networks which have not been pruned, and those pruned through means like L1 and L2 regularization. Their performance can be compared through an analysis of model accuracy, loss, and computations carried out (FLOPS) during training.

5   Conclusion

FL as a learning paradigm is gaining significant attention lately, as it brings numerous benefits to various domains. However, each use case comes with its own specific challenges. This work is focused on optimizing the performance of FL models and the training process itself, based on actions which can be taken during the setup phase and early stages of learning.

By applying the principles outlined in this paper and analysing the environment prior to learning, optimal models can be obtained earlier in the learning process. Furthermore, federated environments with limited resources can benefit from utilizing specific neural network architectures which minimize resource usage while maximizing performance.

In future work, a detailed analysis of the effects which various WS architectures have on FL will be carried out.

Acknowledgments. Mihailo Ilić gratefully acknowledges the financial support of the Ministry of Science, Technological Development and Innovation of the Republic of Serbia (Grants No. 451-03-66/2024-03/200125 & 451-03-65/2024-03/200125).

Disclosure of Interests. The author has no competing interests.

References

1. Alistarh, D., Grubic, D., Li, J., Tomioka, R., Vojnovic, M.: QSGD: communication-efficient SGD via gradient quantization and encoding. In: Advances in Neural Information Processing Systems, vol. 30 (2017)
2. Amor, M.B., Stier, J., Granitzer, M.: Correlation analysis between the robustness of sparse neural networks and their random hidden structural priors. Procedia Comput. Sci. **192**, 4073–4082 (2021)
3. Armacki, A., Bajovic, D., Jakovetic, D., Kar, S.: Personalized federated learning via convex clustering. In: 2022 IEEE International Smart Cities Conference (ISC2), pp. 1–7 (2022). https://doi.org/10.1109/ISC255366.2022.9921863
4. Caldas, S., et al.: LEAF: A benchmark for federated settings (2018). arXiv preprint arXiv:1812.01097
5. Chen, W., Horvath, S., Richtarik, P.: Optimal client sampling for federated learning (2020). arXiv preprint arXiv:2010.13723
6. Cho, Y.J., Wang, J., Joshi, G.: Client selection in federated learning: Convergence analysis and power-of-choice selection strategies (2020). arXiv preprint arXiv:2010.01243
7. Dayan, I., et al.: Federated learning for predicting clinical outcomes in patients with COVID-19. Nat. Med. **27**(10), 1735–1743 (2021)
8. Fallah, A., Mokhtari, A., Ozdaglar, A.: Personalized federated learning with theoretical guarantees: a model-agnostic meta-learning approach. Adv. Neural. Inf. Process. Syst. **33**, 3557–3568 (2020)
9. Hanzely, F., Richtárik, P.: Federated learning of a mixture of global and local models (2020). arXiv preprint arXiv:2002.05516
10. Ilić, M., et al.: The role of federated learning in processing cancer patients' data. In: Savaglio, C., Fortino, G., Zhou, M., Ma, J. (eds.) Device-Edge-Cloud Continuum. Internet of Things. Springer, Cham (2024). https://doi.org/10.1007/978-3-031-42194-5_4
11. Ilić, M., Ivanović, M., Kurbalija, V., Valachis, A.: Towards optimal learning: Investigating the impact of different model updating strategies in federated learning. Expert Syst. Appl. **249**, 123553 (2024)
12. Kairouz, P., et al.: Advances and open problems in federated learning. Foun. Trends® Mach. Learn. **14**(1–2), 1–210 (2021)

13. Koloskova, A., Stich, S., Jaggi, M.: Decentralized stochastic optimization and gossip algorithms with compressed communication. In: International Conference on Machine Learning, pp. 3478–3487. PMLR (2019)
14. Li, L., Fan, Y., Tse, M., Lin, K.Y.: A review of applications in federated learning. Comput. Ind. Eng. **149**, 106854 (2020)
15. Li, T., Sahu, A.K., Zaheer, M., Sanjabi, M., Talwalkar, A., Smith, V.: Federated optimization in heterogeneous networks. Proc. Mach. Learn. Syst. **2**, 429–450 (2020)
16. McMahan, B., Moore, E., Ramage, D., Hampson, S., y Arcas, B.A.: Communication-efficient learning of deep networks from decentralized data. In: Artificial Intelligence and Statistics, pp. 1273–1282. PMLR (2017)
17. Niu, X., Wei, E.: Fedhybrid: a hybrid federated optimization method for heterogeneous clients. IEEE Trans. Signal Process. **71**, 150–163 (2023)
18. Reisizadeh, A., Mokhtari, A., Hassani, H., Jadbabaie, A., Pedarsani, R.: FedPAQ: a communication-efficient federated learning method with periodic averaging and quantization. In: International Conference on Artificial Intelligence and Statistics, pp. 2021–2031. PMLR (2020)
19. Ribero, M., Vikalo, H.: Communication-efficient federated learning via optimal client sampling (2020). arXiv preprint arXiv:2007.15197
20. Sattler, F., Müller, K.R., Samek, W.: Clustered federated learning: model-agnostic distributed multitask optimization under privacy constraints. IEEE Trans. Neural Netw. Learn. Syst. **32**(8), 3710–3722 (2020)
21. Sheller, M.J., et al.: Federated learning in medicine: facilitating multi-institutional collaborations without sharing patient data. Sci. Rep. **10**(1), 1–12 (2020)
22. Sheller, M.J., Reina, G.A., Edwards, B., Martin, J., Bakas, S.: Multi-institutional deep learning modeling without sharing patient data: a feasibility study on brain tumor segmentation. In: Crimi, A., Bakas, S., Kuijf, H., Keyvan, F., Reyes, M., van Walsum, T. (eds.) BrainLes 2018. LNCS, vol. 11383, pp. 92–104. Springer, Cham (2019). https://doi.org/10.1007/978-3-030-11723-8_9
23. Smith, V., Chiang, C.K., Sanjabi, M., Talwalkar, A.S.: Federated multi-task learning. In: Advances in Neural Information Processing Systems, vol. 30 (2017)
24. Stier, J., Granitzer, M.: Structural analysis of sparse neural networks. Procedia Comput. Sci. **159**, 107–116 (2019)
25. Taïk, A., Cherkaoui, S.: Electrical load forecasting using edge computing and federated learning. In: ICC 2020-2020 IEEE International Conference on Communications (ICC), pp. 1–6. IEEE (2020)
26. Traub, T., Nashouqu, M., Gulyás, L.: Efficient sparse networks from watts-strogatz network priors. In: Nguyen, N.T., et al. Computational Collective Intelligence. ICCCI 2023. LNCS(), vol. 14162. Springer, Cham (2023). https://doi.org/10.1007/978-3-031-41456-5_13
27. Wang, K., Mathews, R., Kiddon, C., Eichner, H., Beaufays, F., Ramage, D.: Federated evaluation of on-device personalization (2019). arXiv preprint arXiv:1910.10252

Integrating Pseudo-time Series Analysis Into Telemedicine: Enhancing Real-Time Disease Monitoring and Intervention

Barbara Puccio[✉][iD]

Department of Surgical and Medical Sciences, Magna Graecia University of
Catanzaro, Catanzaro, Italy
barbara.puccio@unicz.it

Abstract. The proposal outlines the development of a new telemedicine
platform that incorporates pseudo time series (PTS) analysis to mimic
the progression of patient health conditions using cross-sectional data.
The platform aims to revolutionize real-time disease trajectory monitoring and intervention, addressing the limitations of existing telehealth
systems that lack dynamic temporal analysis. The preliminary findings
highlight the feasibility of constructing pseudo-temporal trajectories that
effectively mirror disease progression, thereby facilitating timely and personalized medical interventions.

Keywords: Pseudo Time Series · Telemedicine · Healthcare

1   Introduction

The use of advanced information management technologies in medical decision-
making is increasingly important in today's healthcare, especially in light of
challenges like the COVID-19 pandemic [1]. These technologies help in the efficient and secure handling of various types of data such as clinical, demographic,
and environmental information. They are crucial for developing integrated
telemedicine procedures like telemonitoring and telehealth systems [2]. **The
COVID-19 pandemic has highlighted the critical need for telemedicine
systems that are not only responsive but also adaptable to the rapid
changes in healthcare demands, emphasizing the importance of real-
time data processing and dynamic health monitoring.** While traditional
telemedicine platforms offer benefits in terms of accessibility and cost reduction,
they typically cannot analyze the progression of patient health conditions over
time. This limitation is significant, particularly in chronic disease management
and elderly care, where understanding changes over time is crucial. **While existing telemedicine platforms have provided substantial benefits, their
inability to track health condition progressions over time presents a
significant barrier, especially in managing long-term health conditions
that require continuous and dynamic assessment.**

J. Tekli et al. (Eds.): ADBIS 2024, CCIS 2186, pp. 337–342, 2025.
https://doi.org/10.1007/978-3-031-70421-5_30

The reported proposal consists of the creation of a new telemedicine platform that uses pseudo-time series (PTS) analysis to address the patient health condition evolution during time. PTS analysis is a sophisticated statistical tool typically used to deduce dynamic processes from static, cross-sectional datasets [3,4]. By applying PTS to health data, this platform will be able to simulate the progression of diseases over time, providing a more comprehensive view of patient health over time [5].

The platform aims to revolutionize real-time monitoring and intervention of disease trajectories by creating models that replicate the progression of various health conditions [6]. This approach not only promises to enhance the accuracy of telemedicine services but also supports the implementation of more effective and timely medical interventions. Initial studies have shown the feasibility of constructing pseudo-temporal trajectories that effectively mirror real-time disease progression [7]. This capability facilitates personalized patient care, allowing for interventions tailored to the individual's specific health trajectory.

In summary, the project seeks to bridge the gap between static data collection and dynamic health condition monitoring, significantly improving the scope and functionality of telehealth systems. By integrating PTS analysis into telemedicine, this initiative will provide healthcare professionals with a powerful tool to dynamically assess and respond to patient health needs, setting a new standard for data integration in medical practice and potentially transforming patient outcomes across various medical settings. **The challenge addressed by this research lies at the intersection of telemedicine and time series analysis, specifically how to dynamically model disease progression using static, cross-sectional data. Traditional telemedicine platforms lack mechanisms to analyze temporal changes in patient health, a gap that our PTS-based approach seeks to fill. By framing this issue within the context of data science, we address the crucial need for real-time, predictive analytics in healthcare monitoring systems.**

2   Related Works

The use of technology in healthcare, especially through telemedicine, has been crucial in meeting the growing need for remote patient monitoring and decision-making. While cloud computing and high-performance computing have transformed data management in healthcare by allowing storage and analysis of large amounts of clinical, demographic, and environmental data, these systems cannot often the ability to effectively incorporate dynamic temporal analyses necessary for chronic disease management and elderly care [8].

At the same time, pseudo time series (PTS) analysis has become a strong analytical tool, particularly in genetic research and bioinformatics, where it has been used to infer dynamic processes from static, cross-sectional datasets [7]. Originally developed to analyze unordered biological sequences, PTS methodologies reconstruct time-oriented trajectories that simulate the sequential progression inherent in biological and medical phenomena [9]. This approach is particularly useful in situations where collecting longitudinal data is impractical or unavailable due to high costs or logistical challenges [10].

Recent applications of PTS in healthcare research have yielded promising results. For instance, studies have successfully used PTS to model disease progression pathways in chronic conditions such as Alzheimer's and cardiovascular diseases from cross-sectional data. By applying machine learning techniques like clustering and dynamic time warping, researchers have been able to identify distinct phenotypic trajectories that reveal how these diseases evolve over time.

Moreover, the combination of PTS with advanced data analytics has facilitated the development of predictive models that can anticipate disease onset and progression with a higher degree of accuracy than traditional models [11]. These models use complex algorithms to analyze temporal patterns in health data, enabling healthcare providers to make more informed decisions about patient care.

Despite these advancements, the use of PTS in telemedicine is still an open issue. Most telehealth platforms still primarily focus on static data capture and lack the sophistication to integrate PTS-based analyses, which can provide deeper insights into patient health trajectories and enable proactive management of chronic conditions. Current research aims to bridge this gap by developing a telemedicine platform that not only supports high-volume data handling and privacy concerns in modern healthcare, but also incorporates PTS to offer a more nuanced, dynamic view of patient health over time [12].

While existing telemedicine platforms primarily focus on static data capture and management, our research leverages the latest advancements in data science to enable dynamic analysis. This is distinct from prior approaches that have traditionally neglected the temporal dimension of health data. By integrating insights from recent studies that apply PTS in other domains, such as genetics and bioinformatics, we innovate on how telemedicine can utilize similar methodologies to enhance disease monitoring and intervention strategies.

3   Preliminary Results

In my initial investigations, I have demonstrated the utility of pseudo time series (PTS) analysis in transforming static, cross-sectional health data into dynamic, temporal trajectories that accurately mimic the natural progression of diseases. This method, which I previously detailed in my study on aging processes [7], was adapted in my current research to construct realistic health trajectories for patients with cardiovascular diseases, utilizing clinical variables such as blood pressure and cholesterol levels.

Applying clustering techniques, I identified distinct phenotypes of disease progression. This approach allowed me to map out varying pathways of disease advancement among patients of different ages and health conditions. The results have been promising; for instance, the use of PTS enabled me to distinguish between early and advanced stages of cardiovascular conditions, effectively illustrating how these diseases progress over time in a pseudo-temporal order. This analysis not only provided insights into typical progression patterns but also highlighted potential intervention points where medical treatment could be most efficient [7].

Furthermore, clustering of pseudo time series data facilitated a deeper understanding of health trajectories, showing significant correlations between the progression of cardiovascular conditions and age-related changes. This alignment with observed clinical outcomes serves as a strong validation of the PTS method, reinforcing its potential to enhance the accuracy and effectiveness of telemedicine platforms by providing a nuanced view of patient health dynamics.

This utilization of PTS marks a significant advancement in my project, setting a solid foundation for the ongoing development of a telemedicine platform. Moving forward, I aim to expand these analyses to include more chronic diseases and integrate more complex datasets, broadening the applicability and robustness of the PTS approach in real-world healthcare settings.

4 Future Work: Platform Development

In the upcoming stages of this project, next actions regard integrating pseudo time series (PTS) analysis into our telemedicine platform. This integration is crucial as it will allow the platform to process and analyze health data in real-time, which is essential for effective patient monitoring and intervention [12]. Key areas of focus will include:

- Real-Time Data Processing: Implementing technologies that enable immediate analysis of incoming data, crucial for swift response in medical scenarios. **We plan to leverage high-throughput computing frameworks and event-driven architectures to facilitate the instantaneous processing of health data streams.**
- Scalability: Designing the platform to efficiently handle an increase in data and users over time, ensuring consistent performance at any scale. **Our scalability strategy involves the use of cloud-native technologies, enabling the platform to dynamically scale resources according to user demand and data volume.**
- Security Measures: Prioritizing the security of patient data by incorporating advanced encryption and secure communication protocols to protect against unauthorized access. **We will adhere to international standards (GDPR in Europe) to ensure the highest level of data protection through end-to-end encryption and multi-factor authentication mechanisms** [13].
- Interoperability: Developing interfaces that ensure easy integration with existing health IT systems, without disrupting current healthcare workflows. **To enhance interoperability, we will implement FHIR (Fast Healthcare Interoperability Resources) standards, allowing seamless data exchange across different healthcare platforms** [14].

– User-Friendly Design: Creating a simple and intuitive interface that all health-care providers can comfortably use, regardless of their technical skills. **User experience will be continuously refined through iterative testing with healthcare professionals to ensure the interface is not only intuitive but also aligns with their daily operational needs.**

The aim is to build a platform that not only addresses the current needs of telemedicine but is also equipped to adapt to future advancements in healthcare technology. This approach will help bridge the gap between traditional healthcare practices and modern telemedicine capabilities.

5 Conclusions

The state of the project regards the integration of pseudo time series (PTS) analysis into telemedicine platforms with the goal of transforming the real-time monitoring and management of patient health conditions. Preliminary results demonstrated the feasibility of creating dynamic, pseudo-temporal trajectories from static data. The use of PST may significantly improve the way how diseases are monitored and treated and may address new challenges in healthcare management. The potential to enable healthcare professionals to dynamically assess and adapt to evolving patient health needs is particularly promising. Achieving these objectives could redefine the standards for dynamic data analysis in medical practice and revolutionize patient outcomes across various settings.

Acknowledgements. Barbara Puccio PhD fellows is supported by Relatech S.p.A. and by the Next Generation EU: Italian PNRR, Mission 4, Component 2, Investment 1.5, call for the creation and strengthening of 'Innovation Ecosystems', building 'Territorial R&D Leaders' (Directorial Decree n. 2021/3277)-project Tech4You-Technologies for climate change adaptation and quality of life improvement, n. ECS0000009. She spent a visiting period at Brunel University as program of the PhD project. Author is grateful to Allan Tucker for discussion on the topic.

References

1. Guzzi, P.H., di Paola, L., Puccio, B., Lomoio, U., Giuliani, A., Veltri, P.: Computational analysis of the sequence-structure relation in SARS-CoV-2 spike protein using protein contact networks. Sci. Rep. **13**, 2837 (2023). https://doi.org/10.1038/s41598-023-30052-w
2. Gensini, G.F., Alderighi, C., Rasoini, R., Mazzanti, M., Casolo, G.: Value of telemonitoring and telemedicine in heart failure management. Card. Fail. Rev. **3**(2), 116–121 (2017). https://doi.org/10.15420/cfr.2017:6:2
3. Campbell, K.R., Yau, C.: Uncovering pseudotemporal trajectories with covariants from single cell and bulk expression data. Nat. Commun. **9**, 2442 (2018). https://doi.org/10.1038/s4167-018-04696-6
4. Tucker, A., Garway-Heath, D.: The pseudotemporal bootstrap for predicting glaucoma From cross-sectional visual field data. IEEE Trans. IT Biomed. (2010). https://doi.org/10.1109/TITB.2009.2023319

5. Dagliati, A., et al.: Using topological data analysis and pseudo time series to infer temporal phenotypes from electronic health records. Artif. Intell. Med. (2020). https://doi.org/10.1016/j.artmed.2020.101930

6. Boccuto, F., et al.: How patients feel with telemedicine devices as an enabling factor for personalised medicine: a preliminary study. Stud. Health Technol. Inf. **314**, 168–172 (2024)

7. Puccio, B., Tucker, A., Veltri, P.: Clustering Pseudo Time Series: Exploring Trajectories in the Ageing Process. Studies in Health Technology and Informatics **314**, 118–119 (2024)

8. Junaid, S.B., et al.: Recent advancements in emerging technologies for healthcare management systems: a survey. Healthcare (Basel) **10**(10), 1940 (2022). https://doi.org/10.3390/healthcare10101940.

9. Tucker, A., Li, Y., Garway-Heath, D.: Updating Markov models to integrate cross-sectional and longitudinal studies. Artif. Intell. Med. (2017). https://doi.org/10.1016/j.artmed.2017.09.009

10. Sajjadi, S.E., Tucker, A.: Exploiting clinical staging data to constrain pseudo-time modelling of disease progression. In: IEEE 34th International Symposium on Computer-Based Medical Systems (CBMS), Aveiro, vol. 2021, pp. 241–246 (2021). https://doi.org/10.1109/CBMS52027.2021.00026

11. Rahman, M., Murshed, M., Teng, S.W., Paul, M.: FSDR: a novel deep learning-based feature selection algorithm for pseudo time-series data using discrete relaxation. arXiv preprint arXiv:2403.08403 (2024)

12. Haleem, A., Javaid, M., Singh, R.P., Suman, R.: Telemedicine for healthcare: capabilities, features, barriers, and applications. Sens. Int. **2**, 100117 (2021)

13. GDPR (2016/679). Regulation (EU) 2016/679 of the European Parliament and of the Council of 27 April 2016 on the protection of natural persons with regard to the processing of personal data and on the free movement of such data, and repealing Directive 95/46/EC (General Data Protection Regulation). Off. J. Eur. Union L **119**, 1–88 (2016)

14. HL7 FHIR Release 4. FHIR Release 4 (R4) - HL7.org. Health Level Seven International (2019)

Classifying Chest X-Ray Images with Deep Learning Techniques: Challenges and Explainable Analysis

Tommaso Ruga[1,2]([envelope]) [iD]

[1] DIMES, University of Calabria, Rende, CS, Italy
tommaso.ruga@dimes.unical.it
[2] CNR-NANOTEC, Rende, CS, Italy

Abstract. Chest radiography represents an important stage of analysis, allowing the detection of pulmonary opacities, commonly known as infiltrations, which, if detected early, can improve medical follow-up. However, this technique may be limited when the case history under examination represents a complex characterisation of the pathology. Artificial intelligence is proposed as an ideal tool for the analysis of large-scale X-ray images, capable of providing an initial classification of images containing possible infiltrations, providing support to domain experts. The author's Phd activity focuses on providing artificial intelligence solutions in public, administrative and non-administrative fields, such as the medical sector. The aim is to speed up processes and optimise diagnosis, without replacing a domain expert but providing him with valid tools. In this preliminary study, the results produced by some neural networks are analysed with the focus on understanding the nature of the classifications produced in chest x-rays images field. The aim is to highlight typical problems in the domain and increase the explainability of the results, through the use of the Grad-CAM technique. Possible solutions and future developments of the work are therefore discussed.

Keywords: Lung Cancer · Infiltration · Deep Learning · Grad-CAM

1 Introduction

The lungs, essential organs of the human respiratory system, represent a complex network of anatomical and functional structures that enable the vital exchange of oxygen and carbon dioxide [1]. Understanding their anatomy is essential for managing respiratory health and addressing the challenges of pathological conditions that can affect these vital organs. In the field of prevention, a first step of control can be attributed to chest X-rays, which make it possible, in a non-invasive manner, to check the health of the lungs and signal the presence of any abnormalities. Pulmonary infiltrates are a common artifact in chest radiographs [2]. They are characterized by the presence of abnormal substances within the lung parenchyma, outside of the normal air spaces, which appear as opaque and

J. Tekli et al. (Eds.): ADBIS 2024, CCIS 2186, pp. 343–350, 2025.
https://doi.org/10.1007/978-3-031-70421-5_31

whitish areas in the radiographic images. Being able to detect a pulmonary infiltration in its early stages is crucial to prevent the disease from progressing and to treat it promptly. The incidence rate of lung diseases, such as infiltration, has increased in recent years. According to estimates by the World Health Organisation (WHO) and the Global Initiative for Chronic Obstructive Lung Disease (GOLD) [3], chronic lung disease (COPD) alone accounts for 600 million cases per year while 300 million cases per year are attributed to asthma. Lung cancer is one of the most widespread and deadly types of cancer. In 2020, 2.2 million is the estimated number of lung cancer's new cases, accounting for 13% of all new cancer cases according to the AIRC Foundation for Cancer Research [4]. With the focus on this data, coupled with the need to provide an initial diagnosis in a timely manner, recent studies in the field of artificial intelligence (AI) have focused on solutions that can process large amounts of images in a short time, in order to provide a decision-support tool for domain experts. The main problem with the solutions provided is that in many cases the images are processed using deep learning (DL) techniques that analyse the content of X-rays as a black box, increasing scepticism on specialists. In this work, an explainable analysis of a deep convolutional pre-trained network is proposed.

The work is organized as follows: after an overview of the current state of the art, the methodology used is described, analyzing the technical details of the conducted experiment, with a focus on the dataset, the network architectures used, and the Grad-CAM technique for the analysis of the network results, which creates an heatmap per image representing where model points its attention when elaborates it; once the architecture with the best performance is identified, the two products of the classification, namely the predictions and the obtained heatmaps, are then analyzed (Fig. 1); typical domain-related issues and those related to the analyzed architectures are discussed, providing possible solutions and discussing potential future developments.

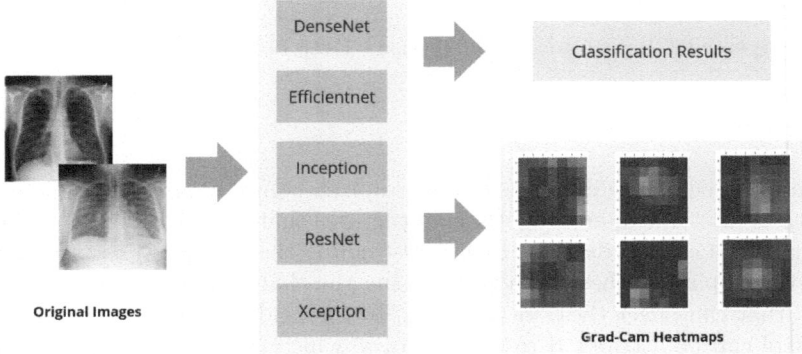

Fig. 1. Experiment Workflow

The main contribution of the work is to provide a preliminary solution for the correct identification of pulmonary infiltration, which is little addressed in the literature, analyzing chest x-ray images, and above all to provide an explicable key to the results, using techniques capable of highlighting on which points of the image the networks are concentrated, reducing the black box nature of networks.

2   Related Works

There are several papers focusing on classification tasks concerning lung disease. However, work often focuses on specific lung diseases, such as pneumonia, and does not directly address lung infiltration. In [5], an ensemble architecture to detect pneumonia is proposed, that use two deep neural network to extract features separately, CheXNet [6] and a VGG-19. Then, the features are concatenated and given as input to a RandomForest Classifier that perform the final classification. Some works focus on the pre-processing steps of the input images, demonstrating the effectiveness of enhancement or segmentation steps to improve the final results. In [7], authors propose a two stage method to segment lung x-ray images from ChestRay-14 dataset and classify them. The first step performs a segmentation using a U-Net, that focuses in global and local branches both, the second step performs a multi-label classification on all 14 dataset classes. Another work that propose a multi-label classification is [8], where an Efficient-NetB4 architecture is used to detect active pulmonary tuberculosis in images. Also in this case, a segmentation step is performed with a U-Net architecture. The authors also show the effectiveness of the proposed solution by validating the results with the Grad-CAM technique, which was used to increase the explainability of the results and to provide visual results to the domain experts who collaborated on the work. This makes it easier to validate the effectiveness of the proposed solution from a medical point of view as well. Instead, in [9] authors highlight importance of image pre-processing step, improving images quality before the learning procedure. Authors apply lightweight techniques that work computing the local variance and mean around each pixel, such as histogram equalization to improve images contrast, and pixel-wise Wiener filter to produce a filtered image, in order to reduce noise and artifacts. The work that currently represents the state of the art in the field of lung infiltration detection is [10], where a framework to classify chest x-ray images, called Slide-Detect, is proposed. In this work a striding window step is used to simulate the image cropping during the training phase. Authors, once again demonstrate, the importance of the pre-processing steps, focusin, in particular, on the image cropping steps to extract the areas of greatest interest, which are involved in the pathology. The cropping areas come from the dataset itself used to validate the model, NIH Chest X-ray [11], the expanded version of the one used in this work, which is correlated with a csv file containing additional information on the patients and information regarding the cropping areas to be performed, validated by the doctors who contributed to the creation of the dataset.

3   Methodology

3.1   Dataset

The dataset used to develop this work is a reduced version of the well-known NIH Chest X-ray Dataset [11], which in its original version contains 112,120 X-ray images with disease labels from 30,805 unique patients. In this reduced version [12], the number of images is 5606, belonging to 15 classes, like in the original version, namely 14 different pathologies and one *No Findings* class representing the category of images without pulmonary problems. Due to the drastic dimension reduction, classes are more sparse in this version, and a data balancing phase has been necessary. In this work, two classes has been used: *Infiltration* (967 images) and *No Findings* (3044 images). However, two aspects of the dataset should be noted: the labelling of the data was obtained by NLP technique, so in some cases, the authors report in the original paper, it may not be correct, however, ensuring that more than 90% of the labels are correct; Moreover, there are relatively few images to have a single class, with the exception of the class No Findings, of course. An image can have several labels at the same time, being categorised among both infiltrations and pneumonias. This aspect is crucial in defining the complexity of the problem and is representative of the reality, as in many cases, a patient who has a pulmonary infiltration also has other diseases attached to it.

3.2   Grad-CAM

Grad-CAM (Gradient-weighted Class Activation Mapping) is a technique used to visualize the regions of an image that are important for a convolutional neural network (CNN) when making a classification decision [18]. Essentially, it is an algorithm made up of simple steps that makes it possible to visualise, given an image as input to the network, the areas of the image where the network is focusing to produce the classification. Grad-CAM has become a standard tool for interpreting CNNs results increasing explainability and improving the debugging phases. Algorithm steps can be summarized as follows: given the image to the network, it produces class scores through its layers; The gradients of those target class score are computed with respect to the feature maps of a convolutional layer, via backpropagation, and then they are averaged over the width and height dimensions to obtain a set of importance weights for each feature map channel; These weights highlights the importance of each feature map for the target class and then are used to perform a weighted combination of the feature maps from the convolutional layer; The product is a coarse localization map highlighting the important regions of the image which becomes the input to a ReLU in order to ensure that only positive influences on the class score are considered; This results in the final class activation map which then is upsampled to the size of the input image and superimposed on the original image; The final product is a heatmap that shows the areas of the image that contributed most to the model's prediction, highlighting any gaps in the model and explaining how the model output was obtained.

4   Results, Discussion and Future Perspective

Several network configuration tests were carried out to evaluate the performance of the best versions and hyperparameters used. All tests were performed on Colab using a Tesla T4 GPU. For the performance evaluation step, several metrics were used, as reported in Table 1.

Table 1. Performances of analyzed models.

Model	Accuracy	Precision	Recall	F1-Score	AUC
DenseNet201 [13]	78%	78%	77%	77%	0.84
EfficientNetB0 [14]	77%	77%	77%	77%	0.85
InceptionV3 [15]	77%	78%	77%	77%	0.85
ResNet50 [16]	**79%**	**79%**	**79%**	**79%**	**0.84**
Xception [17]	75%	75%	75%	75%	0.82

In general, the behaviour of models is homogeneous and rigorous. It must be taken into account, however, that the results were obtained without an image preprocessing step and without a data cleaning step. The aim of the work is not to obtain an ideal model but to analyse the behaviour of the networks net of the best configuration. Anyway, best model in terms of performances is ResNet50 with every metrics equal to 79%, while worst is Xception, although it still provides rigorous results. Results achieved is higher than baselines reported in [10], looking at ResNet50 results show in the paper. Considering that no pre-processing steps, such as enhancing, cropping or image segmentation, were performed, the results can therefore be described as encouraging.

4.1   Grad-CAM Analysis

In order to providing a greater explanation of the results obtained, a qualitative and explainable analysis is now proposed using the Grad-Cam technique regarding the nature of the predictions produced. Selecting the last convolutional layer of the best performing pre-trained network, the heatmaps associated with the images processed by the network were extracted in order to highlight which points in the image contain the most representative features. Given the results obtained, it was easy to think that in the case of some images, the models encountered ambiguities, or a difficulty in focusing on the truly significant features. Thanks to the application of the Grad-CAM algorithm, it was possible to visualise precisely this aspect (Fig. 2). Through an empirical analysis, three case studies were isolated (Fig. 3): a) images containing imperfectly positioned patients, b) images containing medical artefacts (pacemakers, electrodes) or x-ray acquisition machinery artifacts (scale ratios, text), c) misclassified images due to model gaps. In the first case, the images include those where patients were improperly positioned during the X-ray acquisition phase, or X-rays that were not perfectly centered during the scanning phase and conversion to a digital file. In the second case, the images include those that contain any type of artifacts, such as pacemakers or typical objects in X-rays like text inserted by the acquisition machine. In the third case, we find images that are correctly acquired and free of artifacts but are misclassified by the models due to their shortcomings. In the first two cases analyzed, a possible solution would be to develop a custom pre-processing algorithm capable of segmenting the image and indicating the areas of interest before the learning process. Subsequently, the image could be cleaned of any artifacts using techniques such as shown in [9]. In the third case, the predictive quality of the models could be improved by providing more images as input or, as a direct consequence of the first two cases already discussed, by providing more rigorous inputs and highlighting the truly relevant features of the images.

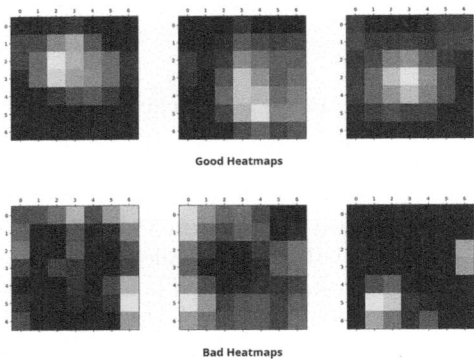

Fig. 2. Examples of Grad-Cam heatmaps.

The aim of the proposed study was not to provide a definitive solution to the problem of classifying images of pulmonary infiltrations. Instead, the aim was to highlight the critical aspects of the field treated, providing a real explanation for the results obtained, as also proposed in [8]. The results obtained are not directly comparable with those proposed in [7,8,10], since a multiclass solution is proposed in these works. Furthermore, the results presented in this work were not obtained downstream of a segmentation process or an advanced pre-processing phase as in [7,9]. None of the works analysed, however, focus on pulmonary infiltration and excluding [8] none of them provide an explanation of the results, although they provide excellent insights for future phases of this work. The contribution of this work is therefore further emphasised as too little has been done yet in the field of explainability of results and in the field of lung infiltration detection.

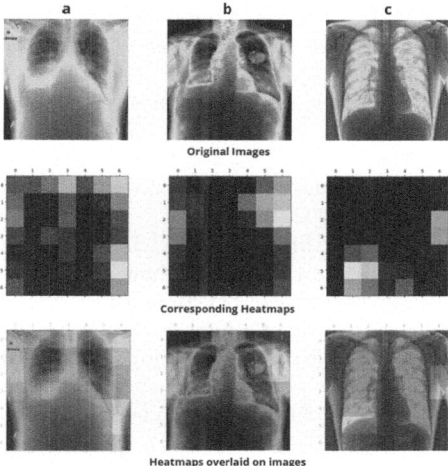

Fig. 3. Examples of bad processed images.

4.2 Future Perspective

The preliminary analyses conducted have highlighted some technical issues typical of the medical domain treated. A problem that directly affects the analysis of chest X-ray images is the difficulty models have in correctly identifying areas of interest due to the presence of the rib cage. Some works present bone suppression imaging techniques that could mitigate this problem [19], while, as already discussed, other works present AI techniques for the removal of other artifacts present in the image [20]. A problem closely related to the domain is the low purity of the classes. Pulmonary infiltrations often occur, as can be observed in the distribution of the dataset classes, in conjunction with other pathologies. In the future, the work will be expanded by increasing the explainable components of the system, through custom machine learning solutions and XAI techniques, proposing solutions that are increasingly aligned with the medical protocols currently in use.

References

1. European Lung. https://europeanlung.org/it/information-hub/keeping-lungs-healthy/anatomia-e-funzioni-del-polmone-sano/. Accessed 21 May 2024
2. Nykamp, S.G., Scrivani, P.V., Dykes, N.L.: I Segni Radiografici Delle Patologie Polmonari: Un Approccio Alternativo. Veterinarian **24**(1) (2002)
3. Lung Cancer 2023 Report. https://goldcopd.org/wp-content/uploads/2023/03/GOLD-2023-ver-1.3-17Feb2023_WMV.pdf. Accessed 22 May 2024
4. AIRC Statistics. https://www.airc.it/news/i-numeri-del-cancro-fotografia-dal-mondo. Accessed 22 May 2024

5. Habib, N., Hasan, M.M., Reza, M.M., Rahman, M.M.: Ensemble of CheXNet and VGG-19 feature extractor with random forest classifier for pediatric pneumonia detection. SN Comput. Sci. **1**(6), 359 (2020)
6. Rajpurkar, P., et al.: CheXNet: radiologist-level pneumonia detection on chest x-rays with deep learning. arXiv preprint arXiv:1711.05225(2017)
7. Chen, B., Zhang, Z., Lin, J., Chen, Y., Lu, G.: Two-stream collaborative network for multi-label chest X-ray Image classification with lung segmentation. Pattern Recogn. Lett. **135**, 221–227 (2020)
8. Devasia, J., Goswami, H., Lakshminarayanan, S., Rajaram, M., Adithan, S.: Deep learning classification of active tuberculosis lung zones wise manifestations using chest X-rays: a multi label approach. Sci. Rep. **13**(1), 887 (2023)
9. Farhan, A.M.Q., Yang, S.: Automatic lung disease classification from the chest X-ray images using hybrid deep learning algorithm. Multimedia Tools Appl. **82**(25), 38561–38587 (2023)
10. Mohamed, A.E., Fayek, M.B., Farouk, M.: Slide-detect: an accurate deep learning diagnosis of lung infiltration. Data Intell. **5**(4), 1048–1062 (2023)
11. Wang, X., Peng, Y., Lu, L., Lu, Z., Bagheri, M., Summers, R.M.: ChestX-Ray8: hospital-scale chest x-ray database and benchmarks on weakly-supervised classification and localization of common thorax diseases. In: Proceedings of the IEEE Conference on Computer Vision and Pattern Recognition, pp. 2097–2106 (2017)
12. NIH Chest X-ray Dataset (Reduced). https://www.kaggle.com/datasets/nih-chest-xrays/sample. Accessed 26 May 2024
13. Huang, G., Liu, Z., Van Der Maaten, L., Weinberger, K.Q.: Densely connected convolutional networks. In: Proceedings of the IEEE Conference on Computer Vision and Pattern Recognition, pp. 4700–4708 (2017)
14. Tan, M., Le, Q.: EfficientNet: rethinking model scaling for convolutional neural networks. In: International Conference on Machine Learning, pp. 6105–6114. PMLR, May 2019
15. Szegedy, C., Vanhoucke, V., Ioffe, S., Shlens, J., Wojna, Z.: Rethinking the inception architecture for computer vision. In: Proceedings of the IEEE Conference on Computer Vision and Pattern Recognition, pp. 2818–2826 (2016)
16. He, K., Zhang, X., Ren, S., Sun, J.: Deep residual learning for image recognition. In: Proceedings of the IEEE Conference on Computer Vision and Pattern Recognition, pp. 770–778 (2016)
17. Chollet, F.: Xception: deep learning with depthwise separable convolutions. In: Proceedings of the IEEE Conference on Computer Vision and Pattern Recognition, pp. 1251–1258 (2017)
18. Selvaraju, R. R., Cogswell, M., Das, A., Vedantam, R., Parikh, D., Batra, D.: Grad-CAM: visual explanations from deep networks via gradient-based localization. In: Proceedings of the IEEE International Conference on Computer Vision, pp. 618–626 (2017)
19. Li, F., Engelmann, R., Pesce, L., Doi, K., Metz, C.E., MacMahon, H.: Small lung cancers: improved detection by use of bone suppression imaging-comparison with dual-energy subtraction chest radiography. Radiology **261**(3), 937–949 (2011)
20. Madesta, F., Sentker, T., Gauer, T., Werner, R.: Deep learning-based conditional inpainting for restoration of artifact-affected 4D CT images. Med. Phys. (2023)

Scalable Deep Learning: Applications in Medicine

Luca Barillaro[✉][iD]

Data Analytics Research Center, Department of Medical and Surgical Sciences,
University "Magna Græcia" of Catanzaro, 88100 Catanzaro, Italy
luca.barillaro@unicz.it

Abstract. This paper aims to introduce my PhD research project, which is focused on scalable deep learning and its applications in the medical context. This project aims to design new DL algorithms or to adapt existing ones, to scalable architectures (e.g., parallel computers, GPUs), to improve the performance of typical ML tasks, such as classifications, and to experiment with them in the analysis of biomedical data, such as bioimages or molecular data. These applications are widely present in current literature thus representing a challenging opportunity. In addition to common DL and HPC approaches, the use of edge devices (i.e., Nvidia Jetson) is being explored, since this may be useful in a medical context using some key features of the edge computing paradigm (such as keeping data near the source) that are significant, i.e., for privacy reasons or legal compliance (i.e., European GPDR). My main investigated applications are on bioimages, such as Computed Tomography (CT) or functional Magnetic resonance imaging (fMRI). Research is being carried out to investigate current methodologies and how to improve them possibly. Moreover, a lot of experiments are carried out to demonstrate the impact of these approaches on traditional medical strategies. This work resulted in some experiments on the classification of medical images (CTs), medical signals (ECG), and gene expression data. My current work's primary focus, boosted by a three-month collaboration with the University of Groningen, involves an extensive project on classifying fMRIs using Machine Learning techniques and concepts from graph theory tailored to exploit High-Performance Computing (HPC) infrastructures.

Keywords: Machine learning · Edge computing · High-Performance Computing · Functional Magnetic Resonance Imaging

1 Background and Objectives

Artificial intelligence resources and techniques are extensively utilized for data analysis in the medical field, offering a superior level of analysis compared to traditional methods. Bio-signals such as ECG (electrocardiogram) and EEG (electroencephalography), as well as bio-images like MRI (magnetic resonance

J. Tekli et al. (Eds.): ADBIS 2024, CCIS 2186, pp. 351–356, 2025.
https://doi.org/10.1007/978-3-031-70421-5_32

imaging), NMR (Nuclear magnetic resonance), or CT (computed tomography), are among the data that benefit from these techniques [7–9].

The precision and ability to overcome natural limitations associated with human-driven procedures make these methods advantageous [10]. Manual analysis is prone to inter-variability when performed by different operators, and intra-variability when the same operator analyzes data under different personal conditions, such as varying fatigue levels.

These techniques, particularly in image analysis, demand substantial computational power due to their reliance on complex data structures and algorithms, resulting in high memory occupancy and computational costs. Consequently, high-performance computing (HPC) has been introduced, leveraging parallel and distributed computers and cloud computing services. This leverages the concept of system scalability.

Scalability in high-performance computing (HPC) refers to the ability of a system, network, or application to efficiently handle an increasing workload by proportionally leveraging additional resources. This characteristic is measured by the system's capacity to maintain or improve its performance as the size and complexity of the computational tasks grow.

A scalable HPC system can effectively manage larger datasets, more complex algorithms, and increased user demands without significant performance or efficiency degradation, ensuring optimal resource utilization across varied computational loads.

This paper aims to introduce my PhD research project, which is focused on scalable deep learning and its applications in the medical context for the reasons previously mentioned.

My PhD project is mainly based on two fields. One deals with bioinformatics data analysis, such as bio-images (i.e., Computed Tomography, CT, functional Magnetic resonance imaging, fMRI) or gene expression data. In this field, research is being carried out to investigate current methodologies and how to improve them possibly and some experiments were conducted.

The other field is real-world experimentation on the edge computing paradigm, to demonstrate the impact of these approaches on medical context using some key features of this paradigm (such as keeping data near the source) that are significant, i.e., for privacy reasons or legal compliance (i.e., European GDPR).

Edge computing is a novel approach to analyzing data near its source, and it offers significant advantages for healthcare organizations by allowing them to keep sensitive data within their network.

This concept has gained traction alongside the proliferation of edge devices, such as IoT devices, which amass substantial amounts of data from diverse origins. Rather than transmitting it to a centralized computing unit, analyzing this data at the edge is often preferable due to factors like communication costs, connection delays (especially in cases where a swift response is imperative), and privacy considerations.

Edge computing presents a robust solution for overcoming these challenges, and the growing computational capabilities of these devices enable them to handle even resource-intensive computations.

Unlike traditional cloud and high-performance computing systems, edge-based models obviate the need to transfer data outside the organization, ensuring rapid responsiveness to computational demands. Additionally, edge models curtail power consumption without compromising computational capabilities, making this feature indispensable for edge environments that are frequently constrained by power limitations.

The research project aims to design new Deep learning algorithms or to adapt existing ones, to scalable architectures (e.g., parallel computers, GPUs), to improve the performance of typical Machine learning tasks, such as classifications, and to experiment with them in the analysis of biomedical data, such as bioimages or molecular data. These applications are widely present in current literature thus representing a challenging opportunity.

In addition, the experience with the edge computing paradigm may open the way to further fields of analysis such as green AI, due to the low power consumption of such devices.

The main current objective is the classification and analysis of fMRIs, thanks to a collaboration with the University of Groningen, the Netherlands, which hosted me for three months in 2023.

This topic requested to acquire a better understanding of network analysis and graph theory because they are currently used [6,14] in fMRI analysis thanks to their expressive power.

2 Material and Methods

The first part of the project was based on literature research to investigate current methodologies and possibly improve them. Many articles were analyzed both in deep learning applications in the medical context and edge computing environments. Some methodologies emerged, and further applications are being explored.

From existing literature, it emerged the need for deep learning applications in the medical field, and how current methodologies could help medical doctors, an example can be found in [12]. Several applications have shown significant results in medical applications, such as in [13] or [11] thus motivating to continue the research in this field.

On the experiment side, during my project, several experiments were conducted to demonstrate the impact of deep learning approaches and edge computing on traditional medical strategies. These experiments were conducted by using common DL frameworks (i.e., Keras [5], TensorFlow [1]), and specific libraries for medical data from the DeepHealth Project.

While the first experiments aimed to perform case studies to explain the validity of an automated method, the subsequent were based on comparisons among different technical solutions on the software side as described above and hardware solutions.

In such experiments, the classification task was mainly stressed, focusing on improving performance scores by using different models or parameter tuning. Several publicly available datasets were used to make experiments and perform comparisons.

Mostly, as introduced above, TensorFlow/Keras was used as the main framework by exploiting the community support and its wide range of adoptions. Models from different fields of application were used for medical data to perform comparisons with existing solutions.

Experiments were conducted mainly on a personal laptop with a NVIDIA GPU, to check the impact of such technology on deep learning applications, compared to CPU-only platforms. Moreover, some experiments also involved using Docker containers to exploit the flexibility of a virtualized environment.

Since the start of my collaboration with the University of Groningen, the dominant data subject of my analyses is functional magnetic resonance imaging (fMRI).

To gain a better understanding of this data type, a study on networks is being conducted, to use them to characterize fMRIs and apply network analysis concepts to perform experiments with fMRI data.

A toolbox, in particular, has been used for network extraction from fMRI, the CONN Toolbox [15] (Matlab based) which has proven to be useful in this context but the not completely open nature of its background could be a limit. In addition, there may be compatibility issues with certain datasets.

The edge computing environment exploration is based on the NVIDIA Jetson platform, which represents a crucial choice in this field, enabling many opportunities. Experiments were conducted on a Jetson AGX Orin Development Kit to explore its capabilities compared to a traditional platform. This actual hardware has been used to study and exploit its advantages and several tests have been conducted.

This edge device is based on a customized version of Ubuntu, and several dedicated containers are available from the manufacturer to perform specific machine-learning tasks.

3   Results

These activities resulted in several papers. At first, they involved an exploration of the field followed by some simple analysis. They aimed to demonstrate the performance gain of a deep learning approach compared to a traditional machine learning one.

In addition, some works explored the application in several fields of bioinformatics, such as CT scan analysis [3] or gene expression data [4]. These papers resulted in some comparatives highlighting the advantages of a different approach to analyzing such data. In such papers we were able to perform classification tasks on different data types, achieving consistent performances especially compared to other works on similar data.

Other works dealt with edge computing, in particular, the aim was to demonstrate its applicability and performance in deep learning-based medical applications, such as ECG analysis [2]. Produced results showed that an edge-based system is capable of consistent performances but leveraging its intrinsic features.

The latter, some works on fMRI analysis were made dealing with a first insight into DL application-based analysis and, more recently with a pipeline that combined graph theory concepts and network analysis to cluster fMRI data.

This approach has proven to be feasible in correctly understanding the data class and opened the way to deeper analysis of the data itself, because, under certain cluster settings, it may reveal potential hidden information.

4 Conclusions and Future Directions

The project is still ongoing, primarily aiming to leverage the experience gained through the collaboration with the University of Groningen to conduct a deeper analysis of fMRI data, ultimately leading to a robust classification of these data.

In future works, in addition to the aforementioned fMRI analysis, I plan to conduct experiments utilizing edge computing technology to further exploit its potential.

Acknowledgement. This work was funded by the Next Generation EU - Italian NRRP, Mission 4, Component 2, Investment 1.5, call for the creation and strengthening of 'Innovation Ecosystems', building 'Territorial R&D Leaders' (Directorial Decree n. 2021/3277) - project Tech4You - Technologies for climate change adaptation and quality of life improvement, n. ECS0000009. This work reflects only the authors' views and opinions, neither the Ministry for University and Research nor the European Commission can be considered responsible for them.

References

1. Abadi, M., et al.: TensorFlow: large-scale machine learning on heterogeneous distributed systems arXiv:1603.04467 (2016). http://arxiv.org/abs/1603.04467
2. Barillaro, L., Agapito, G., Cannataro, M.: Edge-based deep learning in medicine: classification of ECG signals. In: 2022 IEEE International Conference on Bioinformatics and Biomedicine (BIBM), pp. 2169–2174 (Dec 2022). https://doi.org/10.1109/BIBM55620.2022.9995598, https://ieeexplore.ieee.org/abstract/document/9995598
3. Barillaro, L., Agapito, G., Cannataro, M.: Scalable deep learning for healthcare: methods and applications. In: Proceedings of the 13th ACM International Conference on Bioinformatics, Computational Biology and Health Informatics, pp. 1–8. BCB 2022, Association for Computing Machinery, New York, NY, USA (2022). https://doi.org/10.1145/3535508.3545590
4. Barillaro, L., Agapito, G., Cannataro, M.: Using edge-based deep learning model for early detection of cancer. In: 2023 31st Euromicro International Conference on Parallel, Distributed and Network-Based Processing (PDP), pp. 252–257 (Mar 2023). https://doi.org/10.1109/PDP59025.2023.00046, https://ieeexplore.ieee.org/abstract/document/10137014, iSSN: 2377-5750

5. Chollet, F., et al.: Keras (2015). https://github.com/fchollet/keras, publisher: GitHub

6. Farahani, F.V., Karwowski, W., Lighthall, N.R.: Application of graph theory for identifying connectivity patterns in human brain networks: a systematic review. Front. Neurosci. **13** (2019). https://www.frontiersin.org/articles/10.3389/fnins.2019.00585

7. Liu, X., Wang, H., Li, Z., Qin, L.: Deep learning in ECG diagnosis: a review. Knowl.-Based Syst. **227**, 107187 (2021). https://doi.org/10.1016/j.knosys.2021.107187, https://linkinghub.elsevier.com/retrieve/pii/S0950705121004494

8. Lundervold, A.S., Lundervold, A.: An overview of deep learning in medical imaging focusing on MRI. Zeitschrift für Medizinische Physik **29**(2), 102–127 (2019). https://doi.org/10.1016/j.zemedi.2018.11.002, https://linkinghub.elsevier.com/retrieve/pii/S0939388918301181

9. Magadza, T., Viriri, S.: Deep learning for brain tumor segmentation: a survey of state-of-the-art. J. Imaging **7**(2), 19 (2021). https://doi.org/10.3390/jimaging7020019, https://www.mdpi.com/2313-433X/7/2/19

10. Miotto, R., Wang, F., Wang, S., Jiang, X., Dudley, J.T.: Deep learning for healthcare: review, opportunities and challenges. Briefings Bioinf. **19**(6), 1236–1246 (2018). https://doi.org/10.1093/bib/bbx044, https://academic.oup.com/bib/article/19/6/1236/3800524

11. Ningrum, D.N.A., et al.: Deep learning classifier with patient's metadata of dermoscopic images in malignant melanoma detection. J. Multidiscip. Healthc. **14**, 877–885 (2021). https://doi.org/10.2147/JMDH.S306284, https://www.dovepress.com/deep-learning-classifier-with-patientrsquos-metadata-of-dermoscopic-im-peer-reviewed-fulltext-article-JMDH

12. Norgeot, B., Glicksberg, B.S., Butte, A.J.: A call for deep-learning healthcare. Nat. Med. **25**(1), 14–15 (2019). https://doi.org/10.1038/s41591-018-0320-3, http://www.nature.com/articles/s41591-018-0320-3

13. Rajpurkar, P., et al.: CheXNet: radiologist-level pneumonia detection on chest X-rays with deep learning. arXiv:1711.05225 (Dec 2017). http://arxiv.org/abs/1711.05225

14. Wang, J., Zuo, X., He, Y.: Graph-based network analysis of resting-state functional MRI. Front. Syst. Neurosci. **4** (2010). https://www.frontiersin.org/articles/10.3389/fnsys.2010.00016

15. Whitfield-Gabrieli, S., Nieto-Castanon, A.: Conn: a functional connectivity toolbox for correlated and anticorrelated brain networks. Brain Connectivity **2**(3), 125–141 (2012). https://doi.org/10.1089/brain.2012.0073

Towards More Efficient and Improved Federated Learning

Jamsher Bhanbhro$^{(\boxtimes)}$ (ID)

DIMES Department, University of Calabria, Rende, Italy
`jamsher.bhanbhro@dimes.unical.it`

Abstract. The evolution of artificial intelligence models has made safeguarding data privacy more crucial than ever. Federated learning offers a promising solution by enabling collaborative model training without sharing sensitive data. However, challenges arise due to user data differences, unfair aggregation methods, and privacy concerns. This research introduces novel techniques to enhance federated learning in three key areas: handling data heterogeneity, improving personalization, and proposing advanced privacy techniques and fair aggregation methods. By addressing these challenges, our research aims to make federated learning more secure, efficient, and adaptable to the diverse conditions of real-world data.

Keywords: Federated Learning · Data Heterogeneity · Personalization in Federated Learning · Privacy Preserving · Client Clustering · Model Aggregation

1 Introduction

Traditional machine learning (ML) relies on a centralized approach, where data is collected and stored on servers to train predictive models. While this approach has demonstrably succeeded in various applications, such as image recognition and natural language processing, it raises significant privacy and security risks due to the inherent vulnerabilities of centralized data storage. Data breaches and unauthorized access to sensitive information have become increasingly common, emphasizing the need for robust privacy-preserving techniques in artificial intelligence (AI) development.

To address these privacy concerns, federated learning (FL) emerges as a promising technique. It trains local models on individual devices or servers that hold local data, and aggregates the weights using a central global model, thereby keeping the information private and reducing the risk of data leaks. Instead of sharing the raw data itself, only the model updates, called weights, are shared with a central server. This significantly improves privacy preservation, as the raw data never leaves the individual devices. This approach not only preserves data privacy but also minimizes the need for extensive data transfers, which can

© The Author(s), under exclusive license to Springer Nature Switzerland AG 2025
J. Tekli et al. (Eds.): ADBIS 2024, CCIS 2186, pp. 357–365, 2025.
https://doi.org/10.1007/978-3-031-70421-5_33

be resource-intensive. Additionally, it reduces the requirement for high-capacity storage devices.

The concept of FL also aligns well with the principles of data minimization and decentralization, which are increasingly advocated in data protection regulations worldwide, such as the General Data Protection Regulation (GDPR) in the European Union [1]. By ensuring that data remains localized, FL helps in complying with such regulatory requirements while still enabling the development of high-quality AI models. Beyond privacy, FL offers several other significant advantages. It enhances scalability by enabling parallel model training on multiple devices, which can lead to faster training times compared to centralized approaches. This distributed nature also allows for better utilization of computational resources, as each participating device contributes to the overall computation. Furthermore, FL can lead to reduced latency in model updates since local data does not need to be transferred to a central server, making it suitable for real-time applications where quick adaptation to new data is pivotal.

FL has various advantages but it introduces a set of challenges, such as managing data heterogeneity, ensuring fairness in weights aggregation, and preserving privacy when weights are similar. Data heterogeneity arises because data across different clients can be non-independent and identically distributed (non-IID), leading to challenges in model convergence and generalization. Ensuring fairness in weights aggregation is essential to maintain balanced model performance across varied clients. Moreover, privacy concerns become significant when similar weights are shared, potentially revealing sensitive information. Building on the prior discussion of existing FL literature and its limitations discussed in Sect. 2, particularly regarding performance under heterogeneity, aggregation, and privacy, this PhD research aims to address these challenges by developing novel and advanced techniques and methodologies to make FL more efficient.

This research study presents our methodology for addressing data heterogeneity through client clustering and personalization, demonstrating superior performance compared to general FL. Additionally, it proposes innovative ideas for aggregation and privacy enhancements.

2   Literature Review

Recent research has explored advanced techniques to enhance data privacy in centralized ML systems, including encryption methods explored by Smith et al. [2] and blockchain integration investigated by Johnson [3]. However, these methods often introduce computational burdens and implementation complexities, making it difficult for practical deployment. Zhao et al. [4] have discussed differential privacy techniques that add noise to data to protect individual privacy in FL, but this can reduce model accuracy. Similarly, Chen et al. [5] have investigated secure multi-party computation methods to enhance privacy, though these are resource-intensive and can significantly increase computation time.

One of the key issues in both traditional ML and FL is the handling of non-independent and identically distributed (non-i.i.d.) data. Centralized ML systems often assume data is i.i.d., which is rarely the case in real-world scenarios. Kim et al. [6] and Lopez et al. [7] have addressed this by proposing adaptive learning rates and data augmentation techniques, respectively. While these methods improve model robustness, they require constant monitoring and can lead to overfitting if not properly managed. In FL, the challenge of data heterogeneity is even more significant due to the diverse data distributions across different devices. Liu et al. [8] have proposed dynamic clustering algorithms to address this issue, although these add significant computational overhead. Nguyen et al. [9] have suggested personalized FL approaches that balance global and local model performance, but these can complicate the training process and may not generalize well to new data.

Model aggregation is a fundamental aspect of FL, aiming to combine updates from multiple decentralized devices to form a global model effectively. One widely used method is Federated Averaging (FedAvg), introduced by McMahan et al. [10], which averages local model updates weighted by the number of data points on each device. This method ensures fair representation in the global model but may struggle with non-IID data distributions. To address these challenges, Li et al. [11] proposed FedProx, which introduces a regularization term to handle heterogeneity in data distributions across clients. Additionally, Karimireddy et al. [12] developed Scaffold, which uses control variates to reduce the variance in local updates, improving convergence speed and model performance. Wang et al. [13] presented FedNova, a normalization technique that adjusts the contributions of each client based on the number of local updates, further enhancing model aggregation in federated settings. These methods aim to address issues related to data heterogeneity but suffer from fairness and performance problems.

The next section of this research details our methodology and some discussions.

3   Proposed Methodologies

This section discusses techniques to enhance FL efficiency. We present our method for handling data heterogeneity, including results and algorithms, and introduce ideas for dynamic privacy and advanced aggregation methodologies to further improve FL.

3.1   Handling Data Heteroginity with CLient Clustering and Progressive Personalization

The FL environment consists of multiple local clients, each training its own model on local data. These model parameters are then aggregated. See Fig. 1, which shows an FL environment with three clients trained on CIFAR-10 data [14]. Initially, all three clients received the data in identical (IID) form and then in non-IID form. In the IID scenario, clients received the same data; in the non-IID scenario, client one received the first three classes, client two received the next three, and client three received the last four. The figure shows the environment performs better with IID data. When the data is in Non-IID form, even simpler data impacts performance.

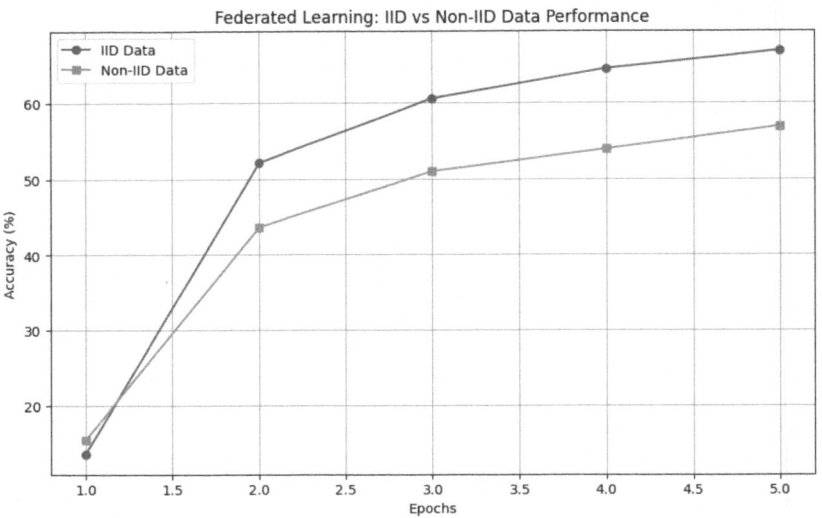

Fig. 1. Data Heterogeneity FL Environments performances

In the FL environment where data is variate and different, client clustering and personalization come to play an interesting role.

We propose a technique that combines client clustering and progressive personalization to reduce the impact of non-IID data. Model personalization in FL is essential for adapting models to specific data distributions of different clients while maintaining overall performance.

To perform a preliminary analysis of the effectiveness of the approach we are designing, we have used the MNIST dataset by dividing it into two non-IID subsets: the first subset contains images of digits 0–4, and the second subset contains images of digits 5–9. These subsets were assigned to two different clients to simulate a non-IID environment. In this experiment, two separate FL environments were trained: a general one and one utilizing the proposed technique. Both environments used the same data. In the general FL approach, each client trains a local model on its respective subset of data. After local training, clients send their model updates to a central server, which performs simple averaging to create a global model by averaging the weights of the client models.

Figure 2 represents the model architecture.

In our methodology, after preprocessing the data, a model was trained and then client clustering and progressive personalization were applied together to achieve better FL performance. Client clustering involves measuring the divergence between the models of different clients to determine how similar their data distributions are. This similarity is assessed by calculating the sum of the

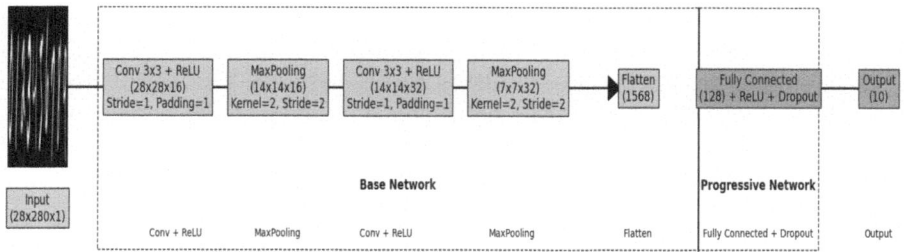

Fig. 2. Model Architecture

squared differences between the parameters of two models, as shown in Eq. 1. If the model divergence between two clients is below a predefined threshold (the experiments in this study use a threshold of 100), their data distributions are considered similar enough to be clustered together (see Eq. 2). For example, if the divergence between client 1 and client 2's models is less than 100, they are placed in the same cluster.

$$D_{ij} = \sum_k \left(\theta_i^{(k)} - \theta_j^{(k)} \right)^2 \tag{1}$$

where $\theta_i^{(k)}$ and $\theta_j^{(k)}$ represent the k-th parameter of the models of clients i and j. Clients are clustered based on this computed divergence. If D_{ij} is less than a predefined threshold T, the clients are grouped into the same cluster:

$$\text{Cluster}_c = \{i, j \mid D_{ij} < T\} \tag{2}$$

Progressive personalization starts with initial local training, where clients perform standard local training on their respective non-IID datasets. After local training, clients within the same cluster share their model updates, which are averaged to create a cluster-specific global model (see the Eq. 3). This cluster-specific global model serves as the starting point for further local training, allowing each client to fine-tune the model to better fit their local data distribution. This iterative process of clustering, model averaging, and additional local training helps the global model in adapting progressively to the unique characteristics of each client's data.

$$\theta_{c,j} = \frac{1}{|Cluster_c|} \sum_{i \in Cluster_c} \theta_{ij} \tag{3}$$

The global model weights θ_{global} are then computed by averaging the weights of all client models:

$$W_{\text{global},j} = \frac{1}{M} \sum_{i=1}^{M} W_{ij} \tag{4}$$

where M is the total number of clients.

We note here that our technique effectively combines client clustering based on model divergence with progressive personalization to handle non-IID data in FL environment. This approach allows the global model to benefit from the strengths of personalized local training while maintaining the robustness of a FL system. In order to practically access the performance of our method, we tested global model using test data (using Eq. 5) The results, as shown in the following Fig. 3, demonstrate a clear advantage over standard FL.

$$\text{Accuracy} = \frac{\text{Correct Digit Predictions}}{\text{Total Test Predictions}} \tag{5}$$

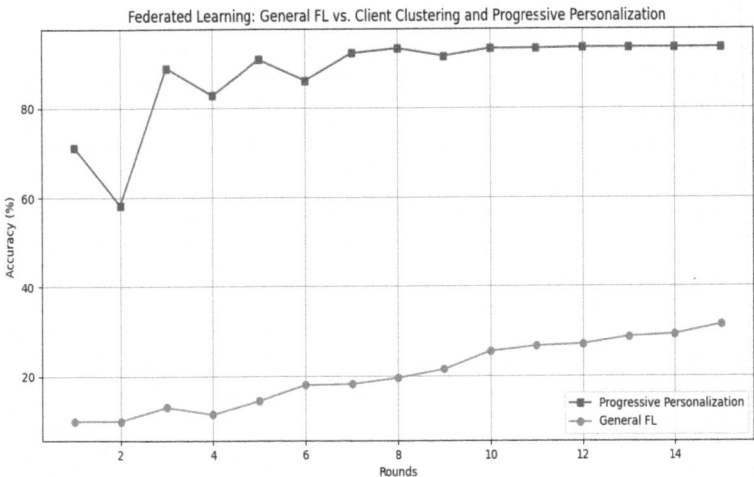

Fig. 3. General FL vs Our Approach

We next discuss some aspects to consider for future work. We are currently in the process of proposing and refining our techniques by experimenting with different datasets and different threshold values to optimize performance. So far, the results presented in this study are our best findings. We are thoroughly examining every aspect to finalize our approach. This is why we have not included testing on other datasets or different combinations in this study. There are a few existing studies addressing Non-IID data in federated learning, including FedProx [11], Scaffold [12], and FedNova [13]. Our preliminary results indicate that our technique outperforms existing methods in terms of environment performance.

3.2   Dynamic Encryption and Decryption in FL

In the FL environment, if the data shared between clients is similar in every training round, the weights will often be similar. If communication protocols are compromised and these weights are seen by unauthorized users, the consistent weights can reveal metadata about the data. Currently available privacy techniques, such as Local Differential Privacy (LDP) and Global Differential Privacy (GDP), secure privacy but significantly impact the performance of the environment due to the noisy weights. For example, Wu et al. [15] proposed a method incorporating LDP to protect privacy, but the added noise adversely affects model accuracy. Similarly, Shen et al. [16] introduced a personalized LDP method, which also suffers from performance degradation due to noise. Additionally, Xu et al. [16] discussed GDP techniques in FL, highlighting the trade-off between privacy and performance.

With our research, we are proposing a new technique for privacy in FL called Dynamic Encryption Key Exchange for FL(DEKE-FL). DEKE-FL will involve the dynamic exchange of encryption keys between the central server and participating edge devices throughout the FL process. Instead of relying on static encryption keys, DEKE-FL will continuously update these keys based on specific triggers such as time intervals, model performance, anomaly detection, and user requests. Initially, each device will generate a unique encryption key pair and securely transmit its public key to the central server. The central server will periodically generate new encryption keys and securely transmit them to all participating devices. Devices and the central server will authenticate each other before exchanging keys using secure communication protocols. Each key will have an expiration timestamp, and keys will be automatically renewed before expiration, with expired keys being securely discarded to prevent unauthorized access. This dynamic approach will enhance security by reducing the window of vulnerability, allow adaptive privacy measures, minimize the attack surface, and help complying with privacy regulations. We are currently working to set up a practical implementation to verify and refine our idea, making it effective and efficient for FL.

3.3   Fair Aggregation Techniques

FL employs various aggregation methods to combine model updates from multiple decentralized devices to form a global model as with aggregation methods mentioned in second last paragraph of Sect. 2. These methods aim to address challenges in FL, including data heterogeneity and communication efficiency, but each has its trade-offs in terms of complexity, fairness, and model performance.

We are currently working towards proposing a fairer and novel technique for aggregation. One idea is to use game theory and mechanism design based methods, which would supposedly ensure effectiveness and fairness by modeling the interactions between clients as a strategic game. Another approach could involve using consensus algorithms, which help in achieving a fair and accurate aggregation by considering the opinions of multiple clients. We aim to utilize these methods to develop an aggregation method that will enhance both performance and fairness in FL.

4 Conclusion and Future Work

In this research, we addressed the issue of data heterogeneity in FL by combining client clustering and personalization techniques. Our approach aims to improve model performance by grouping clients with similar data distributions and personalizing their models accordingly. Our preliminary results indicate that this technique performs well, and we will continue to refine and finalize it to further enhance its performance.

In addition to improving this technique, future work during my PhD will involve developing new aggregation methods and privacy techniques for FL. These efforts will focus on creating more robust aggregation techniques that can better handle diverse data distributions and ensuring stronger privacy guarantees for participants. The research will also define new privacy techniques using dynamic encryption methods as well. Future research will also involve extensive testing of these methods in various real-world scenarios to validate their effectiveness and make FL more efficient and applicable across different domains.

Acknowledgments. I acknowledge the support of the PNRR project FAIR - Future AI Research (PE00000013), Spoke 9 - Green-aware AI, under the NRRP MUR program funded by the NextGenerationEU.

I also want to express my gratitude to Prof. Luigi Palopoli and Dr. Simona Nisticó for their invaluable contributions and support throughout this research.

References

1. Regulation (EU) 2016/679 of the european parliament and of the council of 27 April 2016 on the protection of natural persons with regard to the processing of personal data and on the free movement of such data, and repealing directive 95/46/EC (general data protection regulation) (text with EEA relevance) (May 2016)
2. Smith, A., Zhou, H., Wei, D.: Enhancing data privacy in centralized machine learning. Int. J. Comput. Sci. Inform. Secur. **21**(1), 67–80 (2023)
3. Johnson, R., Lee, S.: Blockchain-integrated machine learning for enhanced data privacy. IEEE Trans. Inf. Forensics Secur. **18**(3), 450–463 (2023)
4. Zhao, R., Lu, Y., Sun, Q.: Differential privacy techniques in federated learning. In: Privacy Enhancing Technologies Symposium 2023, vol. 15, pp. 452–467 (2023)
5. Chen, X., Wang, Y.: Secure multi-party computation in federated learning. Adv. Neural. Inf. Process. Syst. **36**(1), 250–263 (2023)
6. Kim, H., Park, S., Kim, J.: Adaptive learning rates for non-IID data in machine learning. Neural Netw. **140**(4), 205–220 (2023)
7. Lopez, R., Hernandez, F.: Data augmentation techniques for addressing non-IID data in ml. Pattern Recogn. **133**(5), 320–335 (2023)
8. Liu, X., Zhang, T., Huang, J.: Dynamic clustering for heterogeneous data in federated learning. In: Proceedings of the 2023 International Conference on Machine Learning, vol. 145, pp. 2205–2216 (2023)
9. Nguyen, T., Tran, V.: Personalized federated learning for diverse data distributions. IEEE Trans. Neural Netw. Learn. Syst. **34**(7), 785–797 (2023)

10. McMahan, B., Moore, E., Ramage, D., Hampson, S., y Arcas, B.A.: Communication-efficient learning of deep networks from decentralized data. In: Artificial Intelligence and Statistics, pp. 1273–1282. PMLR (2017)
11. Li, T., Sahu, A.K., Zaheer, M., Sanjabi, M., Talwalkar, A., Smith, V.: Federated optimization in heterogeneous networks. In: Proceedings of Machine Learning and Systems 2020, pp. 429–450 (2020)
12. Karimireddy, S.P., Kale, S., Mohri, M., Reddi, S., Stich, S., Suresh, A.T.: Scaffold: stochastic controlled averaging for federated learning. In: Proceedings of the 37th International Conference on Machine Learning, pp. 5132–5143 (2020)
13. Wang, J., Liu, Q., Liang, H., Joshi, G., Poor, H.V.: Tackling the objective inconsistency problem in heterogeneous federated optimization. In: Advances in Neural Information Processing Systems, vol. 33, pp. 7611–7623 (2020)
14. Alex, K.: Learning multiple layers of features from tiny images (2009)
15. Xia, W., Lei, X., Zhu, L.: Local differential privacy-based federated learning under personalized settings. Appl. Sci. **13**(7), 4168 (2023)
16. Shen, X., Jiang, H., Chen, Y., Wang, B., Gao, L.: PLDP-FL: federated learning with personalized local differential privacy. Entropy **25**(3), 485 (2023)

Development of Explainable AI Methods for the Interpretation of Machine Learning Models in Bioinformatics and Medicine

Fedra Rosita Falvo[1,2(✉)] iD

[1] Department of Medical and Surgical Sciences, University "Magna Graecia",
88100 Catanzaro, Italy
[2] Data Analytics Research Center, University "Magna Graecia", 88100 Catanzaro,
Italy
fedrarosita.falvo@unicz.it

Abstract. Developments and adoptions of machine learning in bioinformatics and medicine have led to increasingly complex and accurate models. Machine learning models, while becoming increasingly accurate, often suffer from a lack of transparency and interpretability, hindering the adoption of such models in fields such as bioinformatics and medicine. The field of Explainable AI (XAI) addresses these challenges by aiming to develop methods and techniques to make AI models interpretable, providing understandable explanations of their predictions. The present research aims to develop advanced XAI methods for the interpretation of machine learning models specifically designed for bioinformatics and medicine, in order to provide clear, understandable and reliable explanations of predictions.

Keywords: Explainable AI (XAI) · Bioinformatics · Machine Learning

1 Introduction

In recent decades, artificial intelligence (AI) has profoundly transformed various sectors, including bioinformatics and medicine: rapid advances in machine learning algorithms have enabled the development of increasingly complex and accurate models. The latter are able to analyze large quantities of biological and medical data, discover patterns, make predictions and support decision-making processes that were previously beyond human capabilities; Despite these significant findings, a major challenge persists: the lack of transparency and interpretability of these sophisticated models.

In bioinformatics, machine learning models are widely used to interpret large-scale biological data, aiding in the discovery of critical biological insights and advancing our understanding of complex biological processes and disease mechanisms. For example, models are applied to classify genomic sequences, predict protein structures, analyze gene expression data, and identify biomarkers for various diseases. Similarly, in the medical field, AI models help diagnose diseases,

J. Tekli et al. (Eds.): ADBIS 2024, CCIS 2186, pp. 366–371, 2025.
https://doi.org/10.1007/978-3-031-70421-5_34

predict patient outcomes, and personalize treatment plans based on individual patient data. These applications highlight the central role of artificial intelligence in both research and clinical settings.

However, as machine learning models become more complex, their internal workings often become "opaque", turning them into "black boxes": this opacity poses significant obstacles in fields such as bioinformatics and medicine, where understanding the The logic behind model predictions is critical to building trust and ensuring practical application. The inability to explain how a model arrives at a particular prediction can lead to skepticism among healthcare professionals and researchers, ultimately hindering the adoption of AI technologies in these high-risk sectors. To address these challenges, the field of Explainable AI (XAI) has emerged which focuses on developing methods and techniques that make AI models interpretable and transparent, providing understandable explanations of their predictions. XAI aims to bridge the gap between high model accuracy and user trust, thereby facilitating wider acceptance and integration of AI technologies in critical areas such as healthcare and bioinformatics.

La necessità di spiegabilità nell'intelligenza artificiale è guidata da diversi fattori: in primo luogo, considerazioni etiche e legali richiedono che i sistemi di intelligenza artificiale, in particolare quelli utilizzati nel settore sanitario, siano trasparenti e responsabili. I pazienti e gli operatori sanitari devono comprendere le basi delle decisioni guidate dall'intelligenza artificiale per garantire il consenso informato e mantenere la fiducia; in secondo luogo, i quadri normativi richiedono sempre più trasparenza nelle applicazioni dell'intelligenza artificiale per salvaguardare dai pregiudizi e garantire risultati equi tra i diversi gruppi di pazienti. Infine, l'utilità pratica dell'IA in contesti clinici dipende dalla capacità degli operatori sanitari di interpretare e agire con sicurezza in base alle raccomandazioni dell'IA. Data l'importanza di queste considerazioni, l'obiettivo primario di questa ricerca è sviluppare metodi XAI avanzati adattati specificamente alle esigenze della bioinformatica e della medicina: l'obiettivo è creare tecniche che non solo mantengano un'elevata precisione predittiva ma forniscano anche spiegazioni chiare, comprensibili e affidabili per i loro risultati. Migliorando l'interpretabilità dei modelli di intelligenza artificiale, questa ricerca mira a migliorarne l'usabilità e l'accettazione in contesti clinici e di ricerca.

2 Integration with Broader Doctoral Research

This research is an integral part of a doctoral project aimed at understanding the explainability and usability of artificial intelligence models in critical sectors: the overall objective of the doctoral research is to innovate in the field of artificial intelligence by developing methods that are not only highly accurate but also interpretable and transparent. The focus on bioinformatics and medicine is driven by the need for precise and understandable AI tools in these fields, where the implications of AI decisions are particularly significant: this specific project on XAI is designed to respond to these needs by developing methods that can provide clear explanations for AI predictions, thus fostering greater trust and adoption among professionals and researchers in bioinformatics and medicine.

3   Research Objectives

The main objective of this research is to develop and test advanced Explainable AI (XAI) methods adapted to the specific needs of bioinformatics and medicine: the aim is to provide clear, understandable and reliable explanations for AI predictions, facilitating integration of these models in clinical practice and research. In conclusion, this research seeks to bridge the gap between the accuracy and interpretability of models, ensuring that they are effective and transparent.

3.1   Specific Objectives

1. **Review of State-of-the-Art XAI Techniques**: This objective involves conducting a comprehensive review of current XAI methods, focusing on their application in bioinformatics and medicine: the review will cover techniques such as LIME (Local Interpretable Model-agnostic Explanations), SHAP (SHapley Additive exPlanations), decision trees and systems-based on rules. The goal is to highlight the strengths and weaknesses of these methods, particularly in their application to bioinformatics and medicine, and to identify gaps in the literature and opportunities for innovation. Emphasis will be placed on developing new benchmarks to assess the balance between accuracy, explainability and fairness: an in-depth review of existing literature, case studies and practical applications will provide a comprehensive understanding of the current state of XAI in these fields.

2. **Testing and Development of XAI Methods**: This objective involves in-depth testing of existing XAI methods such as SHAP and LIME to evaluate their performance, identifying their strengths and limitations: based on this evaluation, new XAI methods will be developed to meet specific needs, focusing on balancing accuracy and interpretability for ensure meaningful and reliable explanations. This includes developing new algorithms or improving existing ones to improve their applicability in real-world bioinformatics and medical scenarios.

3. **Application in Bioinformatics and Medicine**: The developed XAI methods will be applied to bioinformatics and medical problems, including genomic sequence classification, protein structure prediction, gene expression data analysis, and biosignal and bioimage classification. To test the methods, datasets from clinical partners will be used, such as the AOU Renato Dulbecco - UOC of Neurology. Detailed explanations will be provided for the AI model's decisions, enabling a deeper understanding of biological processes and strengthening confidence in clinical decision support systems. The effectiveness of these methods will be evaluated through case studies, focusing on improving research outcomes and patient care.

4. **Ethical and Fairness Considerations**: This goal involves ensuring that AI models and their explanations are free from bias and provide fair results across different patient groups: techniques to detect and mitigate bias will be incorporated, and the implications of explainability for patient privacy and

data security. Adherence to the highest standards of ethical development and deployment of AI will be maintained, ensuring fairness and transparency.

5. **Integration and Deployment**: Il progetto di ricerca mira a sviluppare strumenti e interfacce di facile utilizzo per ricercatori e medici per interpretare e utilizzare facilmente le spiegazioni del modello AI: i metodi XAI sviluppati saranno integrati nei quadri biomedici e clinici esistenti, garantendo che siano accessibili e utilizzabili dagli utenti finali. Verranno esplorate future partnership con istituzioni cliniche e aziende per facilitare l'integrazione pratica e l'implementazione di questi metodi.

4 Expected Outcomes

This research aims to explore and develop advanced Explainable AI (XAI) methods adapted to bioinformatics and medicine, which will improve the transparency, reliability and usability of machine learning models in these fields. A key achievement is the creation of XAI techniques that balance predictive accuracy with interpretability: this balance ensures that models provide precise predictions and clear explanations that healthcare professionals and researchers can easily understand. The development of these techniques will involve extensive testing and validation to ensure their effectiveness in real-world applications. The application of these XAI methods to bioinformatics and real-world medical problems is expected to significantly improve biomedical research and clinical practice: clear and understandable explanations for the classification of genomic sequences, prediction of protein structures, analysis of Data on gene expression and interpretation of biosignals and bioimages will improve the quality of research and patient care. This will lead to more informed decision-making and faster progress in understanding complex biological processes and disease mechanisms.

Another important achievement is the definition of new benchmarks for the evaluation of XAI methods: these benchmarks will take into account accuracy, explainability and fairness, providing a comprehensive framework for the evaluation of XAI techniques; this standardization will help compare and improve XAI methods over time. Addressing ethical and fairness considerations is also a crucial expected outcome: developing techniques to detect and mitigate bias in AI models will ensure that the explanations provided are fair and impartial, promoting equitable outcomes for different patient groups. This focus on ethical AI will increase the reliability of models, making them more acceptable to healthcare professionals, researchers and patients.

In addition to these technical advances, the research aims to produce practical tools and interfaces to make the developed XAI methods accessible and usable by end users: these tools will be user-friendly, ensuring that researchers and clinicians can easily interpret and use the explanations of the models artificial intelligence. This will involve creating intuitive interfaces and conducting training sessions and workshops to ease the transition from research to practical application. Furthermore, the research will significantly contribute to bioinformatics and medicine by promoting AI-driven clinical discoveries and applications. The results and tools developed will be shared according to the principles of "Open Science" and "FAIR Data", ensuring broad accessibility and impact: this approach will help make artificial intelligence more transparent, ethical and effective.

Achieving these results will pave the way for more transparent and trustworthy AI applications in healthcare and bioinformatics, which will increase user confidence in AI technologies, promoting wider adoption and integration in these fields, leading to breakthroughs significant in both research and clinical practice.

5 Conclusion and Future Work

The development of advanced XAI methods is fundamental for the broader adoption of AI in bioinformatics and medicine: this research aims to address current challenges by ensuring that AI models are not only accurate but also comprehensible and reliable. By providing clear and dependable explanations for AI predictions, the proposed XAI methods will enhance the usability and acceptance of AI technologies in clinical and research contexts. The successful implementation of this research will result in innovative XAI methods that can be widely adopted, fostering a new era of AI-driven discoveries and clinical applications. The findings and tools developed through this research will be disseminated in accordance with the principles of "Open Science" and "FAIR Data," ensuring broad accessibility and impact, and contributing to ongoing efforts to make AI more transparent, ethical, and effective.

In the long term, this research will not only contribute to the advancement of XAI but also promote the establishment of stronger collaborations between academic institutions, clinical centers, and industry partners. These collaborations will be essential for the continuous development and integration of AI technologies in healthcare, ensuring that the benefits of AI are maximized while maintaining ethical and transparent practices. The anticipated outcomes will set a precedent for future research in XAI and its applications, ultimately leading to improved patient care, more robust scientific discoveries, and a deeper understanding of complex biological systems.

References

1. Goodfellow, I., Bengio, Y., Courville, A.: Deep Learning. MIT Press (2016). https://www.deeplearningbook.org/
2. Tjoa, E., Guan, C.: A survey on Explainable Artificial Intelligence (XAI): towards medical XAI. IEEE Trans. Neural Netw. Learn. Syst. (2020). https://doi.org/10.1109/TNNLS.2020.3027314
3. Arrieta, A.B., et al.: Explainable Artificial Intelligence (XAI): concepts, taxonomies, opportunities and challenges toward responsible AI. Inform. Fusion **58**, 82–115 (2020). https://doi.org/10.1016/j.inffus.2019.12.012

4. Linardatos, P., Papastefanopoulos, V., Kotsiantis, S.: Explainable AI: a review of machine learning interpretability methods. Entropy **23**(1), 18 (2020). https://doi.org/10.3390/e23010018
5. Samek, W., Wiegand, T., Müller, K. R.: Explainable artificial intelligence: understanding, visualizing and interpreting deep learning models. arXiv preprint arXiv:1708.08296 (2017)
6. Confalonieri, R., Coba, L., Wagner, B., Besold, T.R.: A historical perspective of explainable artificial intelligence. Wiley Interdiscip. Rev. Data Min. Knowl. Discov. **11**(1), e1391 (2021). https://doi.org/10.1002/widm.1391

Deep Learning Techniques for Predicting Wildfires in Calabria Italy Using Environmental Parameters

khushal Das$^{(\boxtimes)}$

DIMES, University of Calabria, Rende, (CS), Italy
khushal.das@dimes.unical.it

Abstract. Predicting wildfires has become crucial due to the increasing frequency and intensity of fires driven by climate change. This study focuses on developing advanced predictive models based on the Calabria (Region in Italy) region's environmental characteristics. We employed comprehensive datasets, including meteorological data, atmospheric gas concentrations, and historical fire occurrence records. Using sophisticated deep learning techniques, we implemented Long Short-Term Memory (LSTM) networks, Gated Recurrent Units (GRU), and Bi-Directional LSTM (Bi-LSTM) models. These models demonstrated high accuracy in detecting fires, with testing accuracies reaching up to 93.9%. However, predicting fire distance and direction showed moderate success, indicating areas for further improvement. Our research significantly enhances wildfire preparedness and response strategies in the Calabria region. Future work will involve developing hybrid models that integrate various deep learning architectures, incorporating datasets from different years to improve generalizability, and implementing real-time testing to assess performance in dynamic environments. This study provides a valuable tool for regional fire management and mitigation efforts, leveraging cutting-edge technology to protect communities and natural landscapes from the increasing threat of wildfires.

Keywords: Deep Learning · Wildfire Prediction · Environmental Monitoring · Time Series Analysis · Climate Change

1 Introduction

Wildfires represent a profound threat to both natural ecosystems and human societies, causing extensive damage to vegetation, wildlife habitats, and infrastructure, and posing significant risks to human lives. The Calabria region in Italy, characterized by its unique climate and topography, is particularly vulnerable to such events. In recent years, the frequency and intensity of wildfires have increased globally, a trend closely linked to the escalating impacts of climate change [1]. These fires not only devastate landscapes but also contribute to atmospheric pollution, including the release of significant quantities of greenhouse gases and particulate matter [3].

© The Author(s), under exclusive license to Springer Nature Switzerland AG 2025
J. Tekli et al. (Eds.): ADBIS 2024, CCIS 2186, pp. 372–379, 2025.
https://doi.org/10.1007/978-3-031-70421-5_35

Understanding and predicting wildfires have become critical in climate change and environmental management. Effective prediction models enhance preparedness and response, mitigating wildfire effects [9]. Traditional methods, which rely on static data and simpler statistical models, often fail to capture the complex interactions between environmental factors and fire occurrences [12]. Recent advancements in deep learning have significantly improved the accuracy and reliability of wildfire prediction models. These techniques excel at modeling complex, non-linear relationships and handling large datasets generated by modern environmental monitoring systems, enabling more nuanced analyses of wildfire risk factors. [15].

This research aims to address these challenges by utilizing advanced deep learning techniques to analyze a comprehensive set of environmental parameters. Specifically, this study focuses on the influence of atmospheric gases such as Black Carbon, CO2, and CH4, along with meteorological features including wind direction, wind speed, and air pressure, on fire occurrences in Calabria during 2021. The integration of these diverse data sources allows for a more holistic understanding of the conditions that lead to wildfires.

By using models such as LSTM [7], GRU [4], and Bi-LSTM, this study seeks to develop robust predictive models for detecting fire occurrences, estimating fire direction, and predicting fire distance. Previous studies have demonstrated the efficacy of these models in handling sequential data and time series analysis [6] but their dataset, environmental conditions, and regional conditions differ from our dataset. This research enhances our understanding of environmental factors in fires. The predictive models improve wildfire preparedness and response, applicable to regions affected by climate change. Using deep learning and environmental monitoring, it aims to better predict and manage wildfires. This study has three main objectives:

- **Fire Occurrence Detection:** Developing AI models to detect fires based on Black Carbon, CO2, and CH4, and environmental parameters such as temperature, humidity, wind speed, and precipitation.
- **Fire Direction Estimation:** Determining the direction of a fire from a specific point using AI algorithms.
- **Fire Distance Prediction:** Predicting the distance of a fire from a specific point by analyzing environmental data.

2   Realted Work

In the face of escalating climate change, predicting wildfires has become crucial due to increased fire frequency and intensity. This research aims to build upon existing studies that emphasize the importance of regional characteristics in developing effective prediction models for Calabria. Previous works, such as [14,17], underline the necessity of tailoring models to specific regional traits for enhanced accuracy. [1] highlights the importance of considering specific regional characteristics when developing prediction models for Calabria. These regional characteristics include variations in climate, vegetation, and topography, all of which significantly influence wildfire behavior. Tailored predictive approaches are necessary to account for these unique factors, as generic models may not capture the specific conditions of Calabria. Recent advancements in deep learning have demonstrated significant potential in wildfire prediction.

The studies [8,16] showcase how deep learning models can effectively manage complex relationships between environmental factors and fire occurrences, leading to more precise predictions. Additionally, research [18] highlights the importance of including atmospheric gases like black carbon in predictive models. Studies such as [11] further stress the need for region-specific adaptation strategies, recognizing the intricate interplay between climate change and wildfire risk in the Mediterranean region.

Moreover, [2] illustrates the capabilities of deep learning in real-time wildfire monitoring and prediction. Research like [10] explores the use of deep learning to predict fire direction, which is essential for timely response and resource allocation. Additionally, [13]combines deep learning with physics-informed neural networks, enhancing the accuracy of wildfire spread predictions.

This research advances foundational studies by using deep-learning techniques to analyze atmospheric gases and meteorological features in Calabria. It aims to develop robust models for fire detection, direction estimation, and distance prediction. This work seeks to enhance wildfire preparedness and response strategies, significantly contributing to regional fire management and mitigation efforts.

3   Methodology

The research approach started with collecting comprehensive environmental data specific to the Calabria region. This included meteorological data such as wind speed, wind direction, air temperature, relative humidity, and air pressure, as well as atmospheric gases like black carbon, CO, CO2, and CH4. Additionally, fire occurrence data was compiled, with 300,465 entries indicating fire occurrences and 205,112 entries indicating no fire in 2021. The dataset utilized in this study includes three main labels: Fire Occurrence, Fire Direction, and Fire Distance Range, which provide comprehensive information about fire incidents. The Fire Occurrence label is a binary indicator that signifies whether a fire has taken place within the observed region. A value of 1 denotes the occurrence of a fire, while a value of 0 indicates that no fire has occurred. This binary classification is crucial for identifying and analyzing the presence or absence of fires over time and space. The Fire Direction label categorizes the direction of the fire into one of eight possible values: North (N), South (S), East (E), West (W), Northeast (NE), Northwest (NW), Southeast (SE), and Southwest (SW). This categorical information helps in understanding the directional spread of fires, which is essential for both predicting the movement of active fires and planning effective response strategies. Finally, the Fire Distance Range label represents the distance of the fire from a reference point, measured in increments of 20 km. For instance, a value of 0–20 indicates that the fire is within 20 km, 20–40 signifies a distance of 20 to 40 km, and so on. This continuous variable provides detailed information about the spatial extent of fires, which can be used to assess the potential impact areas and to model the spread dynamics of fires. The methodology process is shown in Fig. 1

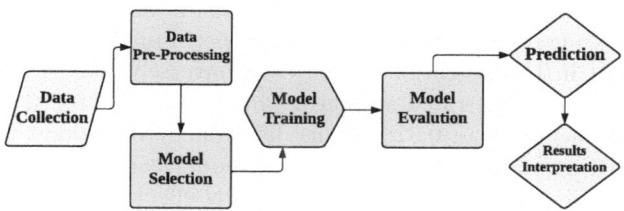

Fig. 1. Research Methodology Process

Together, these labels form a robust framework for analyzing fire occurrences, directions, and distances, enabling the development of predictive models that offer valuable insights into fire dynamics. Following data collection, we cleaned the data to remove inconsistencies, missing values, and outliers, then normalized it to ensure a uniform scale for optimal model performance.

We employed several advanced deep-learning models to predict wildfire occurrences, directions, and spread distances in the model selection and training phase. The chosen models included LSTM networks, GRU, and Bi-LSTM networks.

Table 1. Task Based models and their parameters

Task	Models	Activation Functions	Optimizer	Epochs
Fire Detection	LSTM, GRU, BiLSTM	Relu and Sigmoid	Adam	15
Fire Direction	LSTM, GRU, BiLSTM	Relu and softmax	Adam	15
Fire Distance Range	LSTM, GRU, BiLSTM	Relu and softmax	Adam	15

For fire detection, the models were trained using 15 epochs, with a batch size of 42. The activation functions used were ReLU and Sigmoid, with the Adam optimizer and Binary Cross Entropy as the loss function. For predicting fire distance range and direction, we also used 15 epochs and a batch size of 42, but with ReLU and Softmax activation functions, again utilizing the Adam optimizer and Binary Cross Entropy loss function. As shown in Table 1

4   Results

The models demonstrated varying levels of accuracy across different prediction tasks. To assess their performance, we evaluated the models on key metrics including testing and training accuracy [5]. For fire detection, the models exhibited high accuracy, highlighting their capability to capture the complex relationships within the dataset effectively.

The LSTM network achieved a testing accuracy of 0.9220 and a training accuracy of 0.9269. This consistency between training and testing accuracy indicates a well-generalized model with minimal overfitting. The GRU model showed a slightly higher testing accuracy of 0.9390 while maintaining a training accuracy of 0.9270, suggesting its robustness and superior performance in handling sequential data typical in fire detection scenarios. Conversely, the Bi-LSTM model, although effective, demonstrated a lower testing accuracy of 0.8638 compared to its high training accuracy of 0.9281. This discrepancy may point towards some degree of overfitting, indicating that while the model learns well during training, it might struggle to generalize to unseen data as effectively as the LSTM and GRU models. These performance metrics are summarized in Table 2.

These results highlight the importance of model selection and architecture for optimal performance. The GRU's superior testing accuracy demonstrates its effectiveness in handling sequential data dependencies, crucial for predicting dynamic events like wildfires. Conversely, the Bi-LSTM's potential overfitting suggests the need for regularization techniques, such as dropout or data augmentation, to improve generalizability.

Table 2. Performance Metrics for Fire Prediction Models

Model	Fire Detection		Fire Distance Range		Fire Direction	
	Testing Accuracy	Training Accuracy	Testing Accuracy	Training Accuracy	Testing Accuracy	Training Accuracy
LSTM	0.9220	0.9269	0.4740	0.4370	0.6270	0.6280
GRU	0.9390	0.9270	0.4370	0.4690	0.6270	0.6270
Bi-LSTM	0.8638	0.9281	0.4370	0.4900	0.6270	0.6270

While predicting the fire distance range, the models showed moderate performance, indicating the complexity of accurately estimating this aspect of wildfire behavior. LSTM model achieved a testing accuracy of 0.4740 and a training accuracy of 0.4370, suggesting some generalization capabilities but also pointing to potential challenges in capturing the full dynamics of fire spread.

GRU model displayed a testing accuracy of 0.4370 and a training accuracy of 0.4690. This performance, slightly lower in testing accuracy than the LSTM, might indicate the GRU's limitations in handling the spatial aspects of fire spread, which are inherently more complex than temporal sequences. The Bi-LSTM model showed similar results, with a testing accuracy of 0.4370 and a training accuracy of 0.4900. This slight increase in training accuracy compared to the LSTM and GRU suggests that the Bi-LSTM might be overfitting to the training data, capturing detailed nuances that do not generalize well to unseen data. These performance metrics are summarized in Table 2.

These results highlight the challenges in modeling fire distance range with sequential neural networks. The modest accuracy levels suggest these models may struggle with the complex, multi-dimensional nature of fire spread, which includes temporal, spatial, and environmental factors. Moreover, enhancing training data with diverse and representative samples, possibly through synthetic data generation or data augmentation, could improve generalization and performance in predicting fire distance range.

In predicting fire direction, the accuracy metrics were notably consistent across the different models. The LSTM, GRU, and Bi-LSTM models all achieved a testing accuracy of 0.6270. The training accuracies for these models were similarly close, with the LSTM achieving 0.6280, and both the GRU and Bi-LSTM models achieving 0.6270. This uniformity in performance suggests that each model captures the underlying patterns in the data to a similar extent. These performance metrics are summarized in Table 2.

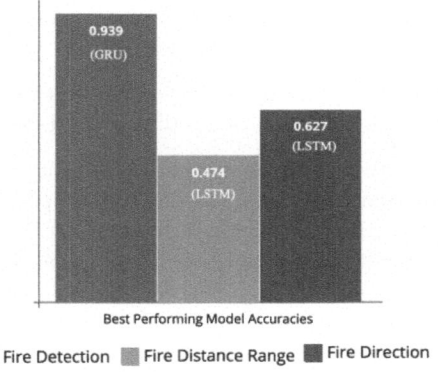

Fig. 2. Testing Accuracies of best-performing models on different tasks

The consistent accuracy across the LSTM, GRU, and Bi-LSTM models indicates they are equally effective at identifying directional patterns due to the inherent design of recurrent neural networks. The slight variation in the LSTM's training accuracy may result from how it processes long sequences. However, this uniformity suggests a performance plateau, indicating that recurrent neural networks alone may not significantly improve fire direction prediction. Future work could enhance performance by integrating additional features, employing ensemble methods, incorporating advanced regularization techniques, or using hybrid models that combine deep learning with domain-specific fire dynamics knowledge. While the consistent performance is promising, it highlights the need for more advanced methodologies to achieve substantial improvements in predicting fire direction accurately.

The best-performing model on fire detection is the GRU model, which achieved an accuracy of 0.9390. Overall, the best-performing model is the LSTM network, which achieved the highest accuracies across all prediction tasks. Specifically, the LSTM model recorded accuracies of 0.922 for fire detection prediction, 0.474 for fire distance range prediction, and 0.627 for fire direction prediction. As shown in Fig. 2.

5   Conclusion and Future Work

By integrating sophisticated deep learning techniques and extensive environmental data, we have developed robust models for fire detection, direction estimation, and distance prediction in the Calabria region. These models demonstrated high accuracy in detecting wildfires, highlighting their potential as valuable tools for enhancing regional fire management and mitigation efforts. Our LSTM networks, GRU, and Bi-LSTM models effectively managed complex relationships between various environmental factors and wildfire occurrences. The results indicated notable success in fire detection, with the LSTM and GRU models achieving testing accuracies of 0.9220 and 0.9390, respectively. However, improvements are needed in predicting fire distance range and direction. Future work will focus on:

Development of Hybrid Models: Combining various deep learning architectures to create more robust predictive systems.

Incorporating Diverse Datasets: Using datasets from different years to capture a broader range of environmental conditions and enhance generalizability.

Real-Time Testing and Deployment: Assessing model performance in dynamic, real-world scenarios to improve responsiveness and accuracy.

In the Future, We aim to develop advanced, reliable wildfire prediction models to enhance understanding and predictive capabilities. These efforts will contribute to comprehensive management strategies applicable across various regions.

Acknowledgments. IR0000032 - ITINERIS, Italian Integrated Environmental Research Infrastructures System (D.D. n. 130/2022 - CUP B53C22002150006) Funded by EU - Next Generation EU PNRR- Mission 4 "Education and Research" - Component 2: "From research to business" - Investment 3.1: "Fund for the realisation of an integrated system of research and innovation infrastructures"

References

1. Abatzoglou, J.T., Williams, A.P.: Impact of anthropogenic climate change on wildfire across western us forests. Proc. Natl. Acad. Sci. **113**(42), 11770–11775 (2016)
2. Barker, L.D., et al.: Scientific challenges and present capabilities in underwater robotic vehicle design and navigation for oceanographic exploration under-ice. Remote Sens. **12**(16), 2588 (2020)
3. Bowman, D.M., et al.: Fire in the earth system. Science **324**(5926), 481–484 (2009)
4. Cho, K., et al.: Learning phrase representations using RNN encoder-decoder for statistical machine translation. arXiv preprint arXiv:1406.1078 (2014)
5. Das, K., et al.: Impact of using e-learning tools on student's psychological health during Covid-19. VFAST Trans. Softw. Eng. **9**(3), 120–127 (2021)
6. Goodfellow, I., Bengio, Y., Courville, A.: Deep Learning. MIT press (2016)
7. Hochreiter, S., Schmidhuber, J.: Long short-term memory. Neural Comput. **9**(8), 1735–1780 (1997)

8. Ji, Y., Wang, D., Li, Q., Liu, T., Bai, Y.: Global wildfire danger predictions based on deep learning taking into account static and dynamic variables. Forests **15**(1), 216 (2024)

9. Jolly, W.M., et al.: Climate-induced variations in global wildfire danger from 1979 to 2013. Nat. Commun. **6**(1), 7537 (2015)

10. Lei, R., et al.: Fossil fuel CO2 emissions over metropolitan areas from space: a multi-model analysis of OCO-2 data over Lahore, Pakistan. Remote Sens. Environ. **264**, 112625 (2021)

11. Molina, J., Herrera, M., y Silva, F.R.: Wildfire-induced reduction in the carbon storage of mediterranean ecosystems: an application to brush and forest fires impacts assessment. Environ. Impact Assess. Rev. **76**, 88–97 (2019)

12. Preisler, H.K., Brillinger, D.R., Burgan, R.E., Benoit, J.: Probability based models for estimation of wildfire risk. Int. J. Wildland Fire **13**(2), 133–142 (2004)

13. Raissi, M., Perdikaris, P., Karniadakis, G.E.: Physics-informed neural networks: a deep learning framework for solving forward and inverse problems involving non-linear partial differential equations. J. Comput. Phys. **378**, 686–707 (2019)

14. Richelmy, T., Re, G.A., Sanna, F., Franca, A., Salis, M., Arca, B.: A spatial analysis of wildfire risk factors in agroforestry areas under climate change: a case study from monte Pisanu, Sardinia (Italy). Environ. Sci. Proc. **17**(1), 97 (2022)

15. Schmidhuber, J.: Deep learning in neural networks: an overview. Neural Netw. **61**, 85–117 (2015)

16. Shadrin, D., et al.: Wildfire spreading prediction using multimodal data and deep neural network approach. Sci. Rep. **14**(1), 1–17 (2024)

17. Tonini, M., D'Andrea, M., Biondi, G., Degli Esposti, S., Trucchia, A., Fiorucci, P.: A machine learning-based approach for wildfire susceptibility mapping. the case study of the Liguria region in Italy. Geosciences **10**(3), 105 (2020)

18. Uecker, T.M., Kaspari, S.D., Musselman, K.N., Skiles, S.M.: The post-wildfire impact of burn severity and age on black carbon snow deposition and implications for snow water resources, cascade range, Washington. J. Hydrometeorol. **21**(8), 1777–1792 (2020)

Tutorials

Data-Driven Analysis for Monitoring Software Evolution

Zheying Zhang[ID] and Kostas Stefanidis[(✉)][ID]

Tampere University, Tampere, Finland
{zheying.zhang,konstantinos.stefanidis}@tuni.fi

Abstract. In the rapidly evolving world of software development, maintaining user satisfaction and adapting to changing needs is critical to the success of software products. This tutorial paper focuses on a data-driven approach to user review and contextual analysis as a means to monitor software maintenance and evolution and ultimately improve user satisfaction. To leverage user feedback to support software evolution, we organize the tutorial in three parts: 1) evaluating collective user expectations and perceived quality; 2) monitoring user satisfaction across software updates; and 3) incorporating user profiles and situational contexts into the analysis. This tutorial aims to provide researchers, IT professionals, and developers with the approaches necessary for adaptive and responsive software evolution.

Keywords: Data Analytics · Monitoring Software · Software Evolution

1 Introduction

Software products, though always being expected to provide satisfactory functionalities and be bug-free, somehow fail to meet the expectations of their users. Thus, software maintenance is inevitable and critical for any software companies who want their products or services to continue profiting [6,16,31]. On the other hand, due to the fierce competitiveness in the contemporary software market, as well as the ease of user churns, monitoring and sustaining the satisfaction of the users is a critical criterion for the long-term success of any software product within the evolution stage. To such an end, continuously understanding and meeting the users' needs and expectations is the key [17,24,27,32], as it is more efficient and effective to allocate maintenance effort accordingly to address the issues raised by users.

On the other hand, accompanied by the rapid development of internet technologies, the volume of user-generated content has been increasing exponentially [2–4]. Among such user-generated content, feedback from the customers, either numeric rating, recommendation, or textual reviews, have been playing an increasingly critical role in product designs in terms of understanding customers' needs [14,15]. Especially for software products that require constant maintenance and are continuously evolving, understanding of users' needs and complaints, as well as the changes in their opinions through time, is of great importance. Additionally, supported by the advance of data mining and machine learning techniques, the effort of knowledge discovery from analyzing such data and specially understanding the behavior of the users shall be largely reduced.

Recently, several studies propose data-driven approaches for feedback analysis, as well as applying such methods to support software maintenance and evolution [25,26]. Many studies focus on the text mining perspective of review analysis towards eliciting users' opinions [5,9,10,13]. Many others focus on the detection and classification of feedback types, e.g., feature requests, bug reports, emotion expression, etc. [1,28,30]. To enhance the effectiveness of software maintenance and evolution practice, an effective approach to the software's perceived user experience and the monitoring of its changes during evolution is required.

To support the practice of software maintenance and evolution targeting enhancing user satisfaction, in this tutorial, we focus on how a systematic data-driven user review analysis can be performed. Specifically, we organize the tutorial into three parts that target at answering the following 3 research questions.

- **(Part I)** How can we analyze users' collective expectations and perceived quality in use with data-driven approaches by exploiting sentiment and topics on a large volume of user review data?
- **(Part II)** How can we monitor user satisfaction over software updates during software evolution using reviews' topics and sentiments? And, how can we identify the problematic updates based on anomalies in review sentiment distribution?
- **(Part III)** How can we analyze users' profiles, software types, and situational contexts as contexts of use that support the analysis of user satisfaction?

Overall, this tutorial studies the application of data-driven end-user review analysis methods that support software maintenance and evolution. We start our discussion by analysing the existing domain knowledge of software maintenance and evolution in terms of taking advantage of the collective intelligence of end-users. In addition, we study various meaningful aspects that contribute to the research on software evolution contexts and lead to interdisciplinary contribution. We present a number of approaches for software maintenance and evolution even in the larger scope of the software industry, and for helping developers ease their efforts in release planning and other decision-making activities. This tutorial also explores the potential of integrating generative AI tools into real-world software development environments to assist requirements management in software maintenance and evolution [35].

2   Tutorial Outline

Typically, the quality and success of software products depends on the needs and satisfaction of the end-users. This is why it is important to study the evaluation and monitor such measures during software evolution. For software evolution, user involvement and support with feedback are critical; however, there are big difficulties in involving a large volume of users and effectively extracting their opinions from low-quality feedback. Towards this direction, data-driven and NLP methods can be used for analyzing large volumes of user reviews and monitoring the changes in users' expectations and satisfaction within software evolution. Furthermore, given that context is considered important in influencing users' perceived quality in use of the interactive software systems, contextual data-driven methods are exploited as well.

In this tutorial, we mostly investigate the application of data mining techniques towards facilitating the continuous monitoring of software evolution status in terms of the users' changing needs and perceived software quality. We focus on data-driven approaches that aim to ease the effort of the developers in continuously monitoring the status of updates fulfilling the needs of their end-users.

2.1 Analyze Users' Collective Expectation and Perceived Quality in Use with Data-Driven Approaches by Exploiting Sentiment and Topics

In this part, we focus on how to use sentiment analysis and topic modeling techniques to evaluate the perceived quality of a software product and to extract the users' collective expectations and needs based on a large volume of user reviews [11, 12, 19, 23]. The purpose of a group of approaches is to effectively support meaningful maintenance and update during the software evolution stage by understanding the needs of the end users' and sustaining their satisfaction, as user satisfaction is formed based on the expectation and the confirmation of such expectation towards the users' perception of performance.

To evaluate the overall perceived quality of a software product based on user reviews, the proposed solutions identify the informative reviews using, for example, a pretrained Bayes classifier after preprocessing the raw reviews. The (informative) reviews are classified into (pre-defined) aspects based on selected quality frameworks using multi-label text classification techniques. Then, for each quality aspect, the average sentiment is calculated based on the classified reviews which reflect the perceived quality of the aspects. Visualization tools can be used to show the perceived quality of the product on diagrams with the quantified evaluation for each aspect visible.

Moreover, the collective expectations and needs of the end-users can be extracted via topic modeling. Following this strategy, positive and negative reviews are separated into subsets using sentiment analysis. For each review subset, topics are then extracted that reflect the collective needs and complaints in terms of each quality aspect from the end-users.

2.2   Monitor User Satisfaction over Software Updates During Software Evolution Using Reviews' Topics and Sentiments

We discuss this part via a two-fold data-driven approach. First, we focus on approaches that observe the trends and identify the changes in the collective feedback of the users, and then on approaches that detect the problematic updates that severely evoke the users' negativity. We use here as input the potential outputs of Part I, namely the positive and negative topic sets extracted at the selected time periods, e.g., between major updates, and the series of quantified perceived quality through software evolution. The core idea is to verify the confirmation of the detected user expectations via the comparison between the topics before and after the updates. For any two adjacent time periods divided by major updates, the latent hypothesis is that the changes in users' opinions between the two review sets reflect the perceived quality of the particular update and that particular version of the software. Therefore, ideally, by tracking such changes through the timelines of the updates in the software evolution stage, the developers shall be able to monitor the users' satisfaction with the software with the specific changes in their needs [21, 22, 33].

More specifically, to monitor user satisfaction during the software evolution stage, one approach is to indicates the users' constant positive opinions on particular features (i.e., topics) that can be reflected by the positive similar topic chains through the updates. On the contrary, the unsolved issues can then be reflected by the negative ones. In addition, the similarities between positive and negative topics reflect the changes in users' opinions on the specific feature, indicating either a positive enhancement or a failed attempt. Other approaches enable as well to detect the abnormal decreases in quantified perceived quality through software evolution timeline using statistical analysis, like for example statistical anomaly detection of distribution. The identified abnormal days are then mapped to a causing former update by comparing the similarity between the top frequent keywords of the reviews on that day and the description of the update.

2.3   Analyze Users' Profiles, Software Types and Situational Contexts as the Contexts of Use that Support the Above Activities

In the third part of the tutorial, we include solutions for 1) situational contexts and ways of interaction analysis towards mobility conflict detection, 2) user types and preference analysis based on user behavior data, and 3) software type and related features analysis based on software feature data.

To analyze the situational contexts of software products, like mobile apps, the existing research uses different concepts of ways of interaction [20]. The hypothesis lies where for each specific software feature the users shall interact with it in one or multiple ideal ways in order to achieve their satisfaction. Therefore, the related approach aims to handle the notion that such conflicts in the expected ways of interaction for the primary software situational contexts and the ideal ways of interaction for software features should be identified and addressed.

On the other hand, alternative approaches take care of different user types and preferences. Specifically, such approaches focus on understanding the latent software user types and the preferences of each type of user [18]. Given an adequate amount of user behavior data, it is important to enable the developers to gain such context information about their users and insights on how to conduct proper maintenance strategy accordingly.

Furthermore, approaches that are based on the software type analysis use different measures, like the centrality measure and a community detection approach from social network analysis, to detect the latent software types within the market with the data of software feature data [17]. Such approaches facilitate the understanding of the software ecology; when knowing the type of a particular software product, the developers easily gain insights into the commonly adopted features within the category and the potential features to be added.

3 Related Tutorials

To the best of our knowledge, exploring data-driven approaches for monitoring software evolution has not been the focus of tutorials at recent conferences in databases, information systems, and data science communities. While our tutorial specifically addresses the application of user reviews and contextual analysis in monitoring software evolution, it is important to acknowledge the related tutorials that potentially enrich our understanding of data analysis methods in software engineering and relevant domains. One closely related tutorial topic discusses the importance of predicting and adapting to uncertainties and changes dynamically in software evolution and adaptation, using formal modelling and verification approaches [8]. While this approach is critical for developers to facilitate continuous evolution, our tutorial topic provides approaches to enabling developers to gain insights into uncertainties of software systems from user reviews. This, in turn, can guide the formal modelling and verification process. Additionally, several other tutorials like [29,34], and [7] have explored data engineering advancements, but in the context of network management and orchestration, falling outside our scope of data-driven analysis for continuous evolution of software systems.

By integrating user feedback analysis with existing software evolution practices, our tutorial offers a novel perspective. We aim to bridge the gap by discussing how user review data can be leveraged to assist decision-making in software development and maintenance.

4 Tutorial Information

Motivation and Target Audience: The tutorial's topic lies in the core of the conference interests. The tutorial aims at researchers and students, as well as IT professionals and developers in data-driven analysis and software evolution, and the general data management community. Researchers and students will get a good introduction to the topic and get inspired by challenging research problems. Furthermore, IT professionals and developers will learn appropriate techniques to promote top-notch methods in their systems. All the materials that will be used for the tutorial will be publicly available.

Prerequisites: The tutorial is carefully structured to accommodate both attendees unfamiliar with the topic and more experienced participants by providing required background knowledge, shared terminology, and a common understanding of the basic concepts related to data analytics and software evolution.

Intended Duration: We are aiming for a 90-minute tutorial.

5   Presenters

Zheying Zhang received her Ph.D degree in computer science and information systems from the University of Jyväskylä, Finland in 2004, and the title of Docent in Software Development at the University of Tampere in 2013. Currently, she is a university lecturer at Tampere University, with over 15 years of research and teaching experience in software engineering, specializing in requirements management. Her research interests include requirements analysis, domain modeling, quality assurance, process improvement, data-driven approach for software development and maintenance, as well as GenerativeAI-assisted software engineering. She has contributed to the software engineering research community with over 60 co-authored peer-reviewed publications.

Kostas Stefanidis is a Professor on Data Science at the Tampere University, Finland. He got his PhD in personalized data management from the Univ. of Ioannina, Greece. His research interests are in the broader area of big data. They lie in the intersection of databases, information retrieval, data mining and the Web, and include personalization and recommender systems, large-scale entity resolution and information integration, and query and data exploration paradigms. His publications include more than 100 papers in peer-reviewed conferences and journals, including SIGMOD, ICDE, and ACM TODS, and a book on entity resolution in the Web of data. He is also actively serving the scientific community. Currently, he is the General co-Chair of the 24th International Conference on Web Engineering (ICWE), and the General co-Chair of ADBIS 2025 and TPDL 2025. He has 9 years experience in teaching.

Prior Tutorials: Recommender Systems [MUMIA Training School'14], Personalization [ICDE'10], Entity Resolution [ICDE'17, ESWC'16, WWW'14, CIKM'13], Fairness in Rankings and Recommendations [ICDE'21, MDM'21, EDBT'20].

References

1. Carreño, L.V.G., Winbladh, K.: Analysis of user comments: an approach for software requirements evolution. In: 2013 35th International Conference on Software Engineering (ICSE), pp. 582–591. IEEE (2013)
2. Christophides, V., Efthymiou, V., Palpanas, T., Papadakis, G., Stefanidis, K.: An overview of end-to-end entity resolution for big data. ACM Comput. Surv. **53**(6), 127:1–127:42 (2021)
3. Christophides, V., Efthymiou, V., Stefanidis, K.: Entity Resolution in the Web of Data. Theory and Technology, Morgan & Claypool Publishers, Synthesis Lectures on the Semantic Web (2015)
4. Efthymiou, V., Papadakis, G., Stefanidis, K., Christophides, V.: Minoaner: schema-agnostic, non-iterative, massively parallel resolution of web entities. In: Advances in Database Technology - 22nd International Conference on Extending Database Technology, EDBT 2019, Lisbon, Portugal, 26-29 March 2019, pp. 373–384 (2019)
5. Feng, L., Chiam, Y.K., Lo, S.K.: Text-mining techniques and tools for systematic literature reviews: a systematic literature review. In: 2017 24th Asia-pacific software engineering conference (APSEC), pp. 41–50. IEEE (2017)
6. Fitzgerald, B., Stol, K.J.: Continuous software engineering: a roadmap and agenda. J. Syst. Softw. **123**, 176–189 (2017)
7. Gao, J., Lei, L., Yu, S.: Big data sensing and service: a tutorial. In: 2015 IEEE First International Conference on Big Data Computing Service and Applications, pp. 79–88 (2015). https://doi.org/10.1109/BigDataService.2015.45
8. Ghezzi, C.: Dependability of adaptable and evolvable distributed systems. formal methods for the quantitative evaluation of collective adaptive systems. 16th International School on Formal Methods for the Design of Computer, Communication, and Software Systems, SFM 2016, Bertinoro, Italy, 20-24 June 2016, Advanced Lectures 16, pp. 36–60 (2016)
9. Guzman, E., Maalej, W.: How do users like this feature? A fine grained sentiment analysis of app reviews. In: 2014 IEEE 22nd International Requirements Engineering Conference (RE), pp. 153–162. IEEE (2014)
10. Iacob, C., Veerappa, V., Harrison, R.: What are you complaining about?: a study of online reviews of mobile applications. In: 27Th International BCS Human Computer Interaction Conference (HCI 2013). BCS Learning & Development (2013)
11. Katsarou, K., Douss, N., Stefanidis, K.: REFORMIST: hierarchical attention networks for multi-domain sentiment classification with active learning. In: Proceedings of the 38th ACM/SIGAPP Symposium on Applied Computing, SAC 2023, Tallinn, Estonia, 27-31 March 2023, pp. 919–928. ACM (2023)
12. Katsarou, K., Jeney, R., Stefanidis, K.: MUTUAL: multi-domain sentiment classification via uncertainty sampling. In: Proceedings of the 38th ACM/SIGAPP Symposium on Applied Computing, SAC 2023, Tallinn, Estonia, 27-31 March 2023, pp. 331–339. ACM (2023)
13. Khalid, H., Shihab, E., Nagappan, M., Hassan, A.E.: What do mobile app users complain about? IEEE Softw. **32**(3), 70–77 (2014)
14. Ko, A.J., et al.: The state of the art in end-user software engineering. ACM Comput. Surv. (CSUR) **43**(3), 1–44 (2011)
15. Kujala, S.: User involvement: a review of the benefits and challenges. Behav. Inform. Technol. **22**(1), 1–16 (2003)
16. Lehman, M.M.: Laws of software evolution revisited. In: Montangero, C. (ed.) EWSPT 1996. LNCS, vol. 1149, pp. 108–124. Springer, Heidelberg (1996). https://doi.org/10.1007/BFb0017737

17. Li, X.: Data-driven analysis towards monitoring software evolution by continuously understanding changes in users' needs (2022)
18. Li, X., Lu, C., Peltonen, J., Zhang, Z.: A statistical analysis of steam user profiles towards personalized gamification. In: 3rd International GamiFIN Conference, GamiFIN 2019. CEUR-WS (2019)
19. Li, X., Zhang, B., Zhang, Z., Stefanidis, K.: A sentiment-statistical approach for identifying problematic mobile app updates based on user reviews. Information 11(3), 152 (2020)
20. Li, X., Zhang, Z.: A user-app interaction reference model for mobility requirements analysis. ICSEA 2015, 170–177 (2015)
21. Li, X., Zhang, Z., Stefanidis, K.: Mobile app evolution analysis based on user reviews. In: New Trends in Intelligent Software Methodologies, Tools and Techniques, pp. 773–786. IOS Press (2018)
22. Li, X., Zhang, Z., Stefanidis, K.: Sentiment-aware analysis of mobile apps user reviews regarding particular updates. ICSEA 2018, 109 (2018)
23. Li, X., Zhang, Z., Stefanidis, K.: A data-driven approach for video game playability analysis based on players' reviews. Information 12(3), 129 (2021)
24. Maalej, W., Pagano, D.: On the socialness of software. In: 2011 IEEE Ninth International Conference on Dependable, Autonomic and Secure Computing, pp. 864–871. IEEE (2011)
25. Mens, T.: Introduction and roadmap: history and challenges of software evolution. In: Software Evolution, pp. 1–11. Springer, Heidelberg (2008). https://doi.org/10.1007/978-3-540-76440-3_1
26. Mens, T., Wermelinger, M., Ducasse, S., Demeyer, S., Hirschfeld, R., Jazayeri, M.: Challenges in software evolution. In: 8th International Workshop on Principles of Software Evolution (IWPSE 2005), 5-7 September 2005, Lisbon, Portugal, pp. 13–22. IEEE Computer Society (2005)
27. Pagano, D., Brügge, B.: User involvement in software evolution practice: a case study. In: Notkin, D., Cheng, B.H.C., Pohl, K. (eds.) 35th International Conference on Software Engineering, ICSE 2013, San Francisco, CA, USA, 18-26 May 2013, pp. 953–962. IEEE Computer Society (2013)
28. Park, D.H., Liu, M., Zhai, C., Wang, H.: Leveraging user reviews to improve accuracy for mobile app retrieval. In: Proceedings of the 38th International ACM SIGIR Conference on Research and Development in Information Retrieval, pp. 533–542 (2015)
29. Rafique, D., Velasco, L.: Machine learning for network automation: overview, architecture, and applications [invited tutorial]. J. Opt. Commun. Netw. 10(10), D126–D143 (2018)
30. Santos, R., Groen, E.C., Villela, K.: A taxonomy for user feedback classifications. In: REFSQ Workshops, vol. 2376 (2019)
31. Swanson, E.B., Beath, C.M.: Departmentalization in software development and maintenance. Commun. ACM 33(6), 658–667 (1990)
32. Szajna, B., Scamell, R.W.: The effects of information system user expectations on their performance and perceptions. MIS Quarterly, 493–516 (1993)
33. Ullah, W., Zhang, Z., Stefanidis, K.: Sentiment analysis of mobile apps using BERT. In: Fujita, H., Wang, Y., Xiao, Y., Moonis, A. (eds.) Advances and Trends in Artificial Intelligence. Theory and Applications - 36th International Conference on Industrial, Engineering and Other Applications of Applied Intelligent Systems, IEA/AIE 2023, Shanghai, China, 19-22 July 2023, Proceedings, Part II. Lecture Notes in Computer Science, vol. 13926, pp. 66–78. Springer, Cham (2023). https://doi.org/10.1007/978-3-031-36822-6_6

34. Zeydan, E., Mangues-Bafalluy, J.: Recent advances in data engineering for networking. IEEE Access **10**, 34449–34496 (2022). https://doi.org/10.1109/ACCESS.2022.3162863

35. Zhang, Z., Rayhan, M., Herda, T., Goisauf, M., Abrahamsson, P.: LLM-based agents for automating the enhancement of user story quality: an early report. arXiv preprint arXiv:2403.09442 (2024)

On Customer Data Deduplication - Research vs. Industrial Perspective:
Lessons Learned from a R&D Project in the Financial Sector

Witold Andrzejewski[1] , Bartosz Bębel[1] , Paweł Boiński[1] ,
and Robert Wrembel[1,2(✉)]

[1] Poznan University of Technology, Poznań, Poland
{witold.andrzejewski,bartosz.bebel,pawel.boinski,
robert.wrembel}@put.poznan.pl
[2] Interdisciplinary Centre for Artificial Intelligence and Cybersecurity,
Poznań, Poland

Abstract. In this tutorial we present the results of researching, designing, implementing, and deploying data deduplication pipelines for customer records in a big financial institution. The tutorial is based on our
experience gained within a R&D project. In the project we developed
two deduplication pipelines. The first one is based on statistical modeling, whereas the second one is based on machine learning. Both pipelines
were extensively tested on a real data set including customer records. The
pipeline based on statistical modeling has already been deployed in the
production system of the financial institution and processes batches of
over 20 million of customer records.

Keywords: data quality · data deduplication · entity resolution ·
entity matching · statistical modeling · machine learning ·
classification · neural networks · graph processing

1 Tutorial Overview

1.1 Data Deduplication Problem and Solutions

Data stored in repositories of companies are often erroneous, typically characterized by missing, wrong, outdated, and misspelled values. On top of it, companies
typically face a problem of multiple database records describing the same physical entity - these multiple records will further be called duplicates. Duplicates
result among others from: (1) using multiple but not synchronized data repositories, (2) using applications that do not check for duplicates while inserting
or updating data (i.e., missing data quality monitoring), and (3) from various
data errors. This problem is of particular importance while dealing with personal
data, e.g., in healthcare, banking, insurance.

The problem of discovering duplicates (i.e., data deduplication) has been
intensively researched and resulted in almost 600 papers within the recent years.

J. Tekli et al. (Eds.): ADBIS 2024, CCIS 2186, pp. 392–400, 2025.
https://doi.org/10.1007/978-3-031-70421-5_37

They were published in top journals (e.g., the VLDB Journal, ACM Computing Surveys, IEEE Transactions on Knowledge and Data Engineering, Information Systems) and conferences (e.g., VLDB, EDBT, PAKDD, CIKM, ADBIS). Figure 1 summarizes the number of papers published on this topic in the period from 2020–2023, based on the dblp.org publication service. The search in this service was run with the following keywords: deduplication, entity resolution, entity reconciliation, entity matching, record linkage, reference reconciliation, data matching, and string matching.

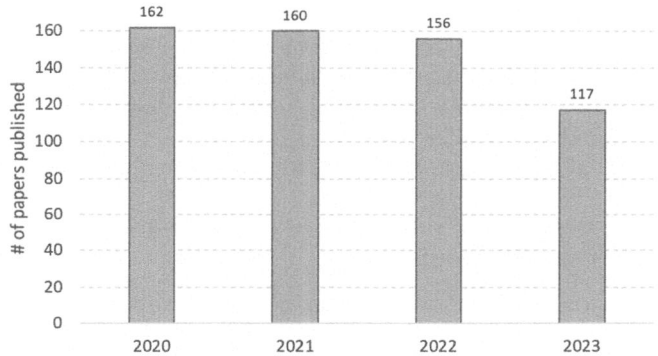

Fig. 1. The number of papers published on: deduplication, entity resolution, entity reconciliation, entity matching, record linkage, reference reconciliation, data matching, and string matching, within period 2020–2023 (available from dblp.org)

Data deduplication is also a technological issue dealt with by companies, since most of them face the problem of duplicated records in their repositories. Duplicates are costly to companies, since they deteriorate their reputation, result in economic loses, increase customer dissatisfaction, consume (human and monetary) resources to periodically run data verification and cleaning campaigns, and distort the results of analyses.

To handle the problem of duplicates, a *base-line data deduplication pipeline* (BLDDP) was developed [8–10,14,20,21]. The pipeline is equipped with multiple complex algorithms, supporting each task in the pipeline. In general, in data deduplication, pairs of records are compared and their similarities are computed. In the naive approach, record comparison has a quadratic computational complexity, which in unacceptable for real applications. Therefore, various techniques for decreasing this complexity were proposed.

While building a deduplication pipeline for a given deduplication problem, a designer has to implement the following tasks and in each task take some decisions, which impact the final result of the deduplication, i.e., groups of records representing the same real world object:

- T1: selecting the most adequate set of attributes A_i^G (i=1, ..., n) for dividing records into smaller groups G_i (to avoid quadratic computational complexity of record comparisons),
- T2: selecting a method for dividing records into groups G_i,
- T3: selecting an algorithm for comparing records in G_i,
- T4: tuning parameters of a given record comparison algorithm,
- T5: selecting the most adequate attributes A_m^C (m=1, ..., k) whose values will be compared,
- T6: selecting the most adequate similarity measures to compare values of A_m^C,
- T7: defining weights of attributes A_m^C, as not every attribute is equally important in the comparison,
- T8: defining an overall similarity formula for a pair of records being compared,
- T9: defining adequate similarity thresholds to distinguish between similarity classes: duplicates, probably duplicates, not duplicates,
- T10: building pairs of similar records,
- T11: building an overall graph of similarities between the compared pairs of records,
- T12: dividing the graph of similarities into sub-graphs, each of which represents groups of similar records.

Even though the BLDDP was developed and made available in multiple research publications, tasks T1, T4, T5, T6, T7, T8, T9 were not researched enough to provide clear and easy ready-to-use methods.

A promising direction in data deduplication is machine learning (ML). Basic deduplication models are based on standard classifiers (e.g., decision trees, SVM), to divide record pairs into classes of duplicates, probably duplicates, not duplicates, e.g., [7,22,24]. More advanced methods propose to apply neural networks, mainly for tasks T1, T2, T5–T9, e.g., [6,12,15,25,26]. However, the main problem faced by the ML approaches in real projects is the lack of learning data sets that would be large enough for a real data set to be deduplicated. Moreover, the deduplication models based on neural networks are typically unexplainable. For these reasons, methods bases on statistical modeling and similarity measures are still commonly used in practice.

1.2   Tutorial Focus

In this tutorial we summarize our experience in researching, designing, implementing, and deploying data deduplication pipelines for customer records in a big financial institution. Within the project we developed two deduplication pipelines. The first one is based on statistical modeling (further called SMP), whereas the second one is based on machine learning (further called MLP). The *SMP* explicitly implement tasks T1–T12, mentioned in Sect. 1, whereas in the *MLP* some tasks are implicitly included in some ML techniques.

In this tutorial we will outline approaches to designing and implementing tasks T1–T12 (mainly in the *SMP*), under technological constraints and other non-functional requirements, which are typically imposed by companies (especially from the financial sector).

Both pipelines were extensively **tested on a real data set of 5 million of customer records**. The pipeline based on statistical modeling has already been **deployed in the production system of the financial institution** and processes batches of over 20 million of customer records.

The tutorial will include two parts. The first part will focus on basics of the base-line data deduplication pipeline. In this part, the fundamental concepts and techniques in this pipeline will be presented. The second part will focus on our experience in building the *SMP*. In particular, we will outline practical methods for handling tasks T1–T12.

This tutorial is based on the following research publications of the authors of this tutorial: [1–5,23]. To the best of our knowledge, it is the only reported R&D project on data deduplication realized for a large company, which was deployed in a production system and runs on such a large data set as over 20 million of records.

1.3   Target Audience and Their Assumed Background

The tutorial will focus on practical aspects of deduplication, rather than on theoretical ones. This makes the tutorial suitable for a broad audience including practitioners (designers and developers) from industry, students, doctoral students, and experienced researchers working in the fields of databases, data integration, data warehouses, data lakehouses, and data quality assurance.

The minimum assumed knowledge of a participant encompasses the following **basics** of: data cleaning, programming languages, similarity measures, algorithm design and complexity, graph structures and processing, as well as machine learning techniques (in particular classification and neural networks).

2   Related Tutorials

Since data deduplication (a.k.a. entity resolution, entity matching, entity reconciliation, record linkage) is being a hot research and technological field, a few tutorials on achievements in this field has been delivered by world class experts. In this section we list the tutorials, which to the best of our knowledge, were delivered on data deduplication within recent ten years.

[11] focuses on the sources that introduce uncertainty in the process of identifying duplicate entities. The authors outline the existing classes of uncertainties, tasks where they appear, and present a few algorithms for handling them. The presented algorithms for entity resolution apply probability theory in various tasks of an entity resolution pipeline.

[17] focuses on novel blocking algorithms designed for deduplicating data from the Web. The authors discuss also performance issues of these algorithms in the context of processing large volumes of data and present two techniques for improving the performance, i.e., meta-blocking and parallel processing. This tutorial was further extended with new blocking algorithms and proposed by the same researchers in [18].

In [13] the authors present and analyze pros and cons of five popular blocking methods supporting data privacy. Next, they present a competing method, which uses locality-sensitive hashing for identifying similar records, previously anonymized by applying an encoding, which is based on the Bloom filter.

[19] addresses challenges of Web data (like heterogeneity, volume, and quality) in the design of a data deduplication pipeline. For such data, schema-agnostic algorithms are needed, which are presented in the tutorial. The tutorial covers: (1) tasks of the BLDDP (mentioned already) and (2) the possibility of improving their performance by means of parallelization.

[16] includes a systematic overview of the evolution of entity resolution pipelines. The pipelines are divided into four generations, each of which focuses on a different challenge posed by data characteristics. Each generation is discussed in details. The last part of the overview focuses on novel approaches to entity resolution, namely deep learning and crowd sourcing.

The aforementioned tutorials present theoretical and practical solutions to data deduplications. Most of the solutions are based on examples that use publicly available data sets. Only [16] addresses deduplication techniques supported by ML.

The tutorial that we propose presents our experience in extending the BLDDP to handle tasks T1-T12. Notice that solutions to some of these challenges have not been discussed in details in the literature. As a consequence, while designing our deduplication pipeline we faced problems for which clear and easy to use solutions did not exist (they still represent research challenges).

The main differences between our tutorial and the aforementioned ones include:

– Our tutorial presents **explicit solutions** to tasks T1, T4, T5, T6, T7, T8, and T9, which to the best of our knowledge, are vaguely described in the literature.
– The tutorial presents **practical solutions** to tasks T1-T12, which have been verified on a large data set in a real project. Notice that: (1) this data set is larger than reported in the related research literature and (2) the quality of deduplicated data was worse than assumed in the related research literature (which introduced yet another challenge to the already difficult problem).
– The tutorial shares our experience that we gained while realizing a **real R&D project for a big financial institution** in Poland. The pipeline based on statistical modeling has already been deployed in the institution.
– The tutorial also **contrasts assumptions** made in research projects with assumptions and requirements made in real R&D project for a big company, which we feel is unique.

3 Scope

The tutorial will include two parts. **Part 1** is based on the following research literature [8–10, 14, 20, 21]. In this part we will present the base-line deduplication pipeline with its main tasks, i.e.,:

- blocking (a.k.a. indexing), which arranges records into groups, such that each group is likely to include duplicates,
- block processing (a.k.a. filtering), which eliminates records that do not have to be compared,
- entity matching (a.k.a. similarity computation), which computes similarity values between record pairs, and
- entity clustering, which creates larger clusters of similar records.

In **Part 2** we will present the *SMP* in details and outline the *MLP*, which we have developed within the R&D project. The *SMP* and *MLP* extend the BLDDP, to serve the particular goals of the project. First, the *SMP* explicitly includes tasks T1-T12, which we found to be crucial for the deduplication process (whereas in the BLDDP, some tasks are either implicit or are not handled). Second, the last two tasks in our pipeline use alternative solutions to further merge groups of similar records into sub-graphs. Third, our pipeline accepts partially dirty data, as in practice (due to a large data volume) it is impossible to perfectly clean all data being deduplicated.

In this part of the tutorial we will outline the solutions, which we have successfully applied in the R&D project. We will also share our experience in implementing the solutions in two environments, namely the standard data engineering (a relational database as storage, SQL and a procedural in-database language for implementation) and the standard data science (csv files as storage, Python standard packages and developed code for implementation). Finally, we will compare the requirements of a real data deduplication project for a financial institution with requirements of the BLDDP assumed by research. As it will be discussed, it turned out that in practice both sets of requirements are often incompatible.

The tutorial is scheduled for 90 min.

4   Presenters

All the authors of this tutorial, in years 2020–2023, were realizing a project on customer data cleaning and deduplication for a big financial institution in Poland. The project successfully ended with the deployment of the solution in the institution.

Witold Andrzejewski (PhD) is an assistant professor of Computer Science at Poznan University of Technology in Poznan, Poland. He received his doctorate from Poznan University of Technology in 2008. His main research interests include optimization of query processing of complex data structures and data mining via auxiliary structures (indices) as well as hardware acceleration and parallelization (GPU based) of database and data mining operations. He has co-authored more than 40 scientific papers mainly related to the above subjects. He has also participated in several research and development projects.

Bartosz Bębel (PhD) is an assistant professor in the Faculty of Computing and Telecommunications, at Poznan University of Technology (Poland). He received his doctorate in 2005, specializing in database systems with a particular focus on data warehousing. He has co-authored more than 50 scientific publications in the area of databases. He is also involved in teaching work on database technologies. Webpage: http://www.cs.put.poznan.pl/bbebel/

Paweł Boiński (PhD) is an assistant professor in the Faculty of Computing and Telecommunications at Poznan University of Technology (Poland). He received his PhD degree in 2015. His research focus on spatial data mining (especially co-location pattern mining), data processing and data integration. Pawel Boiński is the author of dozens of articles published in international journals and conference proceedings. He has participated in several research and development projects mainly related to data processing in bioinformatics and data deduplication.

Robert Wrembel (PhD, Dr. Habil.) is an associate professor in the Faculty of Computing and Telecommunications, at Poznan University of Technology (Poland). In 2008 he received a post-doctoral degree in computer science (habilitation), specializing in database systems and data warehouses. He has been a deputy dean of the Faculty of Computing and Management (2008–2012) and the Faculty of Computing (2012–2016). Since Jan 2023 he is the chair of the Data Processing Technologies group at Poznan University of Technology; since May 2023 he is the leader of the Interdisciplinary Centre for Artificial Intelligence and Cybersecurity in Poznań https://caics.put.poznan.pl/. He was a consultant at software house Rodan Systems (2002–2003) and a lecturer at Oracle Poland (1998–2005). Currently he is an IT consultant in hospital Centrum Medyczne HCP. Within the last 10 years he has realized four R&D projects: for a big financial institution in Poland, one for a company in the energy sector, and two for a corporation in the field of electronics. He cooperates with IBM Software Lab Kraków in Poland. He has led at his University the Erasmus Mundus Joint Doctorate Program - Information Technologies for Business Intelligence - Doctoral College (2013–2020). Robert visited numerous research and education centers, including: Università degli Studi di Milano (Italy), Universitat Politècnica de Catalunya - BarcelonaTech (Catalunya), Université Lyon 2 (France), Universidad de Costa Rica (Costa Rica), Klagenfurt University (Austria), Loyola University (USA), INRIA Paris-Rocquencourt (France), and Université Paris Dauphine (France). In 2012 he graduated from a 2-months innovation and entrepreneurial program at Stanford University. In 2013 he has done an internship in a BI company Targit (USA). His research interests encompass: data integration, data quality, databases, data warehouses, and data lakes. Webpage: http://www.cs.put.poznan.pl/rwrembel/

Acknowledgements. The project is supported by the grant from the National Center for Research and Development no. POIR.01.01.01-00-0287/19.

References

1. Andrzejewski, W., Bębel, B., Boiński, P., Sienkiewicz, M., Wrembel, R.: Text similarity measures in a data deduplication pipeline for customers records. In: International Workshop on Design, Optimization, Languages and Analytical Processing of Big Data (DOLAP) @(EDBT/ICDT), vol. 3369. CEUR Workshop Proceedings, pp. 33–42. CEUR-WS.org (2023)
2. Andrzejewski, W., Bębel, B., Boiński, P., Wrembel, R.: On tuning parameters guiding similarity computations in a data deduplication pipeline for customers records: experience from a R&d project. Inf. Syst. **121**, 102323 (2024)
3. Boiński, P., Andrzejewski, W., Bębel, B., Wrembel, R.: On tuning the sorted neighborhood method for record comparisons in a data deduplicaton pipeline: industrial experience report. In: Strauss, C., Amagasa, T., Kotsis, G., Tjoa, A.M., Khalil, I. (eds.) DEXA 2023. LNCS, vol. 14146, pp. 164–178. Springer, Cham (2023). https://doi.org/10.1007/978-3-031-39847-6_11
4. Boiński, P., Sienkiewicz, M., Bębel, B., Wrembel, R., Gałęzowski, D., Graniszewski, W.: On customer data deduplication: Lessons learned from a R&d project in the financial sector. In: Workshops of the EDBT/ICDT 2022 Joint Conference, vol. 3135. CEUR Workshop Proceedings. CEUR-WS.org (2022)
5. Boiński, P., Sienkiewicz, M., Wrembel, R., Bębel, B., Andrzejewski, W.: On evaluating text similarity measures for customer data deduplication. In: ACM/SIGAPP Symposium on Applied Computing (SAC), pp. 297–300. ACM (2023)
6. Brunner, U., Stockinger, K.: Entity matching with transformer architectures - a step forward in data integration. In: International Conference on Extending Database Technology (EDBT), pp. 463–473. OpenProceedings.org (2020)
7. Chen, X., Xu, Y., Broneske, D., Durand, G.C., Zoun, R., Saake, G.: Heterogeneous committee-based active learning for entity resolution (HeALER). In: Welzer, T., Eder, J., Podgorelec, V., Kamišalić Latifić, A. (eds.) ADBIS 2019. LNCS, vol. 11695, pp. 69–85. Springer, Cham (2019). https://doi.org/10.1007/978-3-030-28730-6_5
8. Christophides, V., Efthymiou, V., Palpanas, T., Papadakis, G., Stefanidis, K.: An overview of end-to-end entity resolution for big data. ACM Comput. Surv. **53**(6), 127:1–127:42 (2021)
9. Colyer, A.: The morning paper on an overview of end-to-end entity resolution for big data (2020). https://blog.acolyer.org/2020/12/14/entity-resolution/
10. Elmagarmid, A.K., Ipeirotis, P.G., Verykios, V.S.: Duplicate record detection: a survey. IEEE Trans. Knowl. Data Eng. **19**(1), 1–16 (2007)
11. Gal, A.: Tutorial: uncertain entity resolution. re-evaluating entity resolution in the big data era. Proc. VLDB Endow. **7**(13), 1711–1712 (2014)
12. Jain, A., Sarawagi, S., Sen, P.: Deep indexed active learning for matching heterogeneous entity representations. VLDB Endow. **15**(1), 31–45 (2021)
13. Karapiperis, D., Verykios, V.S., Katsiri, E., Delis, A.: A tutorial on blocking methods for privacy-preserving record linkage. In: Karydis, I., Sioutas, S., Triantafillou, P., Tsoumakos, D. (eds.) ALGOCLOUD 2015. LNCS, vol. 9511, pp. 3–15. Springer, Cham (2016). https://doi.org/10.1007/978-3-319-29919-8_1
14. Köpcke, H., Rahm, E.: Frameworks for entity matching: a comparison. Data Knowl. Eng. **69**(2), 197–210 (2010)
15. Mudgal, S., et al.: Deep learning for entity matching: a design space exploration. In: SIGMOD International Conference on Management of Data, pp. 19–34. ACM (2018)

16. Papadakis, G., Ioannou, E., Palpanas, T.: Entity resolution: past, present and yet-to-come. In: International Conference on Extending Database Technology (EDBT) (2020)

17. Papadakis, G., Palpanas, T.: Blocking techniques for web-scale entity resolution. In: International Conference on Web Information System Engineering (WISE) (2014)

18. Papadakis, G., Palpanas, T.: Blocking for large-scale entity resolution: challenges, algorithms, and practical examples. In: IEEE International Conference on Data Engineering (ICDE), pp. 1436–1439 (2016)

19. Papadakis, G., Palpanas, T.: Web-scale, schema-agnostic, end-to-end entity resolution. In: The WEB Conference (2018)

20. Papadakis, G., Skoutas, D., Thanos, E., Palpanas, T.: Blocking and filtering techniques for entity resolution: a survey. ACM Comput. Surv. **53**(2), 31:1–31:42 (2020)

21. Papadakis, G., Tsekouras, L., Thanos, E., Giannakopoulos, G., Palpanas, T., Koubarakis, M.: Domain- and structure-agnostic end-to-end entity resolution with JEDAI. SIGMOD Rec. **48**(4), 30–36 (2019)

22. Sarawagi, S., Bhamidipaty, A.: Interactive deduplication using active learning. In: ACM SIGKDD International Conference on Knowledge Discovery and Data Mining (KDD), pp. 269–278. ACM (2002)

23. Sienkiewicz, M., Wrembel, R.: Managing data in a big financial institution: conclusions from a R&D project. In: Workshops of the EDBT/ICDT 2021 Joint Conference, vol. 2841. CEUR Workshop Proceedings. CEUR-WS.org (2021)

24. Silva, J.A., Pereira, D.A.: A multiclass classification approach for incremental entity resolution on short textual data. Int. J. Bus. Intell. Data Min. **18**(2), 218–245 (2021)

25. Thirumuruganathan, S., et al.: Deep learning for blocking in entity matching: a design space exploration. Proc. VLDB Endow. **14**(11), 2459–2472 (2021)

26. Zeakis, A., Papadakis, G., Skoutas, D., Koubarakis, M.: Pre-trained embeddings for entity resolution: an experimental analysis. Proc. VLDB Endow. **16**(9), 2225–2238 (2023)

Author Index

J. Tekli et al. (Eds.): ADBIS 2024, CCIS 2186, pp. 401–402, 2025.
https://doi.org/10.1007/978-3-031-70421-5

GPSR Compliance

The European Union's (EU) General Product Safety Regulation (GPSR) is a set of rules that requires consumer products to be safe and our obligations to ensure this.

If you have any concerns about our products, you can contact us on ProductSafety@springernature.com

In case Publisher is established outside the EU, the EU authorized representative is:

Springer Nature Customer Service Center GmbH
Europaplatz 3
69115 Heidelberg, Germany

The manufacturer's authorised representative in the EU is Springer
Nature Customer Service Centre GmbH, Europaplatz 3, 69115 Heidelberg,
Germany. If you have any concerns regarding our products, please
contact ProductSafety@springernature.com

Printed and bound by CPI Group (UK) Ltd, Croydon, CR0 4YY
29/04/2026
02099532-0010